Educational
Telecommunications

Educational
Telecommunications

Donald N. Wood
California State University, Northridge
Donald G. Wylie
San Diego State University

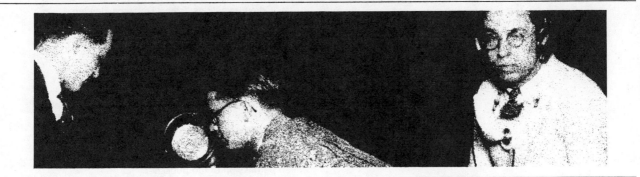

Wadsworth Publishing Company, Inc., Belmont, California

to Marie and Yvonne
. . . for their everlasting patience and gentle
harassment

Communications Editor: Rebecca Hayden
Production Editor: Rebecca Hayden
Designer: Dare Porter
Copy Editor: Ellen Seacat
Technical Illustrator: John Foster

Printed in the United States of America
1 2 3 4 5 6 7 8 9 10—81 80 79 78 77

Library of Congress Cataloging in Publication Data

Wood, Donald N   Date
    Educational telecommunications.

    Includes index.
    1. Educational broadcasting. 2. Television in
education. 3. Telecommunication. I. Wylie,
Donald G., date   joint author. II. Title.
LB1044.8.W66   384.54   76-25585
ISBN 0-534-00494-6

Preface

This book is intended as a textbook and reference work to cover the field of "educational telecommunications." As such, it is a broad look at a comprehensive area. Many books have been written about either public broadcasting or instructional television. We contend, however, that the educational telecommunications (ETC) field can most profitably be studied in an all-inclusive fashion, looking at the interrelationships among different aspects of the total profession—such as the interplay between public television and instructional telecommunications. The distinctions are not always easy to make. Therefore, although we considered writing two separate books (one each for public broadcasting and school TV), we have elected to cover the entire field between two covers.

We have divided the text into four distinct parts. Part One deals with the history and common heritage of both public broadcasting and instructional telecommunications (ITC)—the drama of educational radio and television. Part Two is concerned with an examination of the structure of public broadcasting in the United States today, including a discussion of several of the continuing issues and debates within the professional membership of the field. Theories and organizational approaches to ITC comprise Part Three, opening, we hope, some fresh perspectives in this area. Part Four is concerned with the practice of ETC, primarily from the viewpoint of schooling operations.

In covering this wide scope, obviously we could not get into as much detail as we would have liked. For instance, although we have tried to present a thorough outline of production considerations, we have not written another handbook for the field. For a complete treatise on cameras, lens characteristics, lighting instruments and theories, audio principles, directing techniques, and so forth, the reader is invited to consult any one of a number of good textbooks and handbooks on production, several of which are cited in the notes.

Similarly, we have not tried to write a text-within-a-text on distribution technologies or classroom utilization—although chapters devoted to these topics present the reader with a basic familiar-ization and grasp of these operations. We have tried to present criteria for decision-making in these areas —not a complete how-to-do-it handbook. In a like fashion, we avoided the temptation to include a monograph on instructional theory or a detailed research guidebook. However, the breadth of the book provides a substantial outline and orientation on all of these topics.

Also, we have broken the book down into a number of clearly marked sections that are frequently cross-referenced (with notations such as "See Section 9.7"). Many of these separate sections can stand alone as mini-essays on specific topics. The reader is invited to consider this organizational device as a key to using this book as a reference work.

Primarily, however, we envision this book as a textbook in courses on "educational broadcasting," "public broadcasting," "television in the schools," and related titles. We have kept in mind our own students, who are majoring in radio/television/film—or taking elective courses in schools of education. Many are unsure how they will use their education and skills in telecommunications when they graduate. Many consider going into public broadcasting. Others aim for commercial broadcasting, aware that they may be involved with various school broadcasts that many commercial stations produce and air as a public service. Still others may find employment with schools or industry, producing instructional TV lessons and training materials. Some radio-TV graduates will be associated with independent production centers, producing instructional materials.

In this book we have tried to give these students more than just facts about public and instructional television and radio. We have tried to provide a sense of what it means to be a professional in public broadcasting and school telecommunications. We have tried to communicate a feeling of the excitement and drama of pioneering in a new field such as ETC. And at the same time, we have tried to present honestly some of the frustrations and differences of opinion that the national leadership must face—including many of the current controversies and continuing debates.

No work of this scope is possible without the contributions of numerous individuals—providing materials, suggestions, photographs, editorial assistance, and critical review. Although it is impossible to list everyone who has been an influence and indirect contributor to this work, we would like specifically to single out the following for their very tangible and direct contributions: Jon Anderson, Robert Avery, Charles Clift, James Fellows, Robert Fox, Dennis Harp, Becky Hayden, Thomas Morgan, Richard Phipps, Elinor Richardson, Mollie Robbins, James Sanders, Ellen Seacat, Autumn Stanley, Paul Steen, Gordon Tuell, Brad Warner, John Witherspoon, and Jean Young, as well as many of our students who have reacted to portions of the manuscript.

Despite the generous support of these friends and colleagues, there will undoubtedly be some flaws in the book. The authors assume sole responsibility for the integrity and validity of the final product. For those interested in the specific division of labor: Wood was responsible for most of Part One, all of Part Three, and most of the first three chapters of Part Four, with Wylie writing the bulk of Chapter 17; Wylie was primarily responsible for Part Two, with Wood contributing to Chapter 5.

DNW
DGW

Contents

List of Figures and Tables

one

**Orientation:
Toward a Philosophy
of ETC**

If you have ever watched a public television program, listened to a noncommercial radio station, or viewed a television lesson in a classroom, you have some idea of what this book is about. Educational telecommunications is a half-billion-dollar industry, employing tens of thousands of people, and affecting the lives of millions of viewers/listeners. In schools throughout the nation, millions of students receive part of their formal education from instructional telecommunications.

1.1 Of Promises and Potentials

Educational telecommunications has only begun to realize its potential. Its story has always been one of half-filled promises and untapped strengths. Throughout the years, writers and educators have seen grand visions of what educational broadcasting could become. As early as 1934, one pioneer proclaimed, "Television's place as an effective instrumentality in the educational system of the future is already assured."[1] Writing almost thirty years later, John Walker Powell described the sense of destiny that pervaded the early days of ETV:

> It was the birth of a new institution, of a movement that could transform the older institutions on which it was superimposed. Created by unpredicted leaders, staffed by unheralded talent, watched by unguessed millions, it leaped boldly into the communications picture to meet a social need and to fulfill a social vision: Educational Television, the logical completion of the communications spectrum, and the "intelligence" service of a nation.[2]

[1]E. B. Kurtz, *Pioneering in Educational Television: 1932–1939* (Iowa City: State University of Iowa, 1959), p. 76.
[2]John Walker Powell, *Channels of Learning: The Story of Educational Television* (Washington, D.C.: Public Affairs Press, 1962), p. 4.

Political leaders and decision makers have also been ready to point out educational TV's potential. Frieda Hennock of the Federal Communications Commission (FCC), who almost single-handedly guaranteed that there would be channels reserved for ETV, stated that "educational stations will provide the highest standards of public service."[3] President Lyndon B. Johnson, who signed the bill creating the Corporation for Public Broadcasting, said, "I believe that educational television has an important future in the United States and throughout the world."[4]

Throughout the short history of ETV and educational radio, many other writers, educators, broadcasters, politicians, and public leaders have had numerous occasions to sing the praises of these media. There has been general agreement that the potential of educational telecommunications is, indeed, vast—both for formal schooling purposes and for general at-home viewing. "Through radio or television, thousands of students can be brought into intimate contact with great teachers. The general public need not leave their living rooms to make contact with great minds."[5]

Yet, we also find general agreement that educational TV and radio have failed to reach their potential. Some would say that these media have fallen far short of living up to their promises. Others would point out that there have been substantial accomplishments in the past few decades, but that public broadcasting and instructional telecommunications still have had little impact—compared to what they ultimately can do. To a great extent, this book will be concerned with the story of what these electronic educational media have achieved to date, what possible goals they might reach in the future, and what some of the existing impediments and roadblocks might be.

1.2 Search for a Definition

Before turning to a specific definition of "educational telecommunications," let us look at a term familiar to almost everyone—"educational television" (ETV). Teachers talk about using "educational television." Many TV viewers often watch "educational television." Taxpayers wonder if "educational television" is really any good. So we all know what we mean when we refer to ETV. Right? *Wrong!*

First of all, would everyone agree on exactly what is meant by "educational"? Is a one-shot documentary "educational" by itself, or must it be integrated into a continuing series that is related to a particular curriculum in the schools? Must a teacher follow up on the lesson? Must there be academic credit involved before the program can officially be classified as "educational"? Is the evening news on a commercial station "educational"?

What do we mean by "television"? Is satellite transmission of weather data "television"? Shall we include in our definition cultural programs on commercial television, such as the CBS presentation of the BBC series *America*? What about industrial uses of TV to monitor an assembly-line procedure for training new employees? Should we call this "television" for the purposes of this book? A classroom teacher uses a film that has been recorded onto a video cassette (with proper copyright clearances), using only a cassette player and a single TV monitor. Is this "television"? (Or is it "film"?)

"Educational television" is also a rather awkward and misleading label in the sense that television per se—the medium, the channel—is not "educational" by itself. We do not refer to books in the classroom as "educational print." We think of the typewriter, the pencil, the chalkboard, and other printing resources as *tools used for educational purposes*. The term "educational

[3]Frieda Hennock, quoted in Powell, *ibid.,* pp. 24–25.
[4]Lyndon B. Johnson, quoted in Carnegie Commission, *Public Television: A Program for Action* (New York: Bantam Books, Inc., 1967), p. vii.
[5]Giraud Chester, Garnet R. Garrison, and Edgar E. Willis, *Television and Radio,* 3rd ed. (New York: Appleton-Century-Crofts, 1963), p. 197.

television" makes about as much sense as would the term "educational printing" or "instructional chalkboard"; a more accurate phrase is "television (or telecommunications) used for educational purposes." However, the phrase and its acronym (TUEP) are awkward, and the term "educational television" and its acronym (ETV) are used. Therefore, wherever the term "educational telecommunications" ("ETC") is used in this book, it is to be interpreted as broadly as possible.

The obvious and clear-cut distinction between commercial television and noncommercial television—the only terms with actual legal definitions—is of little value in determining what is actually educational. Much of what is transmitted over commercial television is, without challenge, educational; and some of what is seen and heard on noncommercial television and radio channels has (some critics say) dubious educational value. Others would say that all radio and television is educational to some extent. We learn about places, events, people, politics, crime, conflict, science, and a myriad of other things from radio and television, certainly including commercial media. Former FCC Commissioner Nicholas Johnson phrased it succinctly: "All television is educational television. The only question is, what is it teaching?"[6]

A Diversity of Applications   In referring to television and radio "used for educational purposes," we are talking about:

A means for enriching a school curriculum with occasional programs loosely related to the subject at hand

A series of how-to-do-it programs for the home handyman, cook, or hobbyist

Broadcasts of outstanding symphony or dance performances

An effective medium for reaching an adult audience with informal education in a foreign language, computer programing, or guitar playing

Channels for school administrative communication

A vehicle for combating illiteracy in a backwoods community

A way of informing every salesman in a given company, throughout the country, about the latest modifications in a new product

Shakespeare brought into your living room

A continuing adult education project for a university

Live coverage of a legislative hearing or international debate

A major classroom resource used on a regularly scheduled basis

A channel for distributing conventional audio-visual films throughout a school building

Agricultural educational projects in a sparsely populated rural area

A forum for confrontation between the Establishment and the Rebels, the Left, and the Right, the Haves and the Have-nots

An in-service education teacher training project

A series of evening documentaries on racial tensions

A mirror to allow students to evaluate their own performances

A single overhead camera in a science lab

Relaying pictures back from a distant planet

Educational telecommunications incorporates all this, and much more. ETC is much broader than any single application or approach. It refers to

---

[6]Quoted by Joan Barthel, "Notes in a Viewer's Album," *Life*, September 10, 1971, p. 66.

the all-pervasive, multifaceted, all-encompassing environment/medium/experience/message. "Educational telecommunications" cannot be interpreted in a too-restrictive sense.

Despite the difficulties of finding accurate and precise definitions for some of these related terms, it is necessary to state what this book is about—what we mean, and what we do not mean, by the title.

"Educational Telecommunications" First of all, the term "telecommunications" is the only label that fully covers the scope of the electronic media with which this book is concerned. Use of the label "radio" or the label "television," of course, excludes the other primary medium; and the term "radio" does not include the connotation of *closed-circuit radio,* that is, the use of non-broadcast audio (tape recordings and cassettes). The word "broadcasting" encompasses both media but includes no closed-circuit considerations. Therefore, "telecommunications" is the only word to cover both radio and television—open-circuit and closed-circuit. Also, the term "telecommunications" implies other specialized uses of the electronic media that we want to include in our total framework: slow-scan television, data transmission, facsimile, still pictures, computer communications, Teletype, telephone, and related transmissions.

We will specifically define "educational telecommunications" (ETC) as *noncommercial* (although commercial uses will occasionally be referred to) *television and audio* (and related electronic media) *transmissions of purposeful, broadly educational, communications*—whether for specific classroom objectives or for general public enlightenment.

Related Definitions The field of educational telecommunications has pragmatically been divided into two major categories, which trace their lineage back to the earliest days of educational television: *public broadcasting* and *instructional telecommunications*. Although a primary theme of this book is that these two categories are closely related and intertwined, it is nevertheless convenient—and necessary—to make a theoretical distinction between the two entities.

Public Broadcasting The term "public television" was popularized by the Carnegie Commission Report in 1967, and it was quickly broadened to "public broadcasting." The terms "public radio" and "public television" (PTV) refer specifically to the two media involved. The Carnegie Commission explicitly defined PTV as *"all that is of human interest and importance which is not at the moment appropriate or available for support by advertising, and which is not arranged for formal instruction."*[7] This definition applies generally to noncommercial television and radio stations, and especially to the cultural, informational, public affairs, and informal educational programs that are broadcast for general at-home reception.

Instructional Telecommunications We will be using the term "instructional telecommunications" (ITC) to mean *direct instructional uses of television and related electronic media for specific teaching/learning applications in any formal educational or training institutional situation*. This definition encompasses instructional uses of the media in any formal learning/training setting—not just in schools; these uses include military training, business instruction, industrial applications, law enforcement programs, and so forth.

The concept of instructional telecommunications generally supersedes the older designation of "instructional television" (ITV). Traditionally, ITV has referred loosely to any in-school usage of television—either closed-circuit or, usually, open-circuit. This was a somewhat misleading designation, however, as schools are concerned with much more than just "instruction" per se. Specifically, we will have occasion to use "ITV" in a historical context when referring

---

[7]Carnegie Corporation, *Public Television*, p. 1.

strictly to *direct instructional use of television for specific teaching/learning situations.*

We will also occasionally be using the term *"school television"* (STV) to refer specifically to *any school-related use of television* (or, occasionally, other telecommunications media), *instructional or otherwise.* In addition to direct instructional applications, STV would include observation applications, administrative messages, public relations uses, in-service training, self-evaluation set-ups, and so forth. As a broader term, "School TV" is probably more accurate than the former label, "ITV."

Finally, we find ourselves unable to do away with the term *"educational television."* This much-maligned and overworked phrase has taken on several meanings over the years. Historically, "ETV" has been used to refer both to the broad overall category of noncommercial open-circuit broadcasting and in-school closed-circuit uses (corresponding to "educational telecommunications") and to the more specific programing category of general at-home, evening programs presented in a noncredit, non-course-structured format (today referred to as "public broadcasting"). In various historical discussions, we will be using "ETV" in both of these contexts.

The ETV picture is complicated by the fact that today educational broadcasters are rediscovering a middle ground of programing—one that falls halfway between "public TV" and "instructional TV"—which is often referred to (for lack of a better term) as "educational TV." As it is increasingly being used in this manner, "ETV" could be defined as *sequential, organized series of presentations having a specific body of content, usually designed primarily for noncredit viewing at home but often viewed additionally in the classroom.*[8] Examples of open-circuit

series falling into this classification would include *Sesame Street, America,* and *Ascent of Man.* The use of the term "educational television" in this sense is compounded by the fact that many public broadcasting stations often use "ETV" to refer to their school programing, and many instructional telecommunications practitioners tend to use "ETV" as a synonym for public broadcasting.

To some extent, the above definitions reinforce the concept that there are few hard-and-fast distinctions and clear-cut classifications among the various types of ETC applications, but this working set of definitions should help to guide the reader through the several overlapping categories with which we must be concerned.

These definitions also indicate something about the relative emphasis we will place on television (as compared with radio) in this book. Although we will disappoint our friends in public radio by doing so, we admittedly focus most of our attention on television. Yet, in most instances, when we use such labels as "ETC," "PTV," and "ITC," we refer to both media—at least to some extent. Much of what we say explicitly about television also applies implicitly to audio recording and radio broadcasting.

1.3   Diagrams and Models

For those who want to consider the total realm of educational telecommunications as easy-to-handle compartments, three pictorial models are given below.

ETV Circle of Instruction/Appreciation   In the diagram (1-1), various concentric rings are used to exemplify uses of television from the most basic, intensive, classroom applications of the medium to the least academically oriented, at-home viewing of TV.

The outermost ring, "escapist" programing, comprises the bulk of prime-time programing on commercial stations—that which is farthest removed from what is usually considered "edu-

---

[8]One early proposal involving this revised view of the term "educational television" was presented by James Day, former president of NET, at the San Francisco Convention of the Western Educational Society for Telecommunications, February 28, 1972.

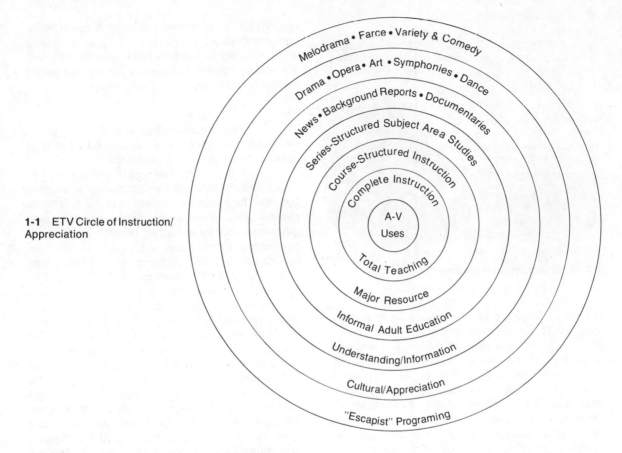

1-1 ETV Circle of Instruction/
Appreciation

cational'' in nature. The next circle, "cultural/
appreciation," covers both commercial and noncommercial "quality" entertainment and cultural
programing—"serious" drama, the arts, and
so forth. The "understanding/information" ring
includes all public affairs and informational applications of television—commercial and noncommercial—such as news reports, documentaries, live coverage of significant events, and
similar items. The next circle, "informal adult
education," could apply to either credit or noncredit uses of the medium for course-structured
materials; this level is primarily concerned with
out-of-school viewing. The "major resource"
circle involves the classroom and implies a sub-

stantial use of television as a major classroom
resource, a learning tool much like a textbook.
The next ring, "total teaching," is more intensive,
incorporating a heavy use of television to carry
virtually the entire instructional burden. Finally,
the innermost circle, "audio-visual uses," includes such items as single-classroom, one-
camera demonstrations, simple closed-circuit
hook-ups, and image magnification purposes—
the use of television as a basic audio-visual teaching tool.

ETV Compartmentalization In the next
diagram (1-2), a tongue-in-cheek attempt is made
to divide the television tube into neat compart-

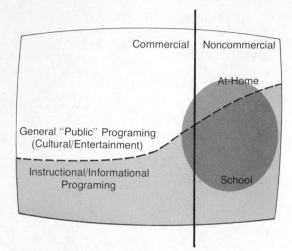

1-2 ETC Compartmentalization

cial, PTV/ITV, at-home/in-school. A much more complex diagram would result if other categories were included: credit/noncredit, specials/sequential series, open-circuit/closed-circuit, and formal education/adult education, among others.

Organizational Chart of ETV The pictorial representation (1-3) of the total ETV scene is based upon a table of organization.

In this chart, we again distinguish between commercial and noncommercial television; but note the overlap in the areas of public affairs, cultural programing, and out-of-school children's programs. We have further tried to divide noncommercial applications into broadcast and nonbroadcast delivery systems, but again, some areas (notably ITV) overlap. In the "non-public broadcasting" area of the model, we have shown the distinction between the "institutional" categories (for example, school TV, medical TV) and the "functional" uses (such as instructional TV, administrative TV, in-service TV, and observation TV). However, even this table of organization is still a gross simplification. Many of the categories overlap in actuality. Various uses cannot easily be assigned to one box or another. For example, an open-circuit program may easily serve more than one of these stated purposes, and closed-circuit equipment almost always serves more than one function.

In trying to arrive at a clear idea of the scope of "educational telecommunications," it may be most fruitful to compare and contrast the three diagrams. Find the inconsistencies; note the overlaps. Note that these figures leave many questions unanswered. How about other types of distribution systems (cassettes, discs, community cable TV, satellites)? Have we accounted for all the distinctions between credit and noncredit programing? How about the interaction of television and other media (including print materials)? Where do we place commercially produced instructional programing? Many other similar questions could be raised.

ments, indicating how the categories of the first diagram (1-1) actually overlap.

Imagine a vertical dividing line that separates the TV screen into two sections—the left-hand section represents all programing broadcast over *commercial* stations, while the right-hand side includes all *noncommercial* broadcasting. Now add an uneven, dotted horizontal dividing line that separates the tube into a top portion, which represents materials of a *"public"* programing nature (cultural and entertainment), and a bottom section, which represents *"instructional/ informational"* programing. Finally, draw a rough circle that includes parts of all four quadrants. Everything inside this circle is appropriate to be viewed *in school,* while all materials outside of the circle should be considered *at-home* programing.

This diagram illustrates how the various categories overlap. Not all in-school materials are noncommercial; some ITV informational materials are designed to be viewed at home; some "public" programs can be used in the classroom.

This simple chart presents only three factors or distinctions: commercial/noncommer-

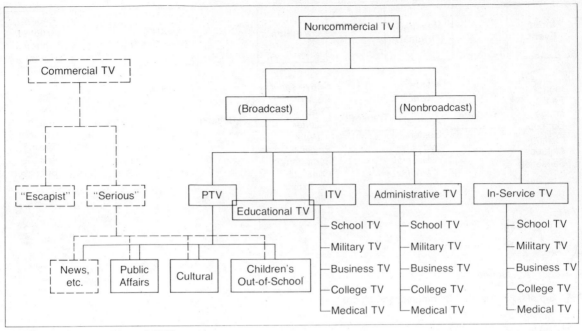

1-3   Organizational Chart of ETV

1.4   A Means of Communication

From a theoretical viewpoint, television and radio (and discs and tapes) are basically just means of communication. The television medium is no more than a means of transforming pictures and sounds into electronic signals, and then transmitting these bits of "information" (electronic signals) over a distance. Television is basically just a means of communication. Of course, as a social force—as a phenomenon among the mass media—television is much more. It is also a message, within itself; it is also a subject for study. This aspect is explored in Section 10.7.

Simplified Communication Model   It might be easier to understand television as a means of communication by examining a simplified one-way communication model (1-4), which is a generalized concept of one such simple model.

As shown in the model, the sender wants to communicate to the receiver some inner event, which may be some idea, a feeling, a fact, or anything else he can conceive in his mind. He first must encode this inner event (or referent) into some sort of message. He may use words—oral or written—or a painting or musical notation or direct action. Of course, he may need to employ machinery merely in order to construct his message (a typewriter, artist's canvas and palette, a piano, a television camera, or whatever). He then must transmit this message through some medium: direct conversation, a symphonic program, an art exhibit, direct physical contact, tape recording—or some version of television, radio, or film. Note that the medium of television can encompass virtually all of the other media—words, print, music, film, and so forth; and some of these other media can be transmitted better by TV than others (for example, television can transmit a

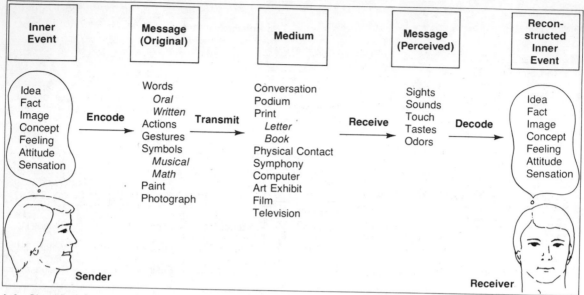

1-4   Simplified One-way Communication Model

close-up of a talking face with more impact than it can convey the totality of a sculpture exhibition).

The encoded and transmitted original inner event is then received by the receiver as a perceived message. Utilizing only his five senses, the receiver should have a perceived message somewhat similar to the original message (words, actions, symbols, or whatever). The receiver then decodes the message and comes up with some sort of reconstructed inner event (idea, fact, emotion, or whatever he reconstructs).

Noise or Interference   The opportunities for communication interference—technically referred to as "noise" (physical, psychological, intellectual)—are manifold. Some break or interference may occur in the communication cycle anywhere along the line. The sender may have aphasia or some other handicap in speaking; he may include some unintentional distracting element in his message (a twitch, shifty-eyed glances, or a sing-song vocal pattern); he may

make mistakes in his encoding (poor typing, out-of-focus camera work, incompetent musical scoring); the medium may distort the message (poor public address system, poor acoustics in the symphony hall, television technical problems); the receiver may make mistakes in perceiving the message (poor eyesight, partial deafness, unusual sensitivity); and, finally, the receiver may decode the message erroneously (mistaken intentions, unfamiliarity with the grammar of the message, inability to interpret symbols, undue emphasis on certain transmitted signals). The reconstructed inner event is seldom an accurate rendition of the inner event that the sender started out with.

Obviously, the role of television as a medium can be extremely complicated. The manner in which the message is constructed, transmitted, and received is dependent upon many complexities of the mechanics of television. The use of the television medium involves numerous steps—any one of which can cause serious problems. For example, suppose that the inner event of the sender is the need to sell Zombat automobiles. Think what is involved in transmission of that inner event, including decisions about what percentage of Zombat Corporation's advertising budget should be invested in a few television commercials; the coordinated labors of various agencies, network and station representatives; the production of several filmed sixty-second spots; the scheduling and transmission of the commercials. What percentage of the receivers are likely to wind up with the reconstructed inner event, "Buy a Zombat"? Imagine the complexities merely in using television to demonstrate the movement of a paramecium's cilia, by means of closed-circuit microscope hook-up, to a biology laboratory session of ten students.

The important concept is simply that television can be used as one link—a complex but efficient link—in a communications chain. Television is not solely *a* program, *a* transmission system, *a* camera, or *a* studio. It is a system of innumerable components that can be used in a variety of ways, as a communication tool, in a specific communication situation.

1.5   Relationship of Communication and Education

The role of television in any educational situation may be approached from a theoretical standpoint by examining the relationship of our simple communication model to simple instructional models.

Feedback/Evaluation   The simplified message transmission presented in the illustration (1-4) is basically a one-way communication model. To complete the communication cycle, the return channel also must be used. The receiver must, in turn, become a sender and send some signal back to the original sender (the viewer goes out and buys a Zombat the following week; the biology student is able to describe the movement of the cilia). Then the communication cycle is complete—feedback has occurred; an evaluation of the original communication has taken place.

This process of completing the communication cycle, or providing feedback, is basically the same as the original process, except that the receiver has become the sender, and the original sender is now the receiver. Once the two-way model is established—once a communication model is constructed with provision for feedback—we theoretically have the basis for a fundamental model of instruction. The teacher lectures (one-way communication); the student responds on a test (feedback communication); and the communication/education model is complete. At this stage, the process is repeated; it becomes, in effect, a continuous process. The teacher modifies the next presentation (based upon the original feedback received from the students) and the students respond a second time—indicating where the communication process fell short the second time.

Many variations can be built upon this basic two-way model: sender-receiver, seller-buyer, stimulus-response, teacher-learner. The principle remains the same: At least at an abstract level, theoretical communication models and theoretical instructional models can be constructed in a virtually identical manner. Whether one is delivering a practice speech in a speech class or presenting a demonstration in a student-teaching situation, the process is similar. Whether the sender is analyzing feedback by a show of hands in a communications class or in an elementary classroom, the process is similar. Whether one is evaluating one's communication performance by votes cast at a ballot box or by the number of errors on a standardized exam, the process is similar.

Education Is Communication   Once the basic similarity of communication models and instructional models can be grasped, the potency of television as an educational tool becomes vivid. *Communication is education. Education is communication*. The process is theoretically identical.

Formal educational situations can be interpreted basically as institutionalized communication proceedings designed for specific individual and societal purposes: to pass on man's heritage of knowledge and culture, to discover new knowledge, to acquire skills for earning a living, to develop appreciation for certain values, and so forth. Television can be a powerful tool for achieving these kinds of communication/education goals.

1.6   Television and Print

Once television is viewed in a theoretical communication context, the similarities between television and print become apparent. Television is basically just a means of communication. Similarly, that is all that print is: a means for transforming (encoding) an inner event (ideas, facts, images, attitudes) into a message (printed symbols) and then conveying (transmitting) this information over a distance through some specific medium (books, handwritten letters, pamphlets, chalkboard) to a receiver (for his decoding).

Analogy to Print   Like television, print is basically just a means of communication (and of education). Too frequently, television is viewed in a narrow context—a teacher may think of classroom TV simply as "that series of twenty-minute illustrated lectures designed as a major resource for fourth-grade English." This is as limiting as to say that "educational print" is textbooks, and nothing else. Is there no other way to use "educational print" than for textbooks? Is there no place for pamphlets, state guides, administrative memoranda, chalkboard improvisations, dittoed class handouts, photocopying services, programed instructional texts, paper and pencil tests, magazines, flash cards, and the like? Similarly, is there no place for the television counterpart to these myriad uses of print communications?

Must we think of television solely as a visual and audio counterpart of the textbook? Cannot television also be used as a pamphlet? As a chalkboard? As a dittoed class handout? Is

there a television equivalent of a photocopying machine? Of a magazine? (These questions are explored further in Section 10.3.) A strong analogy can be drawn between telecommunications media and the printing press. Both are communications channels for getting various types of information from one place to another. This analogy can be extended in many ways, including the following:

Purpose: Like printed materials, television can be used for several different purposes—to entertain, to inform, to persuade.

Audience: Like printed materials, television can be designed for a variety of audiences—from preschoolers to postgraduate students, from slow learners to the gifted, from the economically deprived to the affluent.

Uses: Like printed materials, television can be used for classroom instruction, informal teaching, in-service teacher education, industrial training, administrative messages, adult instruction in the living room—or for any other format.

Producer: Like printed materials, television programing designed for instruction can be produced and distributed by commercial companies, nonprofit organizations, government agencies, or local school districts.

Quality: Like printed materials, television can be used to produce programing of varying quality—ranging from high-quality valuable cultural contributions to low-grade trash.

Distribution: Like printed materials, television can be distributed on various scales—from national coverage to single-classroom dissemination to individual retrieval of stored information.

Utilization: Like printed materials, television can be designed for explicit instructional purposes—generally integrated into the total learning situation as utilized by a competent and well-trained classroom teacher (although sometimes ITC materials, like print materials, are designed to stand alone as reference tools or as "total teaching" tools).

The analogy could be extended in many directions. For every use, perspective, or facet of "educational print," a comparable use/perspective/facet of telecommunications technology can be found.

Of course, the obvious differences between print and telecommunications media (principally TV) cannot be overlooked. Print stresses abstract symbols on a tangible printed page, while television consists of actual sound and pictures in an ephemeral medium. Print is well suited for review and detailed study, while television possesses more impact and dramatic reinforcement.

From a pedagogical and administrative point of view, both the printing press and the television system do share many of the same organizational characteristics. In the broad view, all media must be seen from a similar perspective—having certain functions in common, sharing similar purposes, fulfilling related needs. They all are means to some other end—sending information from one point to another.

This analogy between television and print is a recurring comparison that pervades much of this book. It is an attempt to counteract the unspoken, but ingrained, assumption that print is the accepted instructional medium, while television is something "added on" to the teaching/learning situation, something foreign to the process of education. This analogy has many convenient applications. It helps to place television in perspective as the all-pervasive environment/medium/experience/message that is more than any one single project or limited application.

1.7 Integration of Television and Other Media

Telecommunications must also be examined in another theoretical context—the relationship of television and radio to other nonelectronic media. This is especially true when we consider television used for formal schooling purposes, but it is also true in discussing public broadcasting.

Audio-visual Media in the Schools Television is part of the total media picture in the school. It is a major factor in the "new technology." In some simple local applications, School TV can be seen as "just another" audio-visual tool. Yet, at the same time, television encompasses all other media—such as films, audio tapes, print, overhead projectors, charts, the podium, models, and slides.

As a medium of production, television can integrate all other AV materials and media. As a means of dissemination, it can transmit all other AV materials. Television cannot be considered in isolation. Increasingly, television must be viewed as one approach—an important approach—in a multifaceted, multimedia "systems" educational program.

Even today, it is difficult to draw a valid distinction between School TV, audio-visual projects, and library concerns:

Is the inexpensive portable video recorder an ITV or AV tool?

Is the distribution of an AV film by means of a school closed-circuit system an example of School TV or has the television system become an AV device?

Who should be in charge of the development of audio and video media carrels for individual student retrieval of materials?

Is not the overhead TV camera merely an extension of other AV techniques?

Under whose jurisdiction falls the development of regional storage and retrieval systems using television as a means for disseminating printed information?

With considerations such as these in mind, we can underline the futility of drawing arbitrary distinctions among various media—television, books, films, audio tapes, charts, filmstrips, pictures, pamphlets, chalkboard, magazines, and so forth. In Chapter 12 (specifically, Section 12.8), we further emphasize the importance of considering all public broadcasting, commercial radio and TV, and School TV as one integrated telecommunications picture.

"Functions" Organization In most schools, an administrative distinction divides media support services according to the classification of the media, that is, a school has a library and an audio-visual center and an ITV program. We might call this type of approach a vertical or "media-category" organization.

Perhaps a more realistic educational and administrative organization could be made in terms of the functions or processes of various instructional and media services (rather than among the types of media themselves). Such a horizontal or "functions" approach would include the following categories:

Instructional Planning: The first function would encompass analyzing a particular instructional problem and designing an appropriate solution for it—which might involve any combination of media and services.

Acquisition: This section would be responsible for buying and leasing printed materials, 8-mm film loops, television materials, audio tapes, pictures, slides, records, and all other media.

Production: This unit would be charged with producing those media materials otherwise not obtainable—printed materials, video programs, films, audio recording, and similar items.

Storage and Retrieval: This function would include all the traditional library systems of cataloging and housing all the media listed above.

Distribution: This service would be in charge of getting the materials to where the interaction with the learners occurs—open-circuit and closed-circuit TV systems, film libraries, bookmobiles, resource centers, and the like.

Utilization: This function would directly assist classroom teachers and/or learners in the proper implementation and utilization of all media and materials.

Evaluation: This final function, of course, includes all the pertinent feedback services and more formal research activities.

This type of "functions" organization could lead to the true integration of all "carriers of knowledge" in a "systems" approach to education. The reader—and the potential user of telecommunications media—should begin to think in terms of this type of total integration of the media. Television and radio cannot be viewed in isolation—as separate and distinct entities that can somehow perform educational miracles. (This approach is explored further in Chapter 13.)

This orientation has presented the concept of "telecommunications used for educational purposes" from several different perspectives—finding meaningful definitions and distinctions, a theoretical examination of communication and instructional models, a realistic comparison of educational telecommunications and print, and a functional integration of media technologies.

1.8 Historical Themes and Supporting Groups

Examination of the historical heritage of educational telecommunications reveals that there were several major obstacles and problems. Certain favorable conditions had to be created before any substantial progress could occur. There had to be sources of funding; there had to be acceptance of the media as educational tools; there had to be a legal basis for the educational use of the airwaves; and there had to be general public support. As the story of educational telecommunications unfolds, we encounter several groups of persons who played significant roles in creating the conditions that could overcome these obstacles. We can divide these individuals and groups according to their institutional settings.

Educators To begin with, there had to be tangible support by the educational establishment itself—not only financial assistance, but also acceptance (initially) of the broadcast media as vehicles for educational material. Educators had to feel a need for radio, and later television, before educational broadcasting could become rooted.

Private Foundations Second, there had to be substantial moneys available from private sources other than traditional educational institutions. Someone—other than tax-supported agencies—would have to be willing to invest the initial capital (with no consideration of realizing any immediate results) in order to get a totally new and untried venture off the ground.

Government Educational broadcasting could not develop without some provision for separate channels to be reserved for educational purposes—official authorization for the existence of noncommercial stations that could come only from the federal government. Also, tax dollars—both at the federal level and at state and local levels—would eventually become a primary source of financial support for educational media.

Lay Public Fourth, there had to be general public acceptance—lay support—for the concept of educational broadcasting. Before educational telecommunications could amount to anything more than a feeble source of intellectual debate or an occasional classroom tool, it would need considerable support and encouragement from the public at large.

Professionals in Educational Telecommunications Finally, the story must be told in terms of the educational radio and television professionals who molded and developed the entire educational broadcasting and instructional telecommunications movements. Unlike the four previous groups, of course, the ETC professionals do not exist outside of the realm of public broadcasting and instructional media. Their pro-fessional roles evolved as the four preceding groups emerged: the educational radio broadcasters and the ETV pioneers became the main dramatic characters in the unfolding story of the ETC movement.

In fact, the background and historical material included in the following three chapters can be described as a continuing drama—the first half-century of the educational telecommunications movement. The public TV/instructional TV story is, if nothing else, an exciting tale of conflict, suspense, climaxes, and timely interventions. However, the final resolution has yet to be unveiled. Like a continuing soap opera—or the PTV series, "Upstairs, Downstairs" on *Masterpiece Theatre*—the final scene is never to be written (which is part of the excitement of the field of educational telecommunications).

Continuing Themes Underlying the historical drama of the educational broadcasting movement—and featuring the five groups of players identified above—we could trace seven different themes or dramatic threads that recur throughout the ETC story. With varying emphases, these seven story lines are intertwined throughout the next three chapters (and picked up periodically in the rest of the book).

1. *Use of the Broadcast Media by Educators:* From the earliest days of educational radio we can trace the evolution of instructional uses of the electronic media in formal learning situations. These school-related uses form the basis for much of the development of the ETC movement

2. *Development of the Professional Associations:* From the very outset of educational radio activities, the educational radio operators—and, later, other media specialists—have banded together to form professional organizations and trade associations to further their interests. These professional groups have continually been at the center of all ETC development.

3. *Evolution of the Educational Radio and TV Networks:* In addition to forming various professional groups, the educational broadcasters worked to establish and then to improve the radio and ETV networks. Much of the history of educational broadcasting is revealed in these network developments (which often overlapped with the activities of the professional associations).

4. *Role of the Private Philanthropic Foundations:* Without the early support of private charitable foundations and organizations, educational broadcasting—especially ETV—never could have been established the way it was. Although foundation support plays a less important part now than it once did, the crucial role it played in the formative years cannot be disputed.

5. *Federal Authorization and Regulation:* The federal government plays an important part in the ETV story in at least two ways. The first of these is the role taken by the Federal Communications Commission (and its predecessor, the Federal Radio Commission) in allocating and reserving certain radio and television channels to be used exclusively for noncommercial purposes.

6. *Government Financial Support:* The second way in which the federal government— as well as state and local governments—has played an increasingly important role is with direct financial aid for educational media. This important support is introduced in Chapter 4 and is continued throughout Part Two of the book—as federal support becomes more important (and controversial).

7. *Lay Support of Public Broadcasting:* The last group of persons to assume a major responsibility in the ETC drama was the public itself, the lay leaders who were to assume a vital role in the evolution of educational broadcasting. Although virtually nonexistent in early educational radio, the lay public today is as integral a part of the ETC structure as any of the other elements listed above.

These seven themes—often overlapping, sometimes obscured—are the primary subplots we will be following, then, as we examine the common heritage of public broadcasting and instructional telecommunications.

**Legacies
of Early Educational
Radio and TV**

Today's achievements in educational telecommunications might be traced back through various historical avenues—the development of still and motion picture media, the audio-visual education movement, the successes and crises of the institution of public education itself. However, more than any other historical element, open-circuit radio broadcasting is the most appropriate starting point for the story of "telecommunications used for educational purposes."

2.1   Early Educational Radio Stations

If we were to examine the background of ETC as a historical drama, Act One would be the story of educational radio—an act that (with cutaway scenes to early television activities) would last about thirty years.

In 1917, three years before KDKA (Pittsburgh) and WWJ (Detroit) were to start their conflicting claims as the first radio station on the air, another Midwestern radio station was inaugurated—with little fanfare or champagne. In Madison, the University of Wisconsin started experimental broadcasts over its radio station, 9XM (later to adopt the call letters WHA). However, the first educational institution officially licensed to operate a radio station was the Latter Day Saints' University of Salt Lake City, Utah, "on an unlisted day and month in 1921."[1] Both the University of Wisconsin and the University of Minnesota received their official broadcast licenses on January 13, 1922.[2] Early transmitting and studio facilities are shown in illustrations 2-1 and 2-2. The third picture (2-3) shows what was involved in an early remote broadcast of a non-studio event.

This was the beginning of educational broadcasting—a phenomenon that would employ thousands of persons and cost close to a billion dollars during the next half-century.

[1]S. E. Frost, Jr., *Education's Own Stations: The History of Broadcast Licenses Issued to Education Institutions* (Chicago: University of Chicago Press, 1937), p. 178.
[2]*Ibid.*, p. 464.

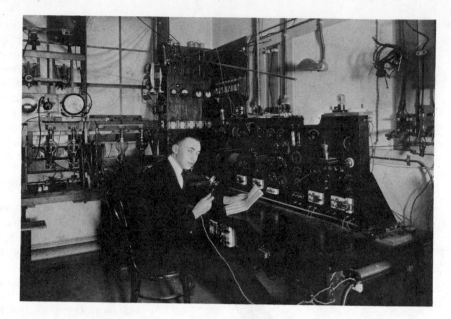

2-1 Transmitting Equipment Used at Station 9XM, University of Wisconsin, Circa 1922 (Courtesy: WHA, University of Wisconsin)

Scores of other early educational radio stations followed these pioneers, many of them activated by electrical engineering departments of prominent universities—designed primarily to experiment with this new gadget, radio. The early flurry of activity was something less than a success story; most of these stations remained on the air less than three years.

The Early Failures   In the mid-1920's, only half of the educational institutions that held broadcast licenses actually had stations on the air. By the mid-1930's, the percentage of active licensees had been still further reduced; up to 1936 a total of 202 broadcast licenses had been issued to educational institutions, but only thirty-eight educational AM radio stations were actually on the air.[3] Not until the late 1940's and early 1950's—with widespread utilization of FM frequencies—did educational radio begin to fulfill its early promise.

Several explanations can be given for the

failure of so many early educational radio experiments—all of which provided valuable lessons for the later television pioneers.

First, many of the early radio stations were faced with insurmountable financial problems. Some of the stations were forced to discontinue their operations even before 1929, but after the Wall Street crash many stations found themselves out of money.

Second, many of the early stations never progressed beyond the point of engineering experiments. Once the experiment proved successful, or once the novelty wore off, the projects were abandoned.

Third, the *commercial* radio interests that began stirring in the mid-1920's provided early educational radio with formidable opposition. In many instances, the educators could not afford, either financially or legally, to contest disputes over broadcast channel assignments—whose profit value was rapidly becoming apparent.

Fourth, educational broadcasters were severely hampered by many rulings of the Federal Radio Commission, established by Congress in

[3]*Ibid.*, pp. 1–5.

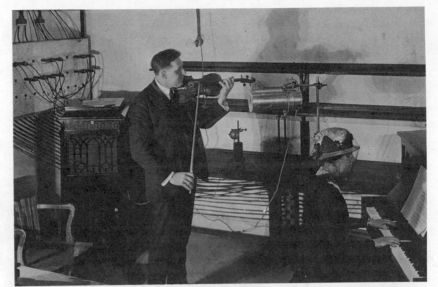

2-2 Studio of Station WHA, University of Wisconsin, 1923 (Courtesy: WHA, University of Wisconsin)

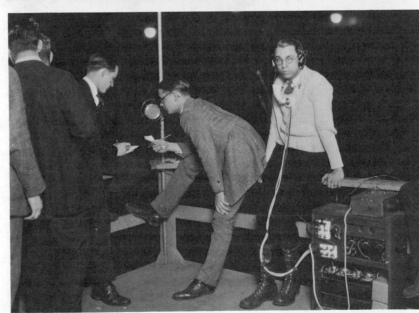

2-3 Remote Broadcast of 1922 Basketball Game, WHA (Courtesy: WHA, University of Wisconsin)

1927. The FRC, charged with assigning channels in the "public interest, convenience, and necessity," was justifiably hesitant to assign valuable air space to educators who were seldom sure that they were going to be able to offer a true program service for listeners (Section 2.6).

2-4 Typical University Radio Studio of Early 1930's (Courtesy: WHA, University of Wisconsin)

Finally—and most basically—the educators themselves were slow to see the real possibilities of the new medium. "Regardless of the explanations offered, it is clear that educators were generally apathetic toward educational broadcasting."[4] They failed to see its educational potential; they failed to grasp its public relations value. As one educational broadcasting pioneer stated more than twenty years later: "The fact was that in almost no instance had these educational radio stations been accepted as major elements in the administrative structures of their universities—the stations were peripheral to the main business of the institution."[5]

Reviewing the situation the early educational broadcasters faced, we can easily see that they had neither the resources, the background, nor the organization to solve many of their problems. Thirty years later the television pioneers were to succeed partially because of the rich experience gained in the early struggling days of radio.

The educators—administrators, teachers, curriculum specialists—were the first group to take its place on the stage for educational radio broadcasting; but the entrance was to be slow and cautious.

The Early Successes   The few early educational radio stations that managed to stay on the air did so because they were fulfilling a well-defined need; each was integrated into an ongoing educational program of a university. These were primarily the stations operated by the Midwestern land-grant colleges—where radio was seen as an obvious means of delivering education to the sparsely populated rural areas

[4]Paul Saettler, *A History of Instructional Technology* (New York: McGraw-Hill Book Company, 1968), p. 205.

[5]Richard B. Hull, "Consider Basic Problems," *AERT Journal*, December, 1956, p. 7

served by the agricultural extension divisions of these universities. Radio was more than just something for electrical engineering departments to experiment with.

During the late 1920's and into the harsh financial climate of the 1930's, many educational radio operations were kept functioning as "schools of the air"—both noncommercial and commercial. One of the first of these was the Ohio School of the Air, founded by Ben H. Darrow, with support from the Payne Study and Experiment Fund of New York. This fund was importantly instrumental during the next few years in the support of both educational radio and other audio-visual projects and associations (Section 2.4).

The Ohio School of the Air, which was maintained under the auspices of the Ohio State Superintendent of Public Instruction, made its debut over commercial station WLW in Cincinnati on January 7, 1929. Although it received some additional outside support, the Ohio School of the Air suffered several cutbacks in state funds and finally ceased operations entirely in 1937. Other early "schools of the air" were established by the Universities of Wisconsin, Kansas, Michigan, Minnesota, and by Oregon State College.

The major commercial networks offered comparable school radio services. The RCA Educational Hour was launched by NBC on October 26, 1928. Labeled the "most ambitious music education program ever undertaken," it featured the famed Walter Damrosch and the New York Symphony Orchestra.[6] The series continued on the air until Damrosch retired in 1942. The comparable offering from CBS was the American School of the Air, which ran from 1930 to 1940. These network programs, plus commercially sponsored series such as *The Standard Oil School of the Air,* were an important part of the early educational radio scene.

2.2 Evolution of the Audio-visual Movement

While the early university-based educational radio stations were struggling for support and recognition, a parallel educational movement had been slowly securing a foothold in the nation's schools.

The use of lantern slides and stereographs, dating back to the 1880's, gave rise to the term "visual instruction" in scattered schoolrooms across the country. By 1910, some schools had begun to adapt films "for regular instructional use."[7] By 1913, Thomas Edison was moved to a somewhat overenthusiastic prediction on behalf of visual instruction and technology:

> Books will soon be obsolete in the schools. Scholars will soon be instructed through the eye. It is possible to teach every branch of human knowledge with the motion picture. Our school systems will be completely changed in ten years.[8]

In the 1920's, "educational films" came from a variety of sources: old theatrical films (retitled and somewhat edited), advertising films, government films, welfare films from large corporations, health films, and a small number of instructional films actually prepared for classroom use. Questions could be raised about the quality of many of these films and about the manner in which they were integrated into classroom activities; nevertheless, the concept of visual aids was gradually growing.

Before the end of the Roaring Twenties, the visual instruction movement could point to several milestones: some colleges and universities were offering credit courses in visual instruction; the first V.I. professional groups were formed; the first professional V.I. journals were

---

[6]Saettler, *Instructional Technology*, p. 199.

[7]*Ibid.*, p. 98.
[8]New York *Dramatic Mirror*, July 9, 1913, quoted in Saettler, *Instructional Technology*, p. 199.

started; the first systematic V.I. research studies were reported; and the earliest V.I. administrative bureaus were established in public schools, colleges, universities, and state departments of education.[9]

2.3   Formation of the Professional Associations

For the next several decades, the educational broadcasting movement and the audio-visual movement developed along similar lines. The former was concerned with open-circuit broadcasting to schools and the at-home public in general; the latter concentrated on classroom uses of all audio and visual media. Each movement developed its own professional associations to further various objectives and goals.

Occasionally the two movements, through their associations, found their paths crossing. They often had common interests and purposes—especially in the area of classroom utilization of broadcast materials—and sometimes found themselves working in tandem (and occasionally at odds).

Although we are primarily concerned here with the story of educational broadcasting, from a historical standpoint it may be well to look first at the development of the professional association of the audio-visual movement.

Development of DAVI/AECT   The National Education Association (NEA) has long been the most prominent organization representing the educational establishment in the United States. It was only natural that it would play a major role in furthering the audio-visual movement. One of the earliest audio-visual groups established was the National Academy of Visual Instruction (NAVI), which was formed in 1920 at one of the NEA meetings. It was to be concerned primarily with standards concerning the use of film in schools.

Another early audio-visual group was formed in 1922 as the Visual Instruction Associa-

tion of America (VIAA). This agency was primarily involved with promoting visual instruction and setting up V.I. demonstrations.

The next year, 1923, the NEA established its own Department of Visual Instruction (DVI) as the official agency for the national association. For several years the three groups engaged in somewhat similar activities—although the VIAA was centered in New York City and the NAVI was identified with "college and university extension divisions from the midwestern region of the United States."[10] (Note the interesting parallel between the NAVI and the activity of early educational radio stations in the same area.)

Finally, in 1932, the two earlier groups merged with DVI to become the one dominant professional voice for the audio-visual movement in the nation. It was not until 1947 that the Department of Visual Instruction added "audio" to its title, becoming the Department of Audio-Visual Instruction (DAVI). At that point, the NEA furnished the DAVI with an executive secretary and a full-time paid staff, providing the agency with even more prestige and national exposure.

In 1970, the organization became independent of the NEA and reformed itself as the Association for Educational Communications and Technology (AECT). The new name and stature reflect a growing concern with the total educational experience from a "systems" or instructional technology viewpoint—as opposed to a limited equipment-oriented perspective connoted by the older term "audio-visual."

With more than 10,000 individual members, the AECT is the professional focal point for formal instructional applications of mediated instruction, programed instruction, computer-assisted instruction, educational systems, and related approaches. It is active in publications, conferences, research projects, placement services, materials cataloging, and similar professional services.

The AECT is organized into special inter-

---

[9]Saettler, *Instructional Technology*, pp. 119–120.

[10]*Ibid.*, p. 129.

est divisions—the oldest of which is the Division of Telecommunications (DOT). The membership of DOT is primarily concerned with classroom uses of television, radio, and other electronic media.

Formation of the ACUBS At the same time that the audio-visual movement was struggling to find its organizational structure, the early educational radio movement was undergoing a parallel process.

As Secretary of Commerce, Herbert Hoover was instrumental in assembling early commercial and university broadcasters for a series of national radio conferences. At the fourth of these, held in Washington, D.C., in 1925, the representatives from the college- and university-owned stations recognized that they had certain problems and needs that were unique to their type of noncommercial operations. So these "educational broadcasters" decided to form their own organization, the Association of College and University Broadcasting Stations (ACUBS).

Despite relatively moderate dues, the association was plagued from the start with membership problems, dropping to twenty-two active members in 1929. Throughout the early 1930's, the ACUBS never listed more than half of the educational radio stations as members.[11] In this early period, the ACUBS was a loosely knit organization, composed of representatives from a few stations (representatives who were primarily concerned with their own stations—not with the association), operating with an inadequate budget, held together only by the determination and foresight of a few leaders who saw the potential advantages of a national organization.

In 1930, the ACUBS held its first annual convention, attended by representatives from twenty-five stations. Out of this convention came

what was probably the first call for the federal government to reserve a few selected broadcast channels for noncommercial (or educational) use. The ACUBS sent a telegram to the annual state governors' conference (which was meeting at the same time), urging their support for congressional legislation assuring "reservation of channels for broadcasting stations owned and operated by the states and by colleges and universities."[12] However, no federal action in that direction was taken for several years (Section 2.6).

Held together by the need for concerted action on such common problems as finances, commercial opposition for radio channels, and lack of sympathy from the Federal Radio Commission (FRC), the small number of ACUBS pioneers also decided to take such steps as appointing an unpaid executive secretary and initiating a regular news bulletin. All these embryonic gropings were important as they foreshadowed later developments and tactics that were to play various roles in the educational broadcasting movement.

Transition to the NAEB In general, however, the situation seemed increasingly bleak for the ACUBS and for all educational broadcasting during the early 1930's. Money problems were increasing, membership was decreasing, the FRC did nothing to encourage noncommercial radio, and educators themselves were turning more frequently to commercial broadcasters for the opportunity to present their programs over adequate facilities.

> These developments were discouraging to some of the ACUBS leaders. They felt commercial broadcasting was becoming more dominant and that educational radio per se was in imminent danger of complete collapse. . . . Officers of the ACUBS felt that the Association should do something to bolster the weakening defenses, but did not know what to do, and they were worried.[13]

---

[11]Harold E. Hill, *The National Association of Educational Broadcasters: A History* (Urbana, Illinois: National Association of Educational Broadcasters, 1954), p. 5. This is an M.A. thesis copyrighted, mimeographed, and distributed by the NAEB.

[12]*Ibid.*, p. 6.   [13]*Ibid.*, p. 13.

At a special meeting called in desperation in the fall of 1934, the ACUBS changed its name to the National Association of Educational Broadcasters (NAEB), and it adopted a new constitution. Both of these changes were designed primarily to attract new members. The new NAEB included not only college- and university-owned stations but also any educational institutions that broadcast regularly over commercial outlets. There also was provision, under the new constitution, for individual memberships for any persons interested in the educational radio field.

Interestingly, this transition to the NAEB was made only two years after the three audiovisual groups had merged into a single association (DVI) in order to consolidate and strengthen the national AV picture. Despite its new face and renewed energy, however, the NAEB still found itself plagued with the same financial, membership, and federal-liaison problems that had haunted the ACUBS for nine years.

Evolution of the "Bicycle" Tape Network
Early in their struggles, the pioneer noncommercial radio men recognized the need to find some way to exchange programing ideas and actual program segments. At first, talk centered around the idea of a script exchange, whereby radio producers could send copies of their more successful scripts to their colleagues at other stations. This attempt, however, met with less than enthusiastic support, and—although it did function for a while—it never did provide a substantial portion of programing support for radio operations.

Still, the educational broadcasters realized that they could never be expected to produce all of their programing at the local level. It simply was too formidable a task for each individual broadcaster to program everything from his own studio. The educational radio broadcasters could not help but cast an envious eye at the programing and networking advantages that their commercial counterparts were enjoying with the Red and Blue Networks of the National Broadcasting Company

and the early success of the Columbia Phonograph Broadcasting System. As early as 1930, the ACUBS members had "envisioned a network of educational stations which would enable the stations to present better programs and to compete on a more even basis with commercial broadcasters."[14] However, they would have to wait almost forty years before that dream could become a reality.

It was obvious from the start that the operation of a live network was prohibitively expensive and educational radio could not afford such a luxury, so most talk began to center around the idea of recording programs and then exchanging the recordings. As early as 1932, consideration was given to the possibility of building a wire recorder that could be used by member stations on a round-robin basis. Such was the kind of early thinking that led eventually to the complex arrangements of tape exchanges, video libraries, regional networks, "bicycle" networks, ITV centers, and similar arrangements that are so much a part of the ETV distribution scene of today.

Nothing was actually accomplished, however, until 1936 when a $500 grant from the National Advisory Council on Radio in Education (Section 2.4) allowed the NAEB to purchase a wire recorder.

Gradual improvements over the years led to tape recording and, finally, to the beginning of the NAEB "bicycle" network in 1949 (so labeled because the individual radio tapes are circulated or "cycled" from one station to another—usually by mail—on a scheduled round-robin basis before being returned to the network center). Seymour Siegel, director of New York's municipal station, WNYC, made five sets of recordings of the 1949 *Herald Tribune Forum* and distributed them to twenty-two NAEB member stations that year. It is generally acknowledged that this was the beginning of a real, viable radio tape duplication and distribution networking operation.

[14]*Ibid.*, p. 9.

2.4   Private Foundations and Other Early Organizations

Up to this point, the narrative of ETC has been largely concerned with the role of college and university educators in operating the first noncommercial stations—and with the emerging leadership of the educational radio movement (primarily in the development of the professional associations). But, as other organizations surface to perform their unique parts, we will also witness the increasing involvement of private philanthropic foundations.

Although the ACUBS was the first association formed to deal specifically with educational broadcasting—and the NAEB remains today as the chief professional association for the field—other organizations, agencies, associations, institutes, councils, committees, centers, corporations, commissions, and the like have been formed over the years. Indeed, observers often comment derisively about the "alphabet soup" of organizations connected with educational broadcasting. However, each of these groups was formed to answer a particular felt need; each organization is important for its contribution to the legacy of educational broadcasting.

One of the earliest of these groups was the Preliminary Committee on Educational Broadcasting (PCEB), which was organized at the 1927 meeting of the NEA department of superintendence, which had set up the National Academy of Visual Instruction seven years earlier. One primary purpose of the PCEB was to determine the feasibility of the concept of a national "school of the air." Several such operations were subsequently established within the next two or three years—usually in conjunction with commercial stations or networks (Section 2.1).

The PCEB also requested the Secretary of the Interior, Ray Lyman Wilbur, to appoint a high-level committee to study the possibilities of instructional radio. Subsequently, in June of 1929, Wilbur established the Advisory Committee on Education by Radio (ACER)—comprised of educators and commercial broadcasters.

The Payne Study and Experiment Fund The ACER was set into operation with financial backing from two philanthropic foundations—the Payne Study and Experiment Fund and the Carnegie Corporation.[15] Note that this was the same year that the Payne Fund also had underwritten the Ohio School of the Air.

The Payne Fund, operating out of New York City, was probably the first major private philanthropic foundation to underwrite various educational media projects. The earliest record of the fund's support for educational media was, perhaps, a 1927 two-year grant of $65,800 to the Committee for the Study of Social Values in Motion Pictures for the purpose of studying theatrical films and making arrangements for excerpts that had positive classroom potential. The Payne Fund was also involved in several other early educational radio activities (as noted below).

Although it remained in existence for only a year, the ACER had a substantial impact in determining the direction of early educational radio. Recognizing that not all educational needs could be met by commercial stations and networks—regardless of how well intentioned they were or how closely they cooperated with educators—the Advisory Committee, in its formal 1930 *Report,* urged that "some air channels should be reserved exclusively for educational purposes."[16] This conclusion undoubtedly had a stimulating effect on the ACUBS which, as noted above, called for such reservations later that year. As might be expected, the commercial

---

[15]The term "corporation" in the title of the Carnegie philanthropic organization may cause some confusion. The word "corporation" in this context simply indicates that the organization has nonprofit *corporate* status (as does almost every similar group). Whether a philanthropic agency is labeled a "foundation," "fund," or "corporation," the purpose, structure, and operations will be similar to other like organizations.

[16]Saettler, *Instructional Technology,* p. 207.

broadcasters did not unanimously concur with this majority report from the ACER. The network representatives especially disagreed with this viewpoint—and the RCA Educational Hour and the CBS American School of the Air were offered as evidence of the great educational service that could be supported only by a viable nationwide commercial network.

The report of the ACER also opened up a split among the educational broadcasters themselves. Some had worked out quite satisfactory relationships with local commercial stations (as well as with the networks) and did not want to jeopardize the status quo, which furnished them with a partial outlet with a minimum of financial investment. Others, representing the spirit of the ACUBS (an association of educational broadcasting *stations*), argued that commercial stations alone could never do the job; the needs of education would become far greater than could ever be met by public service time donated by commercial stations. Also, they argued that if educators were ever frozen entirely out of the stations and operations field (a trend which many felt was being pursued by the Federal Radio Commission), then the commercial stations—unpressured by the threat of any educational stations coming on the air—would gradually lose their incentive to demonstrate any public spirit and their "altruistic" interest in educational programing would soon dissipate.

This split among the educational broadcasters led, in turn, to the creation of other associations.

National Advisory Council on Radio in Education   The idea for the National Advisory Council on Radio in Education (NACRE) grew out of some early recommendations from the American Association for Adult Education. With support from the Secretary of the Interior and William J. Cooper, the U.S. commissioner of education (and with the financial backing from the Carnegie Corporation and from the Rockefeller Foundation), the NACRE was established in July 1930.

Composed of forty individuals—representing the interests of government, educational agencies, commercial broadcasters, and others—the National Advisory Council adopted many generalized goals; but it existed primarily to foster a closer relationship between educators and commercial broadcasters.

Although its most notable success was probably a long series of educational broadcasts carried over the NBC networks, NACRE also was active in many other areas that helped to pull all educational broadcasters in a unified direction: research undertakings, demonstration projects, conferences, and the like. As noted above, the council also was responsible for the grant that enabled the NAEB to purchase its first wire recorder.

The NACRE was disbanded in 1938 when the Rockefeller Foundation withdrew its financial backing. (The disbandment of agencies or projects when sponsoring foundations change their priorities is an occurence that we will see repeated later.)

National Committee on Education by Radio   The National Committee on Education by Radio (NCER) was established partially in reaction to the commercial rapproachement inherent in NACRE. The NCER also was called together by Education Commissioner Cooper, but it was composed of individuals who were concerned with the very survival of educational radio stations—along with other lofty motives. Although cooperation with commercial stations and networks was desirable, the more urgent tasks were guaranteeing the support of noncommercial stations and stimulating educators to get more of their own stations on the air. The National Committee, supported by an initial five-year grant of $200,000 from the Payne Fund, held its first meeting in January 1931.

The resemblance of the NCER to the later Joint Committee on Educational Television is very close (Section 3.4). The first members of the NCER included the ACUBS, the Land Grant Colleges Association, the National Association of State University Presidents, the NEA, the National University Extension Association, the Payne Fund, and other "similar organizations."[17] The composition of this committee reflected the concern with radio's potential as an instrument of rural adult and agricultural education, supported by the Midwestern land-grant colleges and state universities; the NCER was geared primarily toward the traditional educational establishment. Its first chairman was Joy Elmer Morgan, the representative to the committee from the NEA, and the NCER soon found itself housed by the NEA.

The NCER lasted until 1941 as a powerful vehicle for unifying educational radio activities and concerns. Because it was directly connected with schools on a national level (through the NEA affiliation), its policies influenced commercial broadcasters, the Federal Communications Commission, and educators themselves. The commercial broadcasters were goaded by the NCER to develop new forms of school broadcasts on a national scale. The FCC was prodded to live up to the requirements of the Communications Act of 1934, which required the FCC to study the question of assigning "fixed percentages" of radio frequencies to education (Section 2.6). Educators were challenged to produce programs that fit school curriculums and generally to *accept* radio as an important tool for education.

> Educators . . . altered their approach to educational radio. Radio research became an established activity. Colleges and universities began incorporating radio courses as part of their regular program so that the industry could be assured of properly trained personnel in order that teachers might utilize radio more effectively in their instructional procedures. Moreover, public and private schools began to make wider use of radio and many began to apply for their own frequency modulation licenses.[18]

Other Associations Other associations organized during this period contributed materially to the cooperation and the professional growth of the entire field. One of these groups was the Institute for Education by Radio (IER), which held its first session on July 1 and 2, 1930, at The Ohio State University. Since that time, this institute—and its successor, the Institute for Education by Radio and Television (IERT)— has been one of the important annual gatherings in the educational broadcasting field. The IERT program awards given each spring are eagerly sought after by both commercial and noncommercial broadcasters in this country and abroad.

In a move reminiscent of the spirit of the fading NACRE, the FCC, in May of 1935, established the Federal Radio Education Committee (FREC) "to combine forces with the educators on the one hand and the broadcasters on the other, within the present American broadcasting structure."[19] The FREC—consisting of forty individuals from broadcasting and education—was headed by John W. Studebaker, U. S. commissioner of education. The FREC was created, to some extent, as a belated acknowledgment by the FCC that the federal government had not done much to encourage the use of radio as a means of education. "National conferences were held, studies made, reports and newsletters published. However, by the early forties, FREC had become only a relic of an idealistic compromise between educators and the [commercial] radio industry, and it has long since faded away."[20] There was as yet no lasting arrangement that would bind educators with the commercial broadcasters.

One other organization should be mentioned in this listing: the Association for Education by Radio (AER). To some extent, the AER was an outgrowth of the NCER. It represented the same kinds of thinking, the same philosophy, as the National Committee—but it was to be more of a grassroots membership organization. Conceived

---

[17]Hill, *National Association of Educational Broadcasters*, pp. 8–9.
[18]Saettler, *Instructional Technology*, p. 219.

[19]Frost, *Education's Own Stations*, p. vi.
[20]Saettler, *Instructional Technology*, p. 224.

in 1939 and founded at the 1940 IER convention in Columbus, the AER—like the NAEB—was an association of, and for, professionals in educational radio, but the AER was intended to be a much broader organization than the NAEB. It included school personnel, educators concerned with proper media utilization in the classroom, teachers of radio training, and so forth. The two groups, however, obviously overlapped in purpose and membership, and in the mid-1950's the AER-T (with "Television" appended to its name) merged with the older NAEB.

Ford Foundation   Finally, we should take note of one other organization that was established in 1936. In Detroit, with very little fanfare or national notice, was created a relatively small, local philanthropic agency—the Ford Foundation. It was established with an initial endowment of $25,000 in order "to receive and administer funds for scientific, educational, and charitable purposes, all for the public welfare."[21] Its early activities consisted mostly of support for local organizations that were looked upon with favor by the Ford family: the Henry Ford Hospital, family charities, and the like; but its operations were gradually expanded as more Ford Motor Company stock was added to its capital, and by 1948 the foundation was averaging $2 million a year in gifts.

Initially, it had nothing to do with educational broadcasting; there is no evidence that the foundation was even aware of the existence of noncommercial radio for a decade or so. However, once the Ford Foundation entered the picture (Section 3.3), educational radio and television had a benefactor that overshadowed all others—a foundation willing and able to pour hundreds of millions of dollars into educational broadcasting.

2.5   Early ETV Activities

Meanwhile, as the educational radio pioneers were inching their way forward toward some sort of organizational framework and financial base, there were some early tentative attempts to add video to educational audio. Although there had been experimentation with television apparatus dating back to the 1920's (and earlier), it was not until the early 1930's that educational institutions got in on the act. Again, we see the importance of the first group of individuals we identified—the leadership of the educators themselves.

The First Experimental Station   Like many of the earliest educational radio "stations," the first experimentation in educational television was undertaken as part of university engineering projects. Between 1932 and 1934, the State University of Iowa, Purdue University, and Kansas State College (Manhattan, Kansas) all "actually produced experimental teaching programs using the scanning disc method."[22]

The State University of Iowa at Iowa City is usually credited with being the first to actually broadcast ETV programs and to have the most extensive early ETV set-up. The first educational TV program, presented on January 25, 1933, consisted of a violin solo, a dialogue, and a lesson in freehand sketching.[23] In a lecture in March 1933, E. B. Kurtz, head of the electrical engineering department and director of the ETV project, remarked, "Just one word on the place of television in our broadcasting system. In my opinion it is not only going to be a complement to radio but it is going to be radio's 'better half.'"[24] At the IER Convention a year later, Kurtz stated, "Television's place as an effective instrumentality in the educational system of the future is already as-

[21]*Business Week,* October 7, 1950, p. 30, cited in Bobbie Gene Ellis, "An Investigation of the Ford Foundation's Role in the Early-Stage Development of Educational Television in the United States" (unpublished M.A. thesis, University of Houston, 1958), p. 3.

[22]*Television in Education,* U.S. Department of Health, Education and Welfare, Office of Education (Washington, D.C.: U.S. Government Printing Office, 1957), p. vii.

[23]E. B. Kurtz, *Pioneering in Educational Television: 1932–1939* (Iowa City: State University of Iowa, 1959), p. 53.

[24]*Ibid.,* p. 46.

sured."[25] His predictions may have been a little premature, but some forty years later most educators probably would have admitted (perhaps grudgingly) that he was correct.

Between 1932 and 1939, Kurtz's station, W9XK, presented a total of 389 educational broadcasts—primarily on an experimental basis. Gradually, the university's experimental television projects turned to electronic scanning and the university's new station, W9XUI. The last telecast of W9XK was on June 29, 1939.

Educational Use of Commercial Stations
Before educators were ready to plunge fully into initiating their own television operations, there had to be, of necessity, a long period of cautious experimentation with the medium. This most conveniently took the form of extended cooperation with commercial broadcasting stations.

In 1942, one of the earliest pragmatic educational uses of the medium occurred—the training of air raid wardens in New York City, using the experimental RCA station. Shortly after the war, educational programs initiated in both Chicago and New York were produced in cooperation with the boards of education and local commercial stations.[26] In 1947, the Philadelphia public school system, under the immediate leadership of Martha Gable, began a weekly series of classroom instructional programs over three local commercial stations. Within a few years, the service included thirteen programs a week that were used by more than 60,000 Philadelphia elementary and secondary students.[27] Slowly and cautiously, elementary and secondary schools were beginning to take advantage of the opportunities offered by early commercial TV stations.

At the higher-education level, various records indicate that by 1948 there were at least eight colleges and universities involved in ETV production and/or programing: the American University (Washington, D.C.); Iowa State College (now University); the State University of Iowa; Kansas State College (now University); the University of Michigan; the University of Pennsylvania; Creighton University (Omaha, Nebraska); and The Johns Hopkins University (Baltimore).[28] Most of these institutions were involved in programing of an educational and/or public relations nature, generally packaging the programs on campus, then actually producing them at a local commercial station.

2.6   Reserving Radio Channels

Another institution involved in early educational radio (along with schools and private foundations) was the federal government—specifically in playing a role in guaranteeing educators that some broadcasting channels would be reserved for their use. However, this support and authorization was not easy to obtain. When the Federal Radio Commission was established in 1927, it could not afford to be generous with the would-be educational broadcasters.

In an effort to bring order into a situation that, in 1927, was complete broadcasting chaos, the FRC had to impose severe limitations on the broadcasting stations—commercial and educational—then in operation. These included power restrictions, broadcasting time limitations, new frequency assignments, and permanent elimination of many stations. In many of these decisions, the FRC did seemingly discriminate against educational broadcasters. "The FRC frequently restricted educational stations to low power, gave them poor frequencies, or in many cases where

---

[25]*Ibid.*, p. 76.
[26]William B. Levenson and Edward Stasheff, *Teaching Through Radio and Television* (New York: Rinehart and Company, Inc., 1952), p. 44.
[27]Beverly J. Taylor, "The Development of Instructional Television," *in* Allen E. Koenig and Ruane B. Hill (eds.), *The Farther Vision: Educational Television Today.* (Madison: University of Wisconsin Press, 1967), p. 137.

[28]Hill, *National Association of Educational Broadcasters,* p. 37, documents the first five of these examples. The other three are discussed in William Kenneth Cumming, *This is Educational Television* (Ann Arbor, Michigan: Edwards Brothers, Inc., 1954), pp. 14–15.

time was shared with a commercial station, assigned the least desirable operating time to the educational station."[29]

As mentioned above (Section 2.1), the FRC was hesitant to assign valuable spectrum space to college electrical engineering experiments. The first obligation of the commission was to guarantee that the radio spectrum was assigned to solvent broadcasting concerns that could continue over long periods to serve the public with professional programing. Another factor in assigning the radio frequencies to commercial broadcasters was that the commercial operators presented a stronger front to the commission than the educators were able to do; their clamor was louder; their cause was more fervently defended. One typical argument is the statement put forth by commercial station WOW, Omaha, in its application for full-time operation of the channel that was then being shared with educational station WCAJ:

> Our contention is that as a matter of principle educational programs should be given by stations having regular listening audiences . . . not sporadic audiences. It seems that service should be given from a station that has a general and not a specialized value. . . . [WCAJ] spends less in a year than is spent by WOW in one week. WOW desires to take the responsibility of rendering the service to the public.[30]

Educators found it difficult to argue persuasively against this type of presentation.

The relative inability of educators to counteract this commercial pressure and the resulting decisions of the FRC prompted one university extension dean to state that the FRC does not kill educational stations; the commissioners "merely cut off the arms, legs and head

of an educational station and then allow it to die a natural death."[31]

The First Requests for Reservations   However, as it became increasingly apparent over the years that commercial radio stations were not able to perform the kinds of sustained educational services that many felt were desirable, pressure slowly began to mount for some sort of federal action that would guarantee educators a certain share of the broadcast spectrum.

The first recorded official request for reserved educational broadcasting channels came out of the 1930 convention of the ACUBS (Section 2.3). A year later, the first bill was introduced in Congress for the reservation of channels; through the efforts of the NCER, Senator Simeon D. Fess of Ohio asked that 15 percent of all radio channels be reserved for education.[32] (The bill was not passed.)

During congressional debate preceding the establishment of the Federal Communications Commission in 1934, the question of reserving channels for education was again raised. An amendment to the Communications Act of 1934, offered by Senators Robert F. Wagner and Henry D. Hatfield, called for 25 percent of all channels to be set aside for education. Congress failed to approve the amendment (although almost half of the Senate voted for it) but did put a stipulation into the Communications Act requiring the newly formed FCC to investigate assigning nonprofit and/or noncommercial channels to education. The new FCC then initiated a series of hearings in which commercial broadcasters, educators, and the NAEB testified. The hearings provided the NAEB with valuable lobbying experience and helped to establish the need for a representative voice in congressional debate. This was to be a key role for the NAEB for the next four decades.

---

[29]Hill, *National Association of Educational Broadcasters,* p. 11.

[30]*Ibid.,* p. 8.

[31]Quoted in John Walker Powell, *Channels of Learning: The Story of Educational Television* (Washington, D.C.: Public Affairs Press, 1962), p. 31.

[32]Hill, *National Association of Educational Broadcasters,* p. 10.

The FCC was caught in a crossfire in its hearings. It did not want the reputation from its beginning of ignoring education's needs; but neither did it want to deprive the public of the fine commercial programing that was being aired during these "golden years" of commercial radio. Many in the arts sincerely believed that only a commercial system would be able to pay artists for creating and performing. For example, NBC was able to get a respected music educator, Howard Hanson of the Eastman School of Music, to say:

> The practical question must arise as to how radio channels, were they allocated for purely educational purposes, could be used. The proper use of such channels would entail huge subsidies, both for the building of stations and the engagement of artists. Even with such subsidies it is questionable whether facilities could be developed which would compare in any way with the present extensive broadcasting chains.[33]

Faced with strong commercial opposition, the FCC decided against reserving any radio channels strictly for educational broadcasting. It did, however, announce a major conference on educational broadcasting—which led to the FREC (Section 2.4). More importantly, the idea of setting aside a portion of the broadcast spectrum for educational uses had received publicity, and the educational broadcasters' request had become a legitimate cause.

The First Minor Victory   During the next three years, the pressure continued to increase and was applied quite frequently. In 1938, the FCC gave the educators their first taste of victory in the battle for channel reservations. The com-

mission set aside five channels in the UHF (ultra-high-frequency) portion of the broadcast spectrum—later these were moved down to the VHF (very-high-frequency) spectrum. These channels were originally intended for AM broadcasting, although they were in the range where FM would ultimately be developed. It was a relatively easy gesture for the FCC to make since there was little commercial pressure for space in the VHF band; all the action was in the MF (medium frequency) portion of the spectrum where standard AM broadcasting was located. However, this reservation was a precedent and a victory for the educational radio leaders.

At a meeting of the executive committee of the FREC in 1943, FCC Chairman Lawrence Fly issued a prophetic warning that was to become a favorite theme of the FCC commissioners—especially for the next decade. This was a period when the educational broadcasters were asking that more and more channels be reserved for their interests even though they could not very rapidly activate the channels already allocated to them. Chairman Fly warned that the five educational channels already set aside for the exclusive use of noncommercial educational institutions "were not set aside for absentees."[34] This warning reflects back to the earliest concerns of the FRC—fear that the educators would not effectively use the channels reserved for them. This type of warning to the educators had substantial influence upon the first year of educational television, nine years later.

The NAEB responded to Fly's warning by passing a resolution two months later asking for even more FM channel reservations. Its voice was one of many among the constituent groups of the NCER. Accordingly, the FCC announced a complete review of all radio allocations and in September 1944 started its hearings. The NAEB, NEA, the American Council on Education, and the

---

[33]National Broadcasting Company, *Broadcasting, Volume II: Music, Literature, Drama, Art* (New York: NBC, Inc., 1935), p. 51.

[34]Quoted in Anthony William Zaitz, "The History of Educational Television: 1932–1958" (unpublished Ph.D. dissertation, University of Wisconsin, 1960), p. 13.

U.S. Office of Education all played prominent roles in the testimony presented before the FCC.

Educational FM Channel Reservations in the 1970's   These hearings resulted in 1945 in the reservation of a total of twenty FM channels for educational use (including the five previously set aside). These channels, located between 88.1 and 91.9 mHz inclusively, remain the section of the FM spectrum reserved for noncommercial radio.

The FM radio channel reservations are on a block or "cluster" basis; that is, these twenty channels are reserved everywhere across the country—although specific assignments are not made to individual communities. Thus, thirty to forty educational FM stations from Seattle to Tallahassee may be operating on the same frequency, say 89.7 mHz. Nowhere has the FCC specified which cities would use that channel—only that that spot on the dial was reserved for noncommercial educational use. Thus, these twenty channel reservations can serve more than a thousand different stations.

In addition to these specific reservations, at least two important concepts emerged from the 1944–45 hearings and reservations: (1) there should be a nationwide table of allocations; and (2) these allocations should be reserved—some of them for noncommercial use. Both of these concepts were important as precedents when the ETV reservations were to be considered a few years later.

2.7   1948: The Year of the Freeze

So, with television on the horizon and with the first tentative support from the FCC, the NAEB began to show signs of new life. A sense of cautious optimism and reserved excitement began to permeate the meetings of the NAEB members. The first educational FM station, WBOE in Cleveland, had gone on the air in 1941. By the end of 1945, there were six FM educational outlets on the air—although there were still only

a total of twenty-four member stations in the NAEB. Growth was slow, tentative; educational broadcasters were not in a hurry to explore this new VHF region with the recently developed FM transmission. But there was *some* growth. For the first time in the two decades of its existence, NAEB members were confident that at least there would be a future.

The year 1948, in general, marked the awakening of educational broadcasters to the potentials of television. Many of them began to realize that they must move fast or television would be completely dominated by commercial interests. If a lesson was to be learned from their experiences in radio, it was that they had to make their claims early in the development of the medium. The *NAEB Newsletter* expressed this concern in May 1948:

> At the Ohio State [IER] Institute, many educational station operators were wondering "out loud" if their failure to apply for and plan for TV facilities might not find them later in a difficult position. High cost and uncertainty of future developments were given as reasons for delay.[35]

The NAEB in Transition   By 1948, the number of educational radio stations had doubled; there were now fifty stations located in thirty-one states—and there were almost a hundred individual members in the NAEB. FM broadcasting had succeeded in breathing new life into the field—and new blood into the association.

With the rapid growth of any organization or idea or project, there is always the danger that the Old Guard—the very ones who had pushed so hard for success and expansion—will suddenly find themselves in the position of defending the status quo against the onslaught of the newcomers who have recently joined the movement. So it was that in 1948 the NAEB found itself caught

[35]*NAEB Newsletter*, May 31, 1948, quoted in Hill, *National Association of Educational Broadcasters*, p. 5.

up in such a confrontation at its annual convention—which "marked the dawn of a new era for educational broadcasting and the NAEB."[36]

Although the NAEB was ostensibly an organization of stations, the individuals representing those stations were the ones who formulated the policies, waged the arguments, and made the decisions. The new members of the association represented new thinking in the educational broadcasting field; they represented the FM pioneers; they represented school system stations; they represented a wide diversity of geographical interests; they represented people who were thinking about television. The Midwestern, college-oriented, agricultural/extension-division radio leaders who had led the battle for twenty years suddenly found themselves in the minority of their own associaiton. The days of the ACUBS leadership were over.

Thus, *The Billboard* magazine described the 1948 Convention as

a fight within the NAEB, with the one element, the progressive force now in power, favoring the establishment of a network, hiring of an Executive Secretary [on a full-time paid basis], doubling of dues and other aggressive action. Still another group is satisfied with the status quo. . . . First skirmish was won by the progressive element when Dick Hull, head of radio for Iowa State College and its station WOI at Ames, was re-elected president, with the board of directors and membership okaying his expansion plans.[37]

In addition to Richard B. Hull, some of the other more progressive leaders who emerged from the 1948 Convention included Robert Hudson, director of educational broadcasts for CBS

(who would serve almost twenty years as the number-two man with the National Educational Television network), Seymour Siegel of New York (who the following year would initiate the NAEB tape network), and I. Keith Tyler of Ohio State University (who would be instrumental in setting up the Joint Committee on Education Television in a couple of years).

The Request for TV Channel Reservations
There were several areas of contention within the ranks of the NAEB, one of which was the position that the association should take regarding television. The newly identified "progressive" leadership of the group—pushing for reservation of TV channels—would play a significant role in the creation of the ETV structure over the next few years.

Since by 1948 the major VHF channels were already fairly well committed to commercial interests, the UHF band seemed to the NAEB leadership to represent the best opportunity for educational television. Because the educators had waited this long—and the choice VHF spectrum appeared already lost—"fears were voiced that perhaps education once again had done too little and was already too late."[38] Nevertheless, at the 1948 convention, the new NAEB leadership mustered the support for the adoption of a resolution urging the FCC "to reserve certain ultra-high frequency channels for educational and non-profit purposes,"[39] "And the battle over the resolution, though unrecorded, must have further separated the [television] pioneers from the stay-at-homes."[40]

This resolution was based upon the precedents of the 1938 and 1945 FCC reservation of FM channels. Following the pattern of these earlier reservations, the NAEB was thinking

[36]Hill, *National Association of Educational Broadcasters*, p. 39.

[37]*Billboard*, October 23, 1954, quoted in Hill, *National Association of Educational Broadcasters*, p. 39.

[38]Richard B. Hull, "The History Behind ETV," *NAEB Journal*, February, 1958, p. 28.

[39]*NAEB Newsletter*, November 1, 1948, quoted in Zaitz, "History of Educational Television," p. 17.

[40]Powell, *Channels of Learning*, p. 35.

in terms of having a solid chunk of the spectrum reserved, that is, an undivided block or "cluster" of adjacent channels (for example, channels 14 through 29 in the new UHF band of frequencies), rather than individual frequencies dispersed throughout the whole broadcast range.

So, the 1948 scene revealed a small handful of progressive NAEB leaders, looking forward to the potential of TV, crying out, "We've got to move . . . now . . . fast," while in the background the large group of stand-pat educational radio broadcasters was mumbling, "Well, yes . . . but . . . money . . . lack of support from boards of education . . . trustees . . . et cetera."

Freezing the Allocations   The unexpected phenomenal growth of commercial television following the end of World War II had made a shambles of the 1946 FCC Table of Assignments for television. Also, the commission was increasingly aware of the growing concern that education would once again lose out in the battle for limited frequency space. Another major issue facing the FCC was the need for the embryonic television industry to adopt a standard technical format for color television. So the commission, on September 10, 1948, issued the *Report and Order* that was to become known as the "freeze order," suspending all action on station licenses and applications while the FCC restudied its earlier Table of Assignments and debated the question of color TV format.

From the standpoint of the ETV interests, two questions were paramount in the commission's decision to halt all new station activation. First, to what extent should the UHF portion of the spectrum be utilized for additional channels in a revised table of allocations? And, second, what consideration—if any—should be given to education's need to have some channels reserved for its eventual use?

Undoubtedly, this FCC action (more properly, it should be identified as a cessation of action) is what made possible the eventual existence of educational television. Without the

freeze at this particular time, there would have been little opportunity later to come up with the designation of "noncommercial television."

2.8   Educational Radio's Legacies to ETV

By the end of Act One of our metaphoric drama, four of the five groups of players in our cast of characters had made their stage entrances. (And these four institutional groups represented five of the seven themes, or story lines, identified in Section 1.8).

1. *Educators* and educational institutions had become increasingly committed to the idea of noncommercial radio (and, soon, television). They were supporting the educational stations and participating in the decisions that would determine the future directions of educational telecommunications. (This corresponds to the first theme of "use of the broadcast media by educators.")

2. *Private foundations* had begun to open up. The Payne Fund, the Carnegie Corporation, and the Rockefeller Foundation had all made important contributions. Many more were to follow. (The fourth theme was "the role of the private philanthropic foundations.")

3. *The federal government,* represented by the FCC, constituted the third group. (The corresponding theme was identified as "federal authorization and regulation.") The 1938 and 1945 reservation of FM channels for noncommercial radio established the federal basis for saving some TV channels to be used for educational purposes.

4. *Professional educational broadcasters* appeared on the scene. In the early 1920's there were no leaders in educational broadcasting—the phenomenon did not exist. Twenty-five years later, the movement had many experienced leaders, who had learned to organize, to cooperate, to project into the future, and to build upon the

mistakes of others; and the movement had a viable professional association, capable of generating leadership and national exposure. (The second identified theme was "development of the professional associations.")

Finally, one more story line was started by the professional educational broadcasters (the third theme was the "evolution of the educational radio and TV networks"). The idea of program exchange was accepted. There was the growing realization of the need for a network headquarters and some method of program exchange. Although the idea of a (radio) network had received superficial support for a couple of decades, it finally was unanimously recognized that the concept of program exchange was absolutely indispensable for both radio and television.

**The
Formative Fifties:
1948–1962**

Throughout the brief history of educational broadcasting, one can point to certain pivotal years in which several developments converged. Such a year was 1948. This is the way the stage was set for Act Two of our ETC drama:

The National Association of Educational Broadcasters, revitalized and with aggressive new leadership, was pushing ahead with its eyes firmly focused on educational television horizons.

Numerous universities, working with early commercial stations, had compiled considerable experience in ETV programing and production.

The Federal Communications Commission had set a firm precedent by allocating certain radio channels to be reserved for the day when educators could activate them; and the commission had already frozen all new TV station applications in order to study the allocation of video channels.

The NAEB, firmly convinced of the need for networking programs, was on the verge of launching its "bicycle" tape network service to its affiliated radio stations.

Recognizing the value of a consortium of educational agencies such as the National Committee on Education by Radio, the educational broadcasters were looking toward assembling a similar pressure group for television.

One other principal actor was standing in the wings, waiting to make his debut—ready to play what was to be, perhaps, the most dominant role of the formative 1950's. Upon the death of Henry Ford in 1947, $250 million was left to the Ford Foundation. In order to avoid potential inheritance tax complications of mammoth proportions, the Ford family transferred other stock to the foundation; and by the end of 1947 it had a total of more than 3 million shares of nonvoting stock in the Ford Motor Company—more than $1 billion in assets. To determine how best to spend the income from this endowment, a special eight-man study group was set up to undertake an extensive review of the foundation's aims and policies. It would be three years before the resulting study would be ready and the foundation would take its place center stage.

Considering the far-reaching impact and importance of the decisions made during this period, the drama of the next four years or so would be, at times, a ponderous and slow-moving piece of stage action—with much of the activity taking place behind the scenes. It was the most crucial period that ETV has ever faced. Act Two, specifically, lasted from 1948 to 1952.

3.1 Early Freeze Activity

The FCC freeze was issued in September 1948; the following summer, July 1949, the FCC issued a *Notice of Further Proposed Rule Making,* which presented a new Table of Assignments. There were allocations for 2,245 stations in 1,400 different communities, but there were no reservations for any noncommercial educational TV stations.

The Protests of Frieda Hennock Only one of the seven FCC commissioners dissented from this 1949 *Notice*—Frieda Hennock. She vehemently pointed out the need to reserve a portion of the television spectrum for the educators—whether or not they were organized, well financed, or farsighted enough to know what they would do with the channels. Eventually the need would be there—but the channels would be gone, all assigned to commercial broadcasters.

Rallying around Commissioner Hennock, the educational broadcasters—now sensing the seriousness of their situation—were quick to support her position. Many educational organizations formally approached the FCC. In September 1949, Richard Hull, on behalf of the NAEB, filed a petition with the FCC requesting that ten adjacent channels in the *UHF band* be reserved for educational telecasting. This was the follow-

up to the NAEB resolution of the preceding year (Section 2.7).

Other educators and educational broadcasters—such as Harold McCarty, chairman of the NAEB Television Study Committee—felt that educators should also have the opportunity to use some of the *VHF channels*. The National Education Association, the U.S. Office of Education, and Commissioner Hennock were among those who also felt that the educators should receive some portion of the VHF spectrum. This difference of opinion represented a split in the ranks of the educational broadcasters—with the NAEB calling for UHF reservations, and others (NEA, USOE) calling for both UHF *and* some VHF channels.

Allerton House and the NAEB Awakening Meanwhile, coincidental with the action over the FCC consideration of reservations, two historic seminars were held at Allerton Park in Urbana, Illinois, during the summers of 1949 and 1950. These seminars, bringing together the leaders in educational broadcasting, were cosponsored by the Rockefeller Foundation and the University of Illinois Institute of Communications Research. During the 1949 session, a sober sense of the mission they had before them slowly dawned on the participants. "It seemed that suddenly a great truth had been revealed . . . the truth that educational [broadcasting] not only has a job to do, but it is capable of doing it."[1]

Although the Allerton House seminars were primarily directed toward radio development, the motivation and sense of excitement and responsibility instilled at the seminars were carried on into television. As Robert Hudson summed up the significance of the sessions,

> The Allerton House Seminars have been of immeasurable value to the cause of educational broadcasting. They provided stimulus when stimulus was needed, and they pro-

[1] Robert B. Hudson, "Allerton House, 1949, 1950," *Hollywood Quarterly,* Spring 1951, pp. 238–239.

vided a basis for coordinated action when combined strength was needed in seizing the opportunities which beckoned.[2]

Largely as a result of the Allerton House seminars, at the 1950 NAEB convention plans were presented for a permanent headquarters for the association's radio tape network. Dean Wilbur Schramm, of the Illinois Institute of Communications Research, proposed that the tape network be housed at Urbana, with the University of Illinois furnishing the space and some personnel, and the rest of the financing to come from specific grants. Six months later the Kellogg Foundation responded with a $245,000 grant to operate the network for its first five-year period. So the "bicycle" network finally had a home, a staff, and enough money to make it work.

3.2   University Television Activities

Meanwhile—on the television side of the stage—the pioneering universities continued to gather more experience in the new medium. The term "telecourse" was coming into general usage—referring to an educational program for which viewers could register and receive some sort of "certificate of participation." The University of Michigan was one of the earliest institutions to engage in this type of programing, incorporating the telecourses into its *University of Michigan Hour,* starting November 5, 1950, over the Detroit commercial station WWJ-TV.

Many other universities and colleges started offering similar telecourses over local commercial TV stations, following the same general pattern.

Early Credit Courses   Western Reserve University is usually given credit for the nation's first continuous series of full-credit college courses to be taught on television. It started telecasting its *Introductory Psychology* course in September 1951, for college credit over station WEWS in Cleveland. Many other universities followed Western Reserve's example during the next few years.

A USOE survey of the period showed that during the 1952–53 academic year, ninety commercial stations were carrying a total of 256 different educational TV series of one type or another.[3]

Use of Nonreserved Channels   Although there would be no television channels set aside for noncommercial use until 1952 (Section 3.6), there was no legal reason why a college or university (or any other institution) could not operate a station on a nonreserved channel for whatever educational purpose it desired. No one had dictated that a station must be operated for profit. After all, educational radio had been in existence for about two decades—operating on standard nonreserved AM channels—before any radio channels were reserved for noncommercial use.

As early as 1945, President Charles E. Friley of Iowa State College (now University) in Ames saw the advantages of television and applied to the FCC for a construction permit. It was granted; and on February 21, 1950, WOI-TV started regular broadcasts, becoming the "100th television station in the United States and the first non-experimental educational station in the world."[4]

From the station's inauguration, Richard Hull, the first director of WOI-TV (and past president of the NAEB), disagreed with many educators who felt that an educational station necessarily had to be strictly "noncommercial." Hull preferred the term "nonprofit" and—as the station operates on a commercial channel—WOI-TV has carried commercials as a method to help support the broadcasting operation.

---

[2]*Ibid.,* p. 249.

[3]Anthony William Zaitz, "The History of Educational Television: 1932–1958" (unpublished Ph.D. dissertation, University of Wisconsin, 1960), pp. 31–32.

[4]*Ibid.,* pp. 11–12.

During the next few years, many other universities and colleges inaugurated television stations on nonreserved channels. Some of these were St. Norbert College (Wisconsin), Harding College (Tennessee), University of Missouri, Michigan State University, Notre Dame, and Loyola University (New Orleans). Some of these adhere to a rigid noncommercial policy. Others are firmly and successfully commercial. A couple eventually switched to noncommercial channels. One has leased its facilities to a commercial operation.

But the disagreement as to whether educators should ask the FCC for a "nonprofit" designation or whether they should stick to a request for "noncommercial" reservations was one more issue that had to be resolved within the ranks of the NAEB—or some other broadly based ETV organization or coordinating agency.

3.3   The Ford Foundation Enters the Picture

In September 1950, the Ford Foundation made public the 1,400-page study that had been in the making for three years and embarked upon a program that would "ultimately establish it as a new phenomenon of the twentieth century."[5] Essentially, the new program called for broadening the scope of the foundation's activities from that of a local, charitable agency to that of an organization concerned with improving man's conditions and society on a worldwide scale.

During the next two decades, the Ford Foundation was to become the single greatest benefactor of the entire ETV movement. Some of its individual projects may be questioned; some of the research it sponsored may be of doubtful value; some of the experiments it encouraged may not have succeeded. However, there can

be no questioning the fact that educational television, as we know it today, would not have happened were it not for the Ford Foundation—the most visible actor from the cast of philanthropic foundations.

Statements vary as to the total dollar support that the foundation has pumped into all its ETV activities; even records from the foundation come up with different figures. A conservative estimate is that since 1950 the Ford Foundation has provided the ETV movement, directly and indirectly, with about $300 million.

Its modest beginnings, however, did not foretell that extensive an involvement. The leadership of the foundation in the early 1950's favored decentralized, semiautonomous control of its many projects. Thus, in the field of education, the foundation outlined two main areas of concern— *formal in-school instruction* and less formal *general adult education*. Then it created two separate funds, each to work specifically with one area of concern.

Fund for the Advancement of Education One of these specially created organizations was the Fund for the Advancement of Education (FAE), established in April 1951. This agency eventually found itself becoming involved with various programs dealing with the effectiveness of instructional television—or of teaching by television. Some of the classic television teaching and research projects were initially supported by the FAE—the Hagerstown closed-circuit operation, the Chicago TV College, the Midwest Program on Airborne Television Instruction, the National Program in the Use of Television in the Public Schools, and many others (Section 3.9).

Fund for Adult Education As a counterpart to the Fund for the Advancement of Education, the Ford Foundation established the Fund for Adult Education ("the Fund"), which was charged with the responsibility for continuing adult education beyond the formal school years. Like its sister agency, the Fund was set up as a

[5]Bobbie Gene Ellis, "An Investigation of the Ford Foundation's Role in the Early-Stage Development of Educational Television in the United States" (unpublished M.A. thesis, University of Houston, 1958), p. 4.

semi-independent agency with its own board of directors and full autonomy to determine its own policies—although supported entirely with Ford Foundation dollars.

C. Scott Fletcher, former president of Encyclopaedia Britannica Films, was named president of the Fund. For more than twenty years, Fletcher remained actively involved with ETV developments in leadership roles, serving most recently as executive consultant to the NAEB. From the start, Fletcher worked closely with the NAEB—for instance, in suggesting that the association set up an adult education committee that could work closely with the Fund. The resulting NAEB committee consisted of Richard Hull, Harold McCarty, Seymour Siegel, and Robert Hudson, among others. Fletcher relied heavily upon this group for advice and for close NAEB ties.

As soon as the Ford Foundation had announced its expansion plans, educational broadcasters were not hesitant to get into line for handouts. Within a few months, more than $50 million worth of requests had been received. Obviously, the foundation could not immediately hand out these kinds of sums; nevertheless, the "requests fell on sympathetic ears."[6] The study group had been strongly concerned about the role of the mass media, and as far as the leadership of the foundation was concerned, "there was no question in their minds that educational television must be given support."[7]

It became obvious that the primary responsibility for the development of ETV would have to fall to one of the two educational funds. Then, of the two, it appeared that the Fund for Adult Education would be the more appropriate agency to use ETC immediately to attain its own ends.

Obviously educational television would be broader than the area of the Fund's concern. . . . It would span audiences from pre-school to post-retirement and would encompass all serious interests of life. . . . But also it was recognized that the occasion called for a philanthropic institution to consider it as a whole. By agreement with the Ford Foundation the Fund undertook the responsibility.[8]

The implications of this decision are fortunate for the future development of what is now recognized as public TV. For it was determined that ETV would be established by the Fund as a social force, a medium of general adult communication and informal continuing education—that it would not start out merely as an audio-visual aid for the classroom.

Three Goals of the Fund   Upon accepting the responsibility for the development of this nonexistent medium, the Fund for Adult Education found itself faced with a multitude of problems, including, among others, securing reservation of educational channel allocations, establishing stations, training personnel, finding continuing sources of financial support, encouraging local initiative, and program production and exchange. It organized these problems into three general goals:

1. Persuading the FCC to reserve some ETV channels for a sufficiently long period;

2. Stimulating the educators to apply for, build, staff, and support the stations;

3. Creating a "national educational television center for the exchange of programs, ideas, information and services."[9]

---

[6]John Walker Powell, *Channels of Learning: The Story of Educational Television* (Washington, D.C.: Public Affairs Press, 1962), p. 60.

[7]*Ibid.*, p. v.

[8]*Continuing Liberal Education: Report for 1955–57* (White Plains, New York: Fund for Adult Education, 1957), p. 18.

[9]*Ibid.*, p. 19.

The Fund soon determined that there are three ways of implementing a goal—assuming you have the resources (as it did)—once a purpose or direction is established: (1) support those activities and agencies that are moving toward the goal, (2) invest money in other organizations that can be moved to head toward the desired goal, and (3) create new agencies to work toward the established goal.[10]

The story of Fund support of ETV is a study of the utilization of these three approaches. The existing agencies and the new organizations discussed below (JCET, NCCET, and ETRC) can be classified as belonging to one or another of these three implementation categories.

TV-Radio Workshop   One other Ford project should be mentioned at this point, as it was initially administered by the Fund for Adult Education. The Television-Radio Workshop was created in mid-1951 to test the hypothesis that the future of educational TV might well lie in improved cultural and educational programing over *commercial* stations. Basically, the same line of thinking had been advocated for radio by the National Advisory Council on Radio in Education and the Federal Radio Education Committee during the 1930's (Section 2.4).

Four different series were produced by the Workshop, the most durable of which was *Omnibus*. Robert Hudson, however, was one of the leading educational broadcasters who championed the view that commercial broadcasting could seldom, if ever, achieve the goals of a real educational experience. *Omnibus*, for example, had tremendous aesthetic and cultural potential, but as Hudson stated it from the educator's perspective, it was not likely "to persuade anyone to engage in consecutive thinking."[11] He thus pointed out the difference between *education* and *cultural entertainment*—only ETV could

really involve people in a systematic, continuing educational experience. (Some observers in the mid-1970's made the same distinction in discussing programing aired over public TV stations—alleging that programing distributed by the Public Broadcasting Service leaned too much toward *cultural entertainment* at the expense of *educational* materials. Section 8.6). Hudson further characterized the use of commercial television for educational purposes as

> the Advertising Council approach; use radio and television to tell people what to do once it has been decided by law or consensus. I hold that radio and television have much more important functions to perform; they must contribute to the decision-making which leads to law and consensus.[12]

Regardless of its educational validity, the Television-Radio Workshop experiment never attracted enough viewer or sponsor interest to become self-supporting as a commercial undertaking, and it was eventually dropped by the Ford Foundation.

3.4   Joint Committee on Educational Television

Just as there had been several differing types of organizations created to meet specific needs in relation to educational radio, so would a number of separate and specialized councils and associations spring up in relation to educational television. Each, in turn, would have its opportunity to play its role on the ETV stage during the next few years.

Following the 1949 *Notice of Further Proposed Rule Making,* it was apparent that not all of the educators were agreed on what they wanted from the FCC. Two main disagreements needed to be resolved: (1) Should all of the re-

---

[10]Powell, *Channels of Learning,* p. 65.
[11]*Ibid.*, p. 62.

[12]*Ibid.*

served channels be in the UHF spectrum or should some of the more valuable VHF channels also be allocated to educators? (Section 3.1.) (2) Should the channels be reserved for strictly "noncommercial" purposes or would a "nonprofit" designation suffice? (Section 3.2.) It was readily agreed, however, that some sort of coordinating, broadly representative group was necessary— perhaps something like the National Committee on Education by Radio.

In late 1950, therefore, the NAEB was instrumental in establishing a committee of representatives from various educational agencies concerned with the potential of ETV.

> At the suggestion of the Chief of Radio-Television for the U.S. Office of Education, Franklin Dunham, . . . the President of the National Association of Educational Broadcasters [at the time, Seymour Siegel] called together those educational groups which were interested in FCC allocation activities.[13]

This meeting, held on October 16, 1950—at the home of Frieda Hennock—was the first meeting of the ad hoc Joint Committee on Educational Television (JCET).

The original groups that comprised the JCET were the American Council on Education, the Association for Education by Radio (soon to add "and Television" to its title), the Association of Land Grant Colleges, the Association of State University Presidents, the NAEB, the National Association of State Universities, the NEA, and the National Council of Chief State School Officers. The similarity to the composition of the NCER is obvious (Section 2.4).

I. Keith Tyler, director of the Institute for Education by Radio-Television at Ohio State University, was selected as chairman of the JCET. He was successful in enlisting help from many sources, including borrowing some office space

from the NEA in a former bathroom of the old Guggenheim mansion. During its early phase, most of the JCET activities—which were to profoundly shape the profession for years to come— were directed out of "Keith's powder room."

JCET Compromises and FCC Hearings
One of the first accomplishments of the JCET was to reach compromises on the basic differences that separated the various groups interested in ETV. On the UHF-VHF issue, the JCET compromised by calling for the FCC to reserve one VHF channel in every major metropolitan area, plus 20 percent of all UHF channels. As to the classification of the reserved channels, the JCET settled on the designation of "noncommercial," except in one-station markets where "nonprofit" might be more appropriate (this made allowance for WOI-TV, the only existing "nonprofit" station in the country at the time).

The JCET was prominent in the 1950–51 hearings. Through the services of a fund-raising firm, a total of $32,500 was solicited to help finance JCET activities during this ad hoc period.[14] The major activity undertaken with these funds was the coordination of testimony by seventy-six leading educators brought before the FCC. The results of the JCET efforts were so successful that the hearings, which were originally scheduled only for November 1950, had to be extended into January 1951.

Monitoring Study   In conjunction with the NAEB, the JCET also sponsored a commercial television monitoring study that contributed significantly to the JCET testimony. Research specialists were engaged to analyze the total programing output of the seven existing New York commercial television stations. With the help of volunteers, from January 4 to 9, 1951, they moni-

---

[13]William Kenneth Cumming, *This is Educational Television* (Ann Arbor, Michigan: Edwards Brothers, Inc., 1954), pp. 1–2.

[14]Harold E. Hill, *The National Association of Educational Broadcasters: A History* (Urbana, Illinois: National Association of Educational Broadcasters, 1954), p. 43. This is an M.A. thesis copyrighted, mimeographed, and distributed by the NAEB.

tored more than twelve hours of programs and commercials daily on each of the seven stations.

> The results of this ordeal were . . . some very incriminating findings. Among the mass of findings was the significant fact that no time was given to educational programs as such. Most of the so-called children's programs bore little or no relation to the tastes and needs of children. The monitoring study proved to be a milestone in television research.[15]

The Permanent Body   The ad hoc Joint Committee on Educational Television ceased to exist on January 30, 1951, when the FCC hearings were concluded. It soon became apparent, however, that some type of unifying and coordinating organization on a permanent basis was necessary if the young ETV movement was to succeed. So on April 23, 1951, the JCET was reconvened and formed into a permanent body. The "new" committee was immediately involved in obtaining hundreds of written statements of support and need from educators—in response to the FCC's 1951 proposed allocations (Section 3.6).

Perhaps the one outstanding factor that differentiated the permanent JCET from its ad hoc predecessor was a $90,000 grant from the newly formed Fund for Adult Education, which enabled the JCET to establish and maintain a small full-time professional staff. The staff functions of the JCET were essentially a task-force operation: traveling and talking, consulting with would-be station operators, representing ETV generally before the FCC, giving legal and technical advice to everyone interested. As Powell stated in 1962, "There is hardly an ETV station in existence that does not owe its initial progress to the personal advice and help of the JCET staff during the early and critical stages."[16]

In terms of the three implementation categories that the Fund was following, the JCET fell into the first category—an example of an agency that was moving in the right direction—and was therefore supported.

3.5   The National Citizens Committee for ETV

In 1952, the final group of individuals who comprise our cast of players is ready to make its debut. Up to this point, there really had been very little public awareness of or lay support for educational broadcasting (that is, radio). There were probably a couple of reasons for this lack of support. First, there had been relatively little in the background of educational radio to involve the average citizen; most of the programing was aimed at students or at small specialized audiences (such as Midwestern farmers). Second, there had been little need to solicit support and funds from the public at large. With the comparatively modest production budgets needed for radio, most educational radio operations had been able to get by without any additional funds beyond what was available from the educational institutions sponsoring the individual stations.

Television changed the picture, however. It was now obvious that large sums of money would be needed—and that the lay public certainly should be approached for general support. Also, programing should be designed to attract a more general audience. The Fund for Adult Education was especially concerned about reaching as large a segment of the general audience as possible. This, after all, was the basic charge to the Fund—educate the general adult public.

Creating a New Agency   The National Citizens Committee for Educational Television (NCCET) was an example of the third category

---

[15]Dallas W. Smythe and Donald Houston, *New York Television, Jan. 4–9, 1951:* Monitoring Study 1, NAEB, 1951, quoted in Paul Saettler, *A History of Instructional Technology* (New York: McGraw-Hill Book Company, 1968), p. 230.

[16]Powell, *Channels of Learning,* p. 67.

of the Fund's implementation plans (creating new agencies). Designed to complement the academic, legal, and technical services performed by the JCET and other organizations (such as the NAEB and the TV Committee of the American Council on Education), the NCCET was basically the public relations and fund-raising arm of the total station activation movement.

Created in late 1952, the NCCET selected as its executive secretary Robert R. Mullen, who had just completed a successful assignment as public relations chief for Citizens for Eisenhower. He was handed the responsibility of "acquainting business, professional and civic leaders and organizations with the problems and encouraging them to support the educators in their efforts to build and operate the stations."[17]

The Lay Public and Community Stations

The community-sponsored ETV stations—which were a special target of the NCCET—were a phenomenon that had no counterpart in educational radio. The formation of these stations, which are owned and operated by nonprofit corporations composed of private citizens, was due largely to the newly stimulated interest and responsibility of the lay public. Aside from a few municipally owned stations, virtually all educational radio operations were owned either by universities and colleges or by school systems. When faced with the staggering costs of supporting an educational *television* station, many of these institutions had to back off. So the NCCET began identifying groups in a community that could collectively support one ETV station—and the "community" ETV station was created (Section 5.2).

The importance of the role of the lay public in the support of educational broadcasting cannot be overemphasized. Mullen summarized their emergence in an early article for the *AERT Journal:*

There are some new faces in the American broadcasting picture today. And they are growing in number. They are the faces not of educators nor of broadcasters but of a species hitherto much of a stranger to the field—lay persons, members of the general public.

This citizen participation was not there in the history of AM. Nor was it there in the movement to establish stations on the 20 channels reserved for educators in the FM band.

This new fact in broadcasting is peculiar to the rise of educational television and is, I believe, a most hopeful augury of success in this exciting venture.[18]

The validity and permanence of this type of thinking was reaffirmed some fifteen years later by the renewed emphasis on *public* television and the creation of the National Citizens Committee for Public Television—and, later, the National Friends of Public Broadcasting.

3.6   Sixth Report and Order

When the FCC hearings were closed in early 1951, the FCC commissioners were still in a dilemma concerning the need to reserve channels for education. The commission was less than anxious to follow Frieda Hennock's leadership in championing the noncommercial cause; however, "whatever its internal differences of opinion about the likelihood of ETV's success, it could hardly come out against education."[19]

On the other hand, the commission was faced with some strong and convincing arguments against reserving any channels for the educators: (1) Educators cannot afford television, and therefore they will not activate the allocations.

---

[17]"The NCCET—An Enviable Record," *AERT Journal,* February 1956, p. 3.

[18]Robert R. Mullen, "The Citizen and Educational TV," *AERT Journal,* October 1953, quoted in "The NCCET . . . ," p. 4.

[19]Powell, *Channels of Learning,* p. 49.

(2) As far as production is concerned, commercial stations and networks can do a better job of presenting education programs. (3) All that educators really need are closed-circuit systems. These arguments can be traced back twenty years to the earliest days of the Federal Radio Commission. The spotty record of educational radio, the relatively small number of universities then making any use of television, and the early *Omnibus*-type thinking all led the FCC to have doubts about saving spectrum space for the educators.

> The Commission did have to have evidence, and plenty of it, to justify reserving a large part of the spectrum for education when other segments of the society were making strong claims that it was needed for other purposes.[20]

The 1951 Proposed Allocations   In early 1951, the FCC issued its *Third Notice of Further Proposed Rule Making*. As a direct result of the recent hearings, this proposal set aside 209 channels for noncommercial, educational use. This *Notice,* of course, was not a firm ruling—but merely a trial balloon to see what reactions might be stirred up.

Frieda Hennock was the key person to remain stirred up. Not satisfied with the majority opinion of the commission, she called for more ETV reservations. The FCC then asked for "sworn statements and exhibits" to show a further indication of the educators' interest. The educators responded—with prodding from the JCET—and came forth with more than 800 such documents.

End of the Freeze   Over a year later, on April 14, 1952, the FCC released its *Sixth Report and Order,* the document which ended the freeze and gave the country a new Table of Assignments. This called for a total of 2,053 television allocations, with 242 of them reserved for non-commercial purposes. Of the 242 ETV reservations, 162 were in the UHF range and 80 were VHF channels.

As opposed to the earlier "cluster" reservations for educational radio, these were individual channel assignments reserved for each separate community (Channel 2 in Boston, Channel 11 in Chicago, Channel 62 in Tacoma, and so forth).

The effectiveness of the educators' testimony was attested to by part of the *Report* itself:

> There is much evidence in the record concerning the activities of educational organizations in AM and FM broadcasting. It is true, and was to be expected, that education has not utilized these media to the full extent that commercial broadcasters have, in terms of number of stations and number of hours of operation. . . . Furthermore, the justification for an educational station should not, in our view, turn simply on audience size. The public interest will clearly be served if these stations are used to contribute significantly to the educational process of the nation.[21]

The end of the "freeze" brings down the curtain on Act Two of our figurative drama.

3.7   Postfreeze Activity

The decade following the *Sixth Report and Order* (which roughly corresponds to Act Three) was largely a story of steady growth, maturation, and slow evolution. The first non-commercial TV station went on the air in 1953; nine years later there were seventy-four ETV stations broadcasting. Annual spending for all ETV increased from a few thousand dollars during the first few years to about $15 million a year by 1960. The total capital outlay—for plants, facil-

[20]Walter Emery, then on the legal staff of the FCC, quoted in Powell, *Channels of Learning,* p. 49.

[21]Quoted in Powell, *Channels of Learning,* p. 23.

ities, and equipment—for these stations was about $30 million.[22]

The Penn State Institute and a Sense of Urgency   The early activity in ETV circles was characterized by a sense of urgency and a few clouds of uncertainty.

The first burst of activity throughout the country produced a series of conferences and institutes. The earliest and, perhaps, most significant of these sessions—because it triggered so many others—was the Educational Television Programs Institute held at Pennsylvania State College (now University) one week after the *Sixth Report and Order* was made public. Sponsored by the American Council on Education and financed by the Fund for Adult Education and other foundations, the session was hosted by Milton Eisenhower, president of Penn State.

The institute was attended by 116 leaders in education, including eleven college and university presidents. This conference "for the first time posed the mass media problem to senior educational administrators."[23] The theme of the institute dealt with the practical issues of running an ETV movement—building and equipping stations, producing programs, and the need for a national program center.

In one of the sessions Paul Walker, then chairman of the FCC, made one of his first pleas for urgent action—a plea that was to be his favorite educational sermon for the following year: "These precious television assignments cannot be reserved for you indefinitely. They may not even be reserved for you beyond one year unless you can give the Commission concrete, convincing evidence of the validity of your intent."[24]

Behind Walker's urgent call to action was one stipulation of the *Sixth Report and Order,* which stated: "No petitions requesting change in allocations, including reservations, would be entertained until one year from the effective date of the report."[25] The strong *implication* was that these channels might be reserved for educators *for only one year.* After that time, the FCC could consider petitions to make changes in the channel assignments—both commercial and noncommercial.

During 1952–53, Walker addressed many educational groups with the same appeal for urgency, stressing that educators must move fast if they did not want to lose the reservations after the first year. Much of his motivation, undoubtedly, was the desire to get the channels filled as quickly as possible—thus justifying the commission action in reserving the channels for noncommercial use. Otherwise, the continuing commercial pressure for the channels would be hard to ignore.

One of the immediate effects of the Penn State institute—and Walker's urgent prodding—was the stimulation of other regional conferences in states across the country. A sixteen-state conference sponsored by the Southern Regional Educational Board, for example, triggered conferences in eight more southern states. In 1953, a JCET report on ETV activity in all states showed only ten states not engaged in specific activity relating to educational television.[26]

There is little doubt that the cumulative effect of these numerous conferences and feasibility studies helped to a great extent in enlisting the support of top educators throughout the nation in making them aware of their responsibilities in educational television.

Station Activation   Despite—or because of—the frenzied activity during that first year, there was one noncommercial ETV station on the

[22]Lyle M. Nelson, "The Financing of Educational Television," in *Educational Television: The Next Ten Years* (Stanford, California: Institute for Communication Research, 1962), p. 166.

[23]Richard B. Hull, "Consider Basic Problems," *AERT Journal,* December 1956, p. 33.

[24]Paul A. Walker, "The Time to Act Is Now," *in* Carroll V. Newsom (ed.), *A Television Policy for Education* (Washington, D. C.: American Council on Education, 1952), p. 31.

[25]Quoted in Cumming, *This Is Educational Television,* p. 2.
[26]Powell, *Channels of Learning,* pp. 102–103.

air by the end of 1953—KUHT, Houston, which started broadcasting May 25, 1953.[27] During 1954, eight more stations were activated, including major community-owned stations in Pittsburgh, San Francisco, and St. Louis. In 1955, eight more ETV stations entered the arena, including the Boston and Chicago community stations.

By the end of 1958, there were thirty-five ETV stations on the air. Virtually every one of them had received a financial start from the Fund for Adult Education—usually in the form of a 50-50 matching grant. The Fund had spent roughly $3.5 million in station activation.

All five groups of persons were now involved in their roles. The educators and the public were committing themselves to the concept of noncommercial broadcasting as rapidly as their meager resources would allow. The private foundations (primarily Ford) were committed to supporting the young movement. The professional educational broadcasters (station managers and organizations like the NAEB, JCET, NCCET) were gaining valuable experience and national exposure. The FCC was basically satisfied with the "validity of the educators' intent" to the extent that the bulk of the 242 noncommercial reservations were not jeopardized. In fact, additional channels have been periodically added to the list of reserved allocations. By 1965, a total of 632 channels had been set aside for noncommercial use.

For the next several years, the drama of the ETV movement can best be summed up by tracing further growth in station activation, developments with the professional association, and—most importantly for the immediate future— the creation of a networking operation.

---

[27]One other station was activated during 1953, but because of funding problems it discontinued operations the next year. (See Section 14.2.)

3.8 Emergence of the Fourth Network: ETRC to NETRC

Soon after its creation, the Fund for Adult Education saw the need for a national programing operation for the emerging ETV stations; the difficulty of distributing quality program material on a nationwide basis would be the biggest single headache for the new movement. The commercial stations were being connected by the transcontinental coaxial cable; and Uncle Miltie (Berle) reigned supreme. How was ETV to attract any viewers—faced with this kind of commercial/entertainment dominance?

In November 1952, a group of prominent educators and business leaders congregated in Chicago to incorporate the Educational Television and Radio Center (ETRC). For almost twenty years, this organization played one of the key roles in the development of noncommercial ETV— at times being the undisputed kingpin. The formation of the ETRC was the third immediate goal that the Fund had set up for itself shortly after its inception—that of creating a "national educational television center for the exchange of programs, ideas, information and services" (Section 3.3). Also, the Center, like the NCCET, was an example of the third approach that the Fund relied upon—creating new agencies to work toward its established goals.

The First Year of the Center: Son-of-a-Fund The first year of the existence of the ETRC was an interim period of temporary officers, of no permanent office space, but of important policy decisions. It would be impossible to staff the new ETRC with qualified personnel immediately; it would certainly take some time to find the right person with the educational stature and prestige to assume the presidency of the Center. However, it was important that the Center should begin to function as soon as possible. So the ETRC board appointed Fund President C. Scott Fletcher as temporary president of the Center; and with three other part-time officers, the business of the ETRC was conducted out of Fletcher's Pasadena Fund office during this formative first year.

It was obvious that the Center could not match the methods and program budgets of the commercial networks; but a program exchange and duplication service—similar, perhaps, to the NAEB "bicycle" radio tape network—could provide ETV programing that a station could use to supplement its own local programing. In response to this underlying purpose, and in addition to it, five basic policy decisions were formulated at the outset of ETRC operations: (1) The Center would not maintain its own production facilities but would obtain programing from the ETV stations and other sources. (2) The Center would not function as a library but would operate as a network service supplying regularly scheduled programs. (3) The Center would serve as a focal point for production ideas, professional meetings, and other nonprograming activities. (4) The ETRC would also service commercial stations in a peripheral capacity. Finally, (5) the Center would include "radio" in its title and would devote some attention to improving the status of educational radio in the United States.[28] By the late 1960's, virtually all these early policy decisions had been reversed or substantially altered.

The Newburn Era   The Center found its first full-time president in the person of the president of the University of Oregon, Harry K. Newburn—who had the academic stature to give the fledgling ETRC the prestige it needed during its early development. Newburn soon relocated the Center in Ann Arbor, Michigan, and by early 1954 the ETRC had a full-time professional staff of five, plus clerical and technical support positions.

On May 16, 1954, the ETRC programing service was officially launched to the four noncommercial ETV stations then on the air. This service consisted of five hours of programing a week on kinescope film recordings that were circulated, through the mails, to stations on a round-robin basis. By the end of Newburn's administration in 1958, the Center was distributing more than seven hours of programing a week to more than thirty ETV stations. This ETRC programing was divided into several distinct categories in order to guarantee a balance among various subject areas (Section 7.2).

Most of the programing during Newburn's presidency was produced by the affiliated stations (see above). This reflected a definite policy—later to be reversed—that the Center should support local stations by helping them to mature as *producers* of programs (as well as by furnishing them with quality materials to be telecast).

During these first few years, the Center also became involved in several other areas of activity. It supported some educational radio programing during the mid-1950's. The ETRC became increasingly involved with research (due partly to Newburn's academic orientation), sponsoring much of the early research in the field. And the Center took over the responsibility for national promotion of ETV and public information activities on the national level when the Ford Foundation ceased underwriting the activities of the NCCET in the mid-1950's.

Another significant move was the creation of a development department within the Center. From its inception, officials of the Fund for Adult Education and the parent Ford Foundation had insisted that the ETRC would not be supported exclusively by the foundation. They urged the Center to investigate means by which it could find other sources of financial backing. Their insistence was in keeping with the foundation's general policy that "it should not support a given recipient indefinitely."[29] However, although some small grants were received from business corporations and other foundations near the end of

[28]I. Keith Tyler, "The Educational Television and Radio Center," *in* William Yandall Elliot (ed.), *Television's Impact on American Culture* (East Lansing: Michigan State University Press, 1956), pp. 225–230.

[29]*The Ford Foundation in the 1960's* (New York: Ford Foundation, 1962), p. 6.

Newburn's administration, the Center was still dependent almost exclusively upon Fund support.

The White Years: First Phase   When Newburn resigned in 1958, he was succeeded by John F. White, former vice-president of Western Reserve University (where he was instrumental in getting the first credit telecourses on the air in 1951) and, more recently, general manager of community ETV station WQED in Pittsburgh. Most observers agree that, compared with Newburn, White was a more dynamic and "progressive" leader.[30] Under White's guidance, the Center was to enter a new phase of its development.

Two of White's early actions were to change the Center's name to *National* Educational Television and Radio Center and to move the Center's main headquarters to New York City (leaving the technical and distribution departments in Ann Arbor). Within three years, White had tripled the size of the NETRC professional staff. Whereas Newburn had worked with no vice-presidents, White soon had appointed four—one each for programing (Robert Hudson), business, development, and network affairs.

White's administration was early characterized by expansion into several related areas (as well as expansion of staff and space). Concerned with the lack of machinery to strengthen station relations, White soon had established the division of network affairs—which included station relations, distribution, technical affairs, and the NET Washington office. This latter department was essentially the former JCET staff. When the JCET funding expired in 1961, the staff and functions of the JCET operation were transferred to the NETRC (all at the direction of the Ford Foundation—which, of course, furnished the support for both operations). The policy-making functions of the JCET were assumed by the newly constituted Joint Council on Educational Broadcasting (composed of the same constituent groups that made up the former JCET).

Areas of Expansion   The Center also moved into the realm of instructional television in 1962. With support from the U.S. Office of Education, the National Instructional Television Library (NITL) was established under the administrative auspices of the NETRC (Section 4.3). The NITL included the national center (which later was to become the National Center for School and College Television, then the National Instructional Television Center, and finally the Agency for Instructional Television) and two regional centers—one at Boston and one at Lincoln, Nebraska (which eventually was to become the Great Plains National Instructional Television Library).

In the area of radio programing, the Center expanded in two directions. In 1960, the NETRC assimilated the Broadcasting Foundation of America (another Ford-sponsored activity—one that was cofounded by Seymour Siegel, former NAEB president and originator of the NAEB radio network). The BFA functions of international radio programing acquisition and exchange became the international exchange division of the Center. And in 1962, under a special Ford grant, the NETRC assumed responsibility for the Support of the Educational Radio Network—a live, interconnected educational radio network of six FM stations on the East Coast.

But the Center's main area of concern was still general adult educational television programing. While the quantity of programing was increased to ten hours per week to each affiliated station (including reruns and library selections), the primary emphasis was the goal of increasing the quality of NETRC programing. Under White's leadership, "quality became the touchstone of the programing service."[31] Shortly after White became president, this policy was reflected in a popular magazine article that proclaimed, "NET is no ivory-tower operation. It is manned by professional TV men with a flair for showmanship and

---

[30]Powell, *Channels of Learning,* pp. 81–82.

[31]Internal memorandum, National Educational Television and Radio Center, dated July 6, 1962.

an eagerness to prove that cultural programs can be made interesting to large audiences."[32]

In nonprograming areas of NETRC activities, perhaps the most significant field of expansion was in the development effort. With the creation of the position of vice-president for development, White's administration vigorously tackled the problem of fund raising—with the strong encouragement of the Ford Foundation—in an attempt to achieve some degree of autonomy from Ford dollars. Even with this aggressive effort, however, the NETRC was never able to obtain more than about a quarter of its annual operating budget from non-Ford sources. It gradually became apparent that appeals to the public conscience of big business and to the altruistic spirit of other private foundations were not going to furnish the solid financial basis needed for adequate continued support of public ETV. Underwriters were found for some specific series, but such sources would never be enough to support a total network operation. This was a fact of noncommercial business that the Center—and the Ford Foundation—were being forced to face.

Throughout its history of support of the NETRC, the Ford Foundation had always made it clear that it did not propose to be the continued sole support of the ETV movement. It would furnish the initial impetus, the seed money—but continued support would have to be forthcoming from other sources; the Foundation was not in the business of furnishing a never-ending flow of operating funds to any given enterprise.

So, for over a decade the NETRC had been molded in the image of the parent foundation; despite its desire for program autonomy and independent fiscal security, the history of the Center had been intertwined with the policies and operations of the Ford Foundation. With this relationship in mind—and aware of the fact that other sources of financial support for ETV simply had to be uncovered—the foundation undertook

a thorough reexamination of its entire ETV policy in 1963; it wanted to determine exactly what its future position should be relative to educational TV.

This reexamination by the Ford Foundation actually marks the end of Act Three of the ETV story. When the curtain opens on the fourth act (Section 4.1), we will find several new alignments and relationships that soon introduce a whole new cast of actors.

3.9   ITV and the Fund for the Advancement of Education

In the meantime—back in the classroom—significant developments were taking place in several related areas of instructional television. The concept of using television for formal classroom instruction, of course, dates back to the 1940's and the use of cooperating commercial stations (Section 2.5). However, with the establishing of noncommercial educational stations, much more time was available for school-related programing. One consistent factor common to virtually all ETV stations—regardless of ownership or programing pattern—has been that most of the broadcast schedule during school hours has been devoted to instructional telelessons for specific classroom subjects.

A great deal of leadership—and financial support—for the development of ITV came from the Fund for the Advancement of Education (FAE). Just as the Fund for Adult Education was to play the crucial role in establishing ETV stations and promoting the cause of general ETV—what is today labeled "public TV"—so was the Fund for the Advancement of Education to play a corresponding vital role in promoting and facilitating the growth of ITV—or "School TV."

Hagerstown CCTV   Not only were open-circuit ETV stations heavily utilized for school broadcasting, but a great potential was seen in the possibility of installing closed-circuit television (CCTV) systems. One of the first, and most im-

---

[32]John Reddy, "Have You Heard About These Better TV Programs?" *Reader's Digest,* July 1960, pp. 137–140.

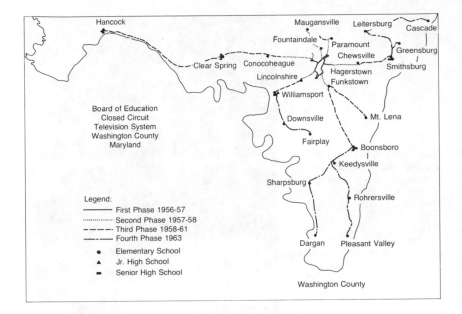

Board of Education
Closed Circuit
Television System
Washington County
Maryland

3-1 Schools Interconnected by Washington County CCTV System, Hagerstown, Maryland

Legend:
—————— First Phase 1956-57
················ Second Phase 1957-58
— — — — Third Phase 1958-61
— · — · — Fourth Phase 1963
● Elementary School
▲ Jr. High School
■ Senior High School

pressive, demonstration projects of a school CCTV system was set up in Hagerstown, Maryland, in 1956. Sponsored cooperatively by the FAE, American Telephone and Telegraph, and the Electronic Industries Association, the system was designed for the Washington County school district.

Eventually every school in the county was wired into an elaborate six-channel closed-circuit system; theoretically six different programs could be simultaneously transmitted to any of the participating schools. To feed this distribution capacity, six production studios were equipped. Originally, virtually alll production was done live; later, however, the introduction of videotape recorders facilitated more efficient use of both the studios and distribution channels.[33] The interconnected CCTV system is shown on the map (3-1).

Initially, the system was designed to help alleviate the teacher shortage (of 352 teachers at the elementary level, 75 held only emergency teaching certificates) and to provide enrichment courses in subjects such as art, music, and foreign languages.[34] Eventually, virtually every major subject in the curriculum was included to some extent in the CCTV project.

Chicago TV College Also in 1956, an even more impressive open-circuit (broadcast) project was started in Chicago at the community college level. The FAE gave a $500,000 grant to the Chicago Board of Education and ETV station WTTW to initiate the "Chicago TV College."

A fully accredited program of the Chicago City Junior College system, the TV College offers a wide variety of courses leading to the Associate of Arts degree. In fact, it is possible to obtain a complete A.A. degree merely by taking courses offered over the open-circuit station. Several

[33]*Washington County Closed-Circuit Television Report* (Hagerstown, Maryland: Washington County Board of Education, 1963).

[34]Saettler, *Instructional Technology*, p. 247.

3-2 Early Spanish Language Series from the Chicago TV College (Courtesy: Great Plains National Instructional Television Library)

hundred students have done so—including many at a nearby state penitentiary. Most of the students taking TV courses, however, also take some campus courses before obtaining the degree.

Most of the courses offered by means of open-circuit TV are supplemented with telephone conferences, occasional open meetings, and the use of programed materials for self-checking. Exhaustive testing and research programs have confirmed repeatedly that the TV students do as well as, or better than, students taking the same courses on campus.

Even more important, [the project] proved that college-credit courses can be televised without compromising course objectives or sacrificing instructional quality. The fact that accrediting agencies, professional associations, colleges and universities promptly agreed to accept credit earned through TV College courses is indication of the instructional quality of the courses.[35]

The illustration (3-2) shows a presentation during a series teaching Spanish and indicates the production quality of the early programs.

The original grant supported the program for its first three years. Since 1959, the project has been entirely self-supporting with an enrollment of more than 10,000 students a year. A TV enrollment of approximately 800 students in one course represents the break-even point.

The National Program   One other significant project of the late 1950's should be mentioned in this quick survey—the National Program in the Use of Television in the Public Schools, also

---

[35]*Eight Years of TV College: A Fourth Report* (Chicago: Chicago Board of Education, 1964), p. 5.

supported by the Fund for the Advancement of Education. What distinguished the National Program was that it was not a single project but a series of loosely related autonomous research endeavors.

Starting in 1957, ten school districts and scores of individual schools from other areas participated in similar research programs. The following year, three more centers were added. By 1959, a total of twenty major centers were participating in various aspects of the program.

Basically, all of the independent research projects were focused on the question of whether or not large-class instruction, using television, can be as effective as conventional teaching situations. In the course of the research projects, many other problems were tackled. The scope of the participating schools was as diverse as possible: large and small, urban and rural, rich and poor, located in every geographical section of the country. Every grade level from primary through secondary schools was included, and virtually every subject matter was tested.

Greatly simplifying the results of such a massively complex research undertaking, one would have to generalize by stating that, in most instances, the TV classes performed better than, or as well as, traditional classes.[36]

Other Projects   In this short overview, we cannot do justice to the many other significant projects and programs that were undertaken during this period. By the end of the decade, sixty or more ETV stations were beaming instructional programs into thousands of classrooms across the country. At least 500 separate closed-circuit systems were operating in schools by the early 1960's.

In 1958, the federal government began making money available directly to schools for equipment, receivers, and television research projects under the National Defense Education Act (Section 4.1). In the same year, a commercial network, NBC, began offering early-morning instructional programing across the nation on *Continental Classroom* —soon to be followed by CBS with its *Sunrise Semester*. In 1959, another Ford Foundation project was initiated, the Midwest Program on Airborne Television Instruction, to test the feasibility of a flying transmitter (the foundation's "wildest scheme," as one critic phrased it).[37] In the 1950's, several southern states built their own networks to upgrade their schooling systems; in 1960, South Carolina initiated an ambitious high-quality statewide *closed-circuit* installation to reach its schools. In American Samoa in 1961, the NAEB started a study that would lead to the world's most ambitious integration of television technology into a schooling system.

The few examples listed above are but a sampling of the many varied ITV programs started during this period (several of these are discussed in Chapter 11). These examples should suffice to illustrate the wide variety of school TV activities that were undertaken during the Formative Fifties, a period of growth that corresponds to Act Three of our ETV drama.

---

[36]*The National Program in the Use of Television in the Public Schools: A Report on the Third Year* (New York: Ford Foundation, 1961).

[37]George N. Gordon, *Classroom Television: New Frontiers in ITV* (New York: Hastings House, Publishers, 1970), p. 23.

Into the Seventies: New Identities

If the extended 1950's could be described as a period of frenzied activity (1948–1952) followed by a time of relatively stable growth (1953–1962), then the 1960's—and into the 1970's—might be labeled a period of regrouping and professional introspection (1962–1966) followed by the emergence of new identities and organizations that would forge new directions for the noncommercial telecommunications movement (from 1967 on). In fact, 1962 should be the date for the beginning of the fourth act of our ETV drama.

4.1 The Educational Television Facilities Act

Act Four begins with a fanfare and an upstage entrance—of direct federal aid to educational broadcasting (one of the seven themes introduced in Section 1.8). This was the first instance of federal money being made available in grants to ETV stations. However, the federal government had previously spoken from the wings in 1958 with the passage of the National Defense Education Act (NDEA), but any aid to open-circuit ETV was indirect.

The Predecessor: NDEA The National Defense Education Act was passed in reaction to Russia's 1957 leap into space with Sputnik I. Congress planned that NDEA would quickly improve the country's educational system—specifically in the areas of mathematics, science, and foreign languages. Title III of NDEA provided funds to enable schools to purchase equipment necessary to beef up instruction in those critical subject areas, with the local schools paying half the cost and the federal funds matching the local dollars. These Title III funds went, for example, to buy language labs, science demonstration equipment—and television reception equipment. Since most open-circuit School TV projects involved some telecourses in math, science, or foreign languages, the TV receivers were justified by the schools in meeting the goals of NDEA, Title III. The acquisition of TV equipment was a very important spur to the entire ETV movement—because the stimulus to purchase TV receivers for the classrooms had a ripple effect that was felt far and wide for many years to come. Schools had an incentive to continue their TV involvement as reception equipment was more easily attainable through the matching federal dollars; and as more classrooms were equipped with receivers, it became more feasible to expand School TV programing and services.

In another part of the act, Title VII, NDEA provided funds for "research, experimentation, and dissemination of information about the uses of television, radio, motion pictures, and related media of communication that might prove useful to educational agencies and institutions."[1] Ultimately, more than $8 million was used by hundreds of schools in acquiring TV equipment to carry out experiments and demonstration projects.

Public Law 87–447   The National Defense Education Act, however, provided little direct assistance to ETV stations themselves. Most of the funds went into school facilities. The station operations were finding it increasingly difficult to update equipment, consider conversion to color production and transmission, and expand facilities. Since the Ford Foundation was becoming less involved in providing "start-up" funds for new ETV stations, groups wanting to put stations on the air were finding it difficult to arrange financing for initial plant and facilities costs.

At the same time, the educational broadcasters did *not* want federal assistance for educational TV or radio *programing*. The specter of federal agencies deciding which ETV programing would get support and which would not represented too much of a threat of government control over programing. Therefore, the repeated request to Congress was for federal dollars to help purchase equipment and facilities.

Finally, in 1962, Congress responded with the Educational Television Facilities Act (Public Law 87–447). This act—and its subsequent renewals—has poured millions of dollars into building and improving physical facilities for educational telecommunications operations across the nation. Providing money on a matching basis, the program—administered by the Department of Health, Education, and Welfare (HEW)—since its inception has partially subsidized the construction and expansion of hundreds of noncommercial radio and television stations. Working with state agencies that helped to establish priorities and process the applications, the federal project initially was a $32-million 50–50 matching program—with the federal government giving one dollar to match every dollar raised by a local station. Later, this was revised upward to a maximum 75–25 matching program (up to three federal dollars for every one local dollar).

Resulting Station Growth   As shown in the table (4-1), "Number of ETV Stations On the Air: 1953–1975," a substantial increase in station activation immediately followed the passage of the Educational Television Facilities Act. Allowing for a two-year lag period (to apply for the grant, raise the local dollars, purchase the equipment, proceed according to the FCC regulations, and actually get on the air), the large increase in the number of stations on the air from the end of 1963 to late 1965 can be attributed, in no small measure, to the impetus from the federal grants. There was about a 40 percent increase in the number of stations during this two-year period.

Undoubtedly, one other major incentive for the surge in station activation was the 1962 all-channel legislation. This amendment to the Communications Act of 1934 authorized the FCC to require all TV receiver manufacturers to equip every new television set to receive all VHF *and* UHF channels; and the commission promptly issued such a regulation which went into effect in mid-1964. Although the "UHF problem" is still far from being equitably resolved, the all-channel

---

[1]John M. Kittross, "Meaningful Research in ETV," in Allen E. Koenig and Ruane B. Hill (eds.), *The Farther Vision: Educational Television Today* (Madison: University of Wisconsin Press, 1967), pp. 214–215.

receiver unquestionably was a large factor in the number of UHF stations—both commercial and noncommercial—that were activated in the mid-1960's. In the table (4-1) note the increase in the number of UHF educational TV stations from 1964 on.

4-1 Number of ETV Stations on the Air: 1953–1975[a]

Year	New VHF stations	New UHF stations	Total stations on the air[b]
1953	1[c]	–	1
1954	5	2	8
1955	7	1	16
1956	4	1	21
1957	5	1	27
1958	6	2	35
1959	5	4	44
1960	3	2	49
1961	7	4	60
1962	5	9	74
1963	3	2	79
1964	7[d]	9	95
1965	6	9	110
1966	6	5	121
1967	4	23	148
1968	4	23	175
1969	1	8	184
1970	6	7	197
1971	5	12	214
1972	2	14	230
1973	1	10	241
1974	3	6	250
1975	2	9	261

[a]Data taken from 1975 Public Telecommunications Directory (Washington, D.C.: NAEB, 1975), pp. 106–120.
[b]Total number of stations operating as of December 31.
[c]Figures reflect only those stations currently on the air.
[d]Six VHF channels assigned to Samoa are included as one station.

4.2   The Transition to NET: Second Phase of the White Regime

By the middle of 1963, there was considerable activity and optimism about ETV. There were approximately eighty ETV stations on the air. Citizen support for community stations was expanding. Many groups and institutions were applying for new station licenses to take advantage of the ETV facilities law. Under John White's leadership, the programing quality of the National Educational Television and Radio Center (NETRC) was improving to the extent that a growing—and influential—audience was increasingly aware of ETV's potential. Instructional TV was being integrated into curricular programs of thousands of schools across the nation.

The Ford Foundation Speaks   Underlying this scene of confidence and progress being acted out upon the main stage, there was an undercurrent of conflict and tension and self-questioning. The problem—as perceived from the Ford Foundation's perspective—was simple: the foundation was furnishing too much of the support for the entire ETV movement.

The Ford Foundation, through the Fund for the Advancement of Education, had furnished the major support to dozens of demonstration and research projects aimed at establishing school TV. The Fund for Adult Education had contributed to the creation of virtually all of the pioneer ETV stations—especially the community stations; and the foundation had already given about $27 million directly to the NETRC.

Not only did this outlay represent a major drain on the resources of the foundation—as vast as they were—but also it had negative connotations for the entire ETV movement. As long as the Ford Foundation continued to be the major support (in some instances, the *sole* support) for various ETV programs and activities, other agencies and institutions were less inclined to pick up a share of the tab. "Let Ford do it" was a common feeling. Educational TV was becoming too much identified as a Ford Foundation cause. The question could be raised, "If ETV really has the potential that the Ford Foundation reports say it has, then why aren't others helping to foot the bill?"

Indeed, this was a question raised by the

foundation itself. The foundation's position had always been that it would never become the sole continuing source of financial support for any single institution or organization or program. It would provide seed money; it would help a project get started. But then the operation—if it had any merit—would be able to find other sources to furnish continuing support. This continuing support might be tax dollars, private contributions, business support, other charitable organizations, or whatever; but it should not continue to be exclusively the Ford Foundation.

Deemphasis on Nonprograming Activities
The Foundation therefore undertook a thorough review of its ETV commitment and specifically its support of NETRC (Section 3.8). This resulted in the decision, in mid-1963, that the Ford Foundation could not continue to be the sole backer of a major all-inclusive *network* operation. Many of the areas into which White had expanded NETRC operations would have to be curtailed (since other outside sources of support were not readily available).[2]

In July, the Center announced that it was discontinuing all its radio services (the live Educational Radio Network and the international activities of the Broadcasting Foundation of America). Shortly thereafter, the name was changed to National Educational Television (NET), Incorporated, dropping "Radio" from its title. NET followed this announcement by dropping or greatly curtailing activities in other areas. Instructional television (the NITL operations) was eventually severed from NET ties to become independent (Section 4.3).

The Washington office was eliminated, with the former JCET activities being picked up by its successor, the Joint Council on Educational Broadcasting—later to be renamed the Joint Council on Educational Telecommunications (back to the JCET again). NET endeavors in development (fund raising), network affairs, and station relations were also substantially cut back—as was the extended services operation (supplying NET programs to commercial stations in communities where no ETV station existed).

These actions resulted in a major reorganization of the NET staff, with some of the former activities—on a reduced scale—being absorbed into new departments. Two of the vice-presidents resigned (to become ETV station managers in Los Angeles and Philadelphia). The program department was overhauled to place more emphasis on public affairs and cultural programing. These changes marked the beginning of a new phase of the center's history—a phase characterized by a reduction of nontelevision activities—and a corresponding concentration on quality ETV programing.

Continued Commitment to Programing
The Ford Foundation's decision to limit the scope of NET's activities in effect resulted in a substantial increase in support for its program service. This move reflected the parent organization's conclusion that "noncommercial television is unlikely to obtain a national program service of higher quality without the support of the Foundation."[3] Also, the foundation indicated—for the first time—that future financial backing would be considered as a continuing part of its program in educational television: "Subsequent grants may be made, if justified, on the basis of a review each year."[4] This, of course, represented a new approach for the foundation, which had heretofore insisted that ultimately NET would have to be able to function independently as a completely autonomous organization.

After 1963, NET continued in the capacity of a primary producer of noncommercial television programing. And the Ford grants—aimed specif-

---

[2]This is not to imply that White and the Ford Foundation were pulling in different directions. Quite to the contrary, White enjoyed close personal ties with the leadership of the foundation. The decision to cut back nonprograming activities was a mutual agreement.

[3]Ford Foundation news release, dated October 2, 1963.
[4]*Ibid.*

ically for programing support—continued to grow in size. Into the 1970's, the foundation was still contributing $10 million to $12 million a year to the ETV programing effort.[5]

4.3   Development of the ITV Libraries

Also during the early 1960's, another federally supported project had tremendous impact on the development of ETC in the schools. Just as it was not feasible to think in terms of a general public ETV movement without some sort of national networking/programing service, so would it be inconceivable to proceed very far into School TV considerations without coming to grips with the necessity of nationwide sharing of the better ITV materials provided by various schools and agencies.

The McBride-Meierhenry Study   Funded by an NDEA Title VII grant, a 1960–61 milestone study was undertaken by two University of Nebraska educators, Jack McBride (Director of Television) and W. C. Meierhenry (assistant dean of the Teachers College). The authors stated the need for their survey, "A Study of the Use of In-School Telecast Materials Leading to Recommendations as to Their Distribution and Exchange," as follows:

During these past years, instructional programming has grown from an ever increasing number of ETV stations, closed-circuit, and production-center studios. . . . Each year the amount of local instructional production increases and the amount of recorded instructional programming continues to grow. Correspondingly, the need continues unabated for a mechanism to facilitate the

acquisition, for local use, of instructional programming produced elsewhere.[6]

After an exhaustive study—involving numerous interviews and questionnaires—one of the findings of the study was that "it is evident that there is a sufficient quantity of recorded televised instruction available to warrant the establishment of pilot regional centers and/or a national center for distribution."[7] Therefore, the first recommendation of the study was that "a non-profit center for recorded televised instruction be established."[8]

Establishment of the NITL   Subsequently, in early 1962, the National Instructional Television Library (NITL) was created with another Title VII NDEA grant from the U. S. Office of Education (USOE). Although it had a separate national advisory board of educators, NITL was organized administratively as a unit of the NETRC. The library represented the first attempt at the organization of a national agency that would evaluate and provide means for cataloging and distributing the mass of STV materials that had been recorded throughout the country.

Specifically, the NITL had three major services: (1) to publish an annual "Guide to Films, Kinescopes, and Videotapes Available for Televised Use," (2) to provide sample lessons of courses listed in the Guide, and (3) to act as the distribution agency for a few of the courses. In general, however, the library served just as the national cataloging and information hub; interested users had to make arrangements for the use of most of the materials directly with the producers.

Regional Libraries   In addition to the national library, two regional libraries were also established, in Boston and in Lincoln, Nebraska.

---

[5]For a more detailed account of the history of NET, see Donald N. Wood, "The First Fifteen Years of the 'Fourth Network,'" *Journal of Broadcasting,* Spring 1969, pp. 131–144.

[6]Jack McBride and W. C. Meierhenry, *Final Report of the Study of the Use of In-School Telecast Materials Leading to Recommendations as to Their Distribution and Exchange* (Lincoln, Nebraska: U.S. Office of Education, 1961), p. 20.

[7]*Ibid.,* p. 50.

[8]*Ibid.,* p. 66.

Theoretically, the two regional libraries were established to help facilitate local acceptance of materials, to effect certain economies, and to encourage consideration of programing of a regional nature. As a practical matter, both of these two areas had the leadership and the desire for a larger role in the development of ITV library operations.

The *21-inch Classroom* in the Boston area was already serving 160 school systems in a four-state area with local/regional programs at the time of the USOE study. In Lincoln, under McBride's initial leadership, there was to develop perhaps the largest, most integrated ETV/ITV complex of coordinated operations to be found anywhere in the country.

NCSCT/ NITC/ AIT When NET disengaged itself from all ITV activities shortly after 1963, the NITL library functions were split into separate operations. It may be noted that the tripartite arrangement had never quite worked out in practice as it had been designed on paper. In effect, each of the libraries (National, Northeastern, and Great Plains) was serving users in overlapping geographical locations throughout the United States.

The Northeastern Regional Instructional Television Library has become integrated into the Eastern Educational Television Network (Section 5.6) and as such still operates out of a regional organization—but with a national clientele.

The National Instructional Television Library itself, severed from NET ties, moved to Bloomington, Indiana, and became the National Center for School and College Television (NCSCT).[9] The NCSCT continued to assess,

acquire, duplicate, and distribute programing from various sources, later changing its name to the National Instructional Television Center (NITC). As it became increasingly involved in an in-depth program of assessing needs in various subject areas, it turned to the production, through consortium arrangements, of outstanding materials in these various areas (Section 11.7). By the mid-1970's, the NITC had become international in scope and had been transformed into the Agency for Instructional Television (AIT).

Great Plains NITL The third library operation was initially titled the Great Plains Regional Instructional Television Library. As it split from the original triumvirate, it gradually grew—by several stages—into the Great Plains National Instructional Television Library (GPNITL).

The Lincoln library has always been the largest operation as measured by quantity of materials available for distribution. Starting with one series for distribution in 1962, the Great Plains Library was self-supporting by 1966 (when the original USOE grant expired). By the mid-1970's, GPNITL was distributing more than 150 various ITV courses—ranging from kindergarten to college and in-service teacher training materials (Section 11.7).

4.4   **The Role of the ETS Division**

Another important actor now makes his entrance upon the ETV stage. Just as the federal government helped to fill a *financial* void (with the ETV facilities law) when the Ford Foundation pulled back in certain funding areas, so did another agency step in to fill a *professional* void (when NET pulled back in some of its activities).

The NAEB and the Professional/Service Functions Actually, there had been some confusion (some would say conflict, others would say competition) between the roles of the NAEB and NET for several years—during the late 1950's and early 1960's. The NAEB was essentially the pro-

---

[9]Bloomington, Indiana, had become the distribution center for several ETV/ITV agencies: the old NET library and Extended Services Plan, the ETS Program Service (now the Public Television Library), and the NCSCT (now the AIT). Users have found it convenient to contract for space and certain administrative services through the Indiana University Foundation—partially due to the administrative and practical experience of the renowned Indiana University Audiovisual Services.

fessional organization, consisting of individuals working to further their profession, and NET was the *service* agency, providing a networking/programing operation for the stations. However, the distinction was not always that clear. To some extent, both organizations were serving as trade associations. The NAEB provided a programing *service* (going back to the tradition of the "bicycle" radio network); and NET was involved in many *professional* activities (research, station development, professional meetings, legal and technical consultation, and the like).

(To some extent, this organizational blur continues as, for example, the NAEB and the Corporation for Public Broadcasting discuss which agencies should have what responsibilities for the further development of instructional telecommunications, Section 8.6.)

So it was only natural for the NAEB to move into the professional vacuum when NET discontinued its various nonprograming services (Section 4.2).

Establishment of the ETS Division   At its 1963 convention, the NAEB membership—in a move reminiscent of its reorganization of 1934 and its metamorphosis of 1948—approved a new constitution for the association. The reorganization resulted in the creation of a new division—the Educational Television Stations division (ETS) of the NAEB.

To head up the operation, the NAEB called on an old friend and leader, C. Scott Fletcher—former president of the Fund for Adult Education and acting president of the ETRC during its first year. When the Fund was formally disbanded in 1961, Fletcher had remained active in educational broadcasting circles, serving in various consultancy capacities. In March of 1964, he accepted the position of acting president of the ETS division of the NAEB. He later stepped down from that time-consuming situation and remained active for many years as executive consultant to ETS.

When it was created, ETS assumed responsibilities in several areas—picking up some of the service functions previously assumed by NET and adding some of its own. Specifically, ETS spelled out six areas in which it would function: (1) activation and development of new ETV stations, (2) representation of stations before government and private agencies, (3) compilation of data regarding fund-raising activities (ETS would not itself get directly into fund-raising activities), (4) facilitating personnel training and placement programs, (5) holding regional and national conferences, and (6) establishment of an ETV program library service.

ETS Program Service   One of the first, and most important, of these activities fell into the sixth category—establishment of the ETS program service, operating out of Bloomington, Indiana. With grants from the W. K. Kellogg Foundation and the National Home Library Foundation, the ETS library service was initiated in 1965. It served as sort of a counterpoint to the NET scheduled "networking" operation. The Educational TV Stations division was the library/duplicating center that the ETRC tried hard not to be from its very start. Not a centralized distribution organization, ETS maintained no scheduled delivery service. It merely accepted the best ETV programing that stations wanted to offer, then duplicated these series and made them available to other stations, as requested. By furnishing this kind of library service, ETS made available throughout the country the kind of systematized, structured, educational series that NET had been abandoning over the years (Section 7.2). The ETS program service has since become the Public Television Library, operated as a department of the Public Broadcasting Service, located in Washington, D.C. (Section 7.4).

But it was in the area of long-range financing that the ETS division was to make what would be, undoubtedly, the most significant contribution of its early history. Before pursuing the

ETS contributions, however, we should note the developments taking place upon the ITV portion of the stage.

4.5   New School TV Patterns and Technologies

It can be pointed out that progress made in the realm of instructional telecommunications is related closely to distribution technologies. The earliest ITV—in the 1940's—made use of *open-circuit commercial television* channels (Section 2.5). As *noncommercial TV* channels came into use—in the early 1950's—this distribution medium made School TV available to thousands of schools. In the late 1950's, *closed-circuit distribution* developments made another medium available to additional thousands of classrooms (Section 3.9). In the 1960's, three new major hardware distribution breakthroughs were to make possible various other school applications of educational telecommunications.

Instructional Television Fixed Service Shortly after the first ETV stations came on the air, it became apparent that single-channel open-circuit broadcasting channels could not meet all of education's needs. In order to be able to make maximum use of television's potential, any ambitious STV system would need to be able to schedule more than one channel at a time. Multichannel closed-circuit systems provided the solution for many school systems, but would prove impractical or too expensive for other schools.

In 1961, an experiment was set up with the Plainedge public schools in Long Island, New York, to test the feasibility of a new limited-broadcast technique—using the upper reaches of the UHF spectrum (above 2,500 mHz). Far above the ordinary broadcasting ranges, this portion of the electromagnetic spectrum is used for—among other purposes—microwave transmission. The characteristics of this portion of the spectrum are such that the telecommunication

signals can be transmitted only when there is direct line-of-sight.

The Plainedge experiment proved satisfactory, and in July 1963, the FCC authorized establishment of the Instructional Television Fixed Service (ITFS) class of broadcast station in the 2,500–2,690 mHz range of the electromagnetic spectrum. The ITFS reservations included thirty-one different channels (each channel, as with standard VHF and UHF open-circuit transmission, is 6 mHz wide), limited strictly to educational users. (Three of these channels have since been deleted by the FCC.) The primary advantage to the ITFS system is that any single licensee can apply for up to four different channels—thus making a multichannel service possible.

4-2 ITFS Transmitter Repeater (Courtesy: Patrick Loughboro)

Further restrictions limit transmitter power; the effective range of transmission is at maximum about twenty miles. It is a point-to-point service from the transmitter to one or more specified receiver locations; each receiving antenna must be licensed with the FCC. The system requires a special receiving antenna and down-converter so that the transmitted 2,500-mHz signal can be converted down to a standard VHF channel and can be displayed on ordinary receivers. Therefore, the ITFS system is not feasible for ordinary home reception—or for any widespread *broad*-casting applications.

However, the advantages of the multi-channel capability, plus the relatively lower transmitting equipment costs (price of a single ITFS transmitter can be only 5 to 10 percent of the cost of a high-powered VHF or UHF transmitter), make it ideal for many educational systems.

The first licensed systems went on the air in late 1964: Mineola (Long Island) and Parma, Ohio, school districts. By the early 1970's, more than 100 different licensees were operating about 300 channels.

Slant-Track Video Recorders The history of hardware advances in radio and television, as in so many other fields, has largely been a record of technical breakthroughs that have resulted in miniaturization and greater quality at lower costs. Such a breakthrough came in the mid-1960's with the development of the helical-scan or slant-track video recorders.

When the first video recorders were developed in the mid-1950's, they were cumbersome and expensive to purchase and to use. With a complex system of four recording heads, these recorders were labeled "transverse," or "quadruplex," recorders. Using 2-inch wide videotape, the four heads are rotated rapidly at a right angle (transverse) to the path of the tape. These professional recorders are still the standard of the broadcast industry.

After some experimentation and development, a new simplified video recording system was introduced about a decade later. This was called the "helical-scan," or "slant-track," process because the revolving head scans the video-tape in a long diagonal curve or helical pattern. (Either one or two recording heads are commonly used in slant-track formats.) The resulting recording normally does not meet the technical quality of quad-head machines, but the purchase price and operating costs are so much lower that individual schools could think in terms of buying their own video recorders. Many models were available for less than $1,000.

One problem with the slant-track recorders was the early proliferation of formats and models. By the late 1960's, more than a dozen manufacturers were advertising almost forty different formats and models. Two of these models are pictured (4-3). Tape widths varied from ¼ inch to 2 inches, and at least eight or nine different tape speeds were employed. This early lack of standardization certainly frustrated attempts at circulation and distribution of materials—but in the free marketplace of competition, the better standards and formats eventually would win out. (Early standardization would have tended to freeze the state of the art at a less sophisticated stage.) A similar situation existed in the middle 1970's relative to development of video disc systems, with several early models and formats competing for acceptance and standardization (Section 15.1).

The big advantages of the slant-track recorders centered around the fact that they were relatively inexpensive and easy to use. School districts and individual schools were now able to set up inexpensive closed-circuit systems. Teachers could easily record demonstrations or portions of a lesson for later use. Administrators could use the video recorders to record messages or sample materials for repeated or distant showings. It was now feasible to involve students in the production and recording of television programing (Section 10.7). Schools could now use TV technology for self-evaluation purposes—recording a performance and then playing it back for analysis. Perhaps one of the biggest advantages was that—even without any studio equipment—schools could now record open-circuit programs off the air and play them back later at the convenience of the classroom teacher, in those instances where permission was granted and there were no copyright problems. The slant-track recorder had, indeed, opened up a whole new category of applications for ETC in the schools—and in industrial training programs.

4-3 Representative Helical-Scan Slant-Track Video Recorders: ½-inch Model (left) and 1-inch Model (right) (Courtesy: JVC Industries, Inc., and International Video Corporation)

Random-Access Systems One other specific advance should be mentioned in connection with the technical developments of the 1960's. This was the random-access retrieval system (also referred to as "dial-access," "information retrieval," and other terms).

The heritage of random-access information retrieval systems can be traced back to the original language laboratories in the armed services in World War II. As remote-controlled audio information retrieval technologies were developed by schools—typically in conjunction with foreign language instruction—the possibility of including video retrieval was explored. During the 1960's many such systems were installed.

Basically, the random-access video information retrieval system consists of a number of individual learning stations—or carrels—each furnished with a chair, small table or desk (usually built-in and partitioned from neighboring carrels), small TV monitor, headsets, and an intercommunication system (telephone dial or pushbuttons), plus other supplementary accessories needed for the particular installation—ancillary materials, talk-back capability, audio recorder, slide projector, typewriter, or whatever. The learning carrel is connected to some remote control/projection center that actually houses all the stored video (and audio) materials. Then, as the student requests a given item of information or program—by dialing, pushing buttons, or direct voice contact—an operator threads up the correct material and transmits it to the appropriate learning station through one of the many channels in the system. A typical carrel installation is shown in the illustration (4-4).

Such a system is expensive to install and operate; but if the learning situation could justify the price for the resulting individualized system, then the technology was available. The primary significance of the random-access systems is that they did represent an opportunity to use the telecommunications hardware to facilitate individualized instruction. Television did not have to be used solely as a means of *mass* education and large-group instruction.

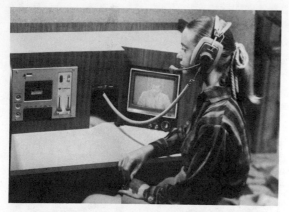

4-4 Student Using Learning Carrel Equipped with TV Receiver and Audio Cassette Player/Recorder (Courtesy: Califone International, Inc.)

As we shall see in Chapter 15, some of these advances and technologies shall themselves become outmoded or too expensive as newer developments and innovations (cassettes, discs, satellites, community cable TV hook-ups, and the like) become feasible for STV applications.

4.6   1967: A Major Turning Point

Earlier we pinpointed 1948 as a crucial year for ETV, and several other pivotal years have been identified: 1952, the year of the *Sixth Report and Order;* 1962, when Congress passed the Educational Television Facilities Act and the Ford Foundation began the transformation of NET; and 1967 must be listed as one of the most crucial turning points of all. In fact, this year is clearly the beginning of Act Five of the ETV melodrama. Several major developments reached fruition during that year. The most significant study ever made of ETV came forth with far-reaching recommendations that would dominate the industry for the next decade or two. A new corporation to oversee all of noncommercial broadcasting was created by federal action— and with federal financing. The first nationwide full-scale excursion into live, interconnected,

coast-to-coast ETV programing was undertaken. There was the first serious discussion of the possibility of the use of satellites for noncommercial broadcasting.

These events did not "just happen," however; they were the results of several years of less dramatic, but equally important, behind-the-scenes groundwork.

The First National Conference on Long-Range Financing   In December 1964, C. Scott Fletcher—on behalf of the ETS division of the NAEB—convened the First National Conference on the Long-Range Financing of Educational Television Stations. After his long tenure as head of the Fund for Adult Education, no one better understood the frustration of trying to find long-range financing security. Also, he understood, better than anyone else in the industry, the necessity for finding other sources of support—outside of the Ford Foundation—if noncommercial television was ever to achieve its full potential.

One of the key elements of this First National Conference was the call for a "Presidential Commission or National Commission with Presidential Approval." A three-page document, outlining the need and plans for such a commission, was presented to the White House in June 1965. Five months later, the Carnegie Commission on Educational Television was announced. That this prestigious commission had Presidential support is evidenced by the following message sent to James Killian, chairman of the Commission, from President Lyndon Johnson:

> I am happy to learn of the plan for a Commission financed by the Carnegie Corporation to study the role of noncommercial educational television in our society.... From our beginnings as a nation we have recognized that our security depends upon the enlightenment of our people; that our freedom depends on the communication of many ideas through many channels. I believe that educational television has an

important future in the United States and throughout the world. . . . I look forward with great interest to the judgments which this Commission will offer.[10]

It is significant that this influential and highly publicized commission was underwritten by the same philanthropic corporation that earlier in the century had almost single-handedly guaranteed the establishment of the concept of free public libraries throughout the country, that the Carnegie Corporation at this time was headed by John W. Gardner (soon to be named Secretary of HEW by President Johnson), and that the obvious analogy between free public libraries and free noncommercial television was to be emphasized by the adoption of the term "public television."

After fourteen months of serious and in-depth investigation and study, on January 26, 1967, the commission published its report, *Public Television: A Program for Action*.[11] It is no exaggeration to state that, as of the late 1970's, this report has been the most influential private document on noncommercial broadcasting released.

The Report of the Carnegie Commission The essence of the Carnegie Commission report was a series of twelve far-reaching recommendations, covering the establishment of a "Corporation for Public Television"; increased federal, state, and local financing; the creation of two major national production centers; live interconnection; uses of improved technology; increased research; federally guaranteed support of the corporation; more federal funds for ETV facilities; more attention for uses of television in formal education; and various avenues of support for local ETV stations.[12] Several of these proposals are discussed below, including the idea of creating a federally chartered corporation.

Aside from the impact of the specific recommendations coming from the commission, perhaps the most important influence of the report was the mere existence of the report. Here was an extremely prestigious, influential, highly regarded commission—funded by the independent and respected Carnegie Corporation (with no axe to grind)—saying many of the same things that the Ford Foundation had been saying for the past fifteen years: the idea of noncommercial educational television has a tremendous potential in this country; it is the obligation of the federal government to take some leadership in guaranteeing support of the resource; ETV can no longer be considered a private philanthropic cause of the giant Ford Foundation; and so on. The important thing was that someone else—not the Ford Foundation—was saying it this time. Even more importantly, there were persons in Washington—both in the White House and in the halls of Congress—who were listening.

The Public Broadcast Laboratory As the Carnegie Commission was drafting its momentous report, the Ford Foundation was turning its attention to still other ways to promote the cause of noncommercial television. In anticipation of the Carnegie Commission's serious call for network interconnection of ETV stations, the Ford Foundation decided to underwrite a demonstration project showing what actual interconnection could accomplish.

The Public Broadcast Laboratory was funded by Ford as a special NET project. More than half of the NET affiliates were connected by network lines for the 1967–68 season. The network was in operation six days a week, eight hours a day—transmitting not only public broadcast

[10]Warren L. Wade and Serena E. Wade (eds.), *A Report on the Second National Conference on the Long-Range Financing of Educational Television Stations* (Washington, D.C.: National Association of Educational Broadcasters, 1967), p. iii.

[11]Published and sold as a paperback book by Bantam Books, Inc., New York.

[12]The twelve Carnegie Commission proposals are reprinted in Appendix B.

materials, but also program previews, promotion materials, and in-service programing.

Additionally, the project included a live program every Sunday evening—in a demonstration of the value of immediate, current, live programing to the ETV movement. This series further developed the use of the "magazine-type" format for public affairs presentations—later successfully used by commercial public affairs programs such as CBS's *60 Minutes*.

The First Satellite Proposals While demonstrating the desirability of live interconnection, the Ford Foundation was also promoting the use of even more advanced technology. The FCC had issued a *Notice of Inquiry* in 1966 for interested parties to express their views and comments "in the matter of the establishment of domestic non-common carrier communication-satellite facilities by non-governmental entities."[13] The Ford Foundation was the first agency to respond. It presented its preliminary proposal in late 1966 and followed up with supplemental comments and a legal brief in early 1967.

The Ford plan called for the establishment of the "Broadcasters' Non-Profit Satellite Service." This plan would have established six commercial channels and five noncommercial channels in each of the four continental time zones. One of the ETV channels would be for what was soon to be labeled public TV, and the other four noncommercial channels would be for instructional materials. The costs of the operation of the satellite services, including the PTV/ITV transmissions, would be paid by the commercial broadcasters who would benefit because it would be less expensive for them to use the transcontinental satellite service than the existing long-lines network interconnection.

This proposal, if adopted, would have represented one way that the Ford Foundation

could gracefully get out from under its continuing obligation to support ETV so extensively. Although the plan was not adopted, it was important in that it was the first comprehensive domestic satellite proposal submitted to the FCC (many others were to follow from commercial networks, communications industries, aerospace concerns, and the like) and in that it established the precedent for including noncommercial broadcasting in any satellite considerations.

The Public Broadcasting Act of 1967 Seldom have the gears of federal government turned quite so expeditiously in churning out legislation for a completely new program—especially one that was not a response to a national crisis. The Carnegie report was issued in late January 1967. President Johnson submitted his proposed Public Broadcast Act to Congress in March. The Senate passed the bill; the House passed its version; a compromise version was worked out and passed by both houses; and President Johnson signed the bill into law on November 7, 1967—a little more than nine months after the report was first issued.

The Public Broadcasting Act of 1967 had three parts. The first part was an extension of the Educational Television Facilities Act of 1962 (Section 4.1); it authorized expenditures of $38 million for educational TV and radio facilities during the next three years.

Part Three authorized $500,000 for "a comprehensive study of instructional television and radio . . . and such other aspects thereof as may be of assistance in determining whether and what Federal aid should be provided for instructional radio and television."[14] This was to be a companion study (for instructional telecommunications) to the Carnegie Commission report. A special Commission on Instructional Technology was established, chaired by Sterling M. McMurrin (dean of the Graduate School, University of Utah, and former commissioner of education, USOE),

---

[13]*Comments of the Ford Foundation in Response to the Commission's Notice of Inquiry of March 2, 1966* (New York: Ford Foundation, 1966).

[14]Reprinted in Koenig and Hill, *The Farther Vision,* p. 380.

and the scope of the study was broadened to include all of instructional technology. The commission's report, *To Improve Learning,* was submitted to Congress in March 1970, with six suggested recommendations. Despite the scope and solid background of the report (it involved a comprehensive study of the status of instructional media in the United States), it failed to excite the public imagination or have the same legislative impact as did the Carnegie report on public broadcasting.[15]

The crux of the Public Broadcasting Act, of course, was the middle part, which established the Corporation for Public Broadcasting (CPB). (The Carnegie Commission had called for the establishment of a corporation for public television, but during the congressional hearings the proposed corporation was expanded to include radio.) The CPB was created as a nonprofit publicly chartered corporation, "which will not be an agency or establishment of the United States Government."[16]

The fifteen members of the board of directors of the corporation were to be appointed by the President of the United States (with the advice and consent of the Senate), with not more than eight to be members of the same political party. The act repeatedly affirms that the CPB is not to be a governmental agency; the members of the board are not to be regarded as government employees.

The CPB is charged generally with the responsibility of assisting new stations in getting on the air, establishing one or more systems of interconnection, obtaining grants from federal and other sources, providing funds to support program production, making grants to stations to support local programing, and conducting research and training projects. However, the corporation is specifically prohibited from owning or operating

any station, system of interconnection, community antenna TV (CATV) cable system, or production facility. (The structure and functions of the CPB are discussed further in Section 6.3).

Federal funding of $9 million was given to the corporation for its first year of operation. Additional support was pledged by the Ford Foundation, Carnegie Corporation, CBS, and others.

4.7   The Corporation Takes Over

Although the CPB was authorized by the end of 1967, it was still a year or so before it could become operational and establish the other entities that would fulfill the promise of the Public Broadcasting Act of 1967.

The Creation of the Corporation   Most of 1968 elapsed before the board of directors was complete. As chairman of the board, President Johnson appointed Frank Pace, former Secretary of the Army and later head of the International Executive Service Corps.

In February 1969, the CPB board selected John W. Macy to serve as its president. Macy had been head of the Civil Service Commission for the Johnson administration. With additional officers appointed to head up specific operational divisions—television activities, radio activities, public information, research and development, and so on—the CPB was functioning in its Washington offices by mid-1969. The corporation was then ready to turn to the specific problems of providing for an organization to run the interconnection service.

The Public Broadcasting Service   Since the corporation was prohibited from owning or operating the network service, involvement of another agency was necessary. Although it was conceivable that some existing organization might meet the need (NET would have been the most obvious choice), the CPB and the station representatives chose instead to create a new entity to serve as the television network—the Public

---

[15]*To Improve Learning: A Report to the President and Congress of the United States,* Commission on Instructional Technology (Washington, D.C.: U.S. Government Printing Office, March 1970).

[16]See Appendix C, page 350.

Broadcasting Service (PBS). The essence of the PBS operation—which distinguished it from NET—was that it would be a station-operated interconnection, a true "mutual" network. The station managers elected their own representatives to serve as the PBS board of directors; and Hartford N. Gunn, Jr., general manager of the prestigious Boston public television station, WGBH-TV, was chosen as president of PBS.

The Public Broadcasting Service was established solely as the networking agency. Its functions were to select, schedule, and distribute network programing over the interconnection; PBS was not to be a production center. It is interesting to note that the scope of PBS's activities and policies were parallel to the charges given to the fledgling ETRC nearly two decades earlier (Section 3.8).

Under Gunn's leadership, the first PBS network package was initiated in October 1970. The interconnected fourth network had finally become a permanent reality—not a "bicycle" network, not a temporary demonstration project, but an actual nationwide television interconnection. It took forty years from the time when the early educational broadcasting pioneers first dreamed of a real network operation; and, ironically, it came to television first—although radio was not far behind.

National Public Radio As the scope of the corporation included radio, it was apparent that a parallel radio network should also be established. Although the name of the Public *Broadcasting* Service would seem to imply that the networking agency could encompass both television and radio, it was felt that a separate entity should be established to handle radio networking activities. Just as NET was passed over in favor of creating a new organization for television networking, so was the National Educational Radio (NER) division of the NAEB (which ran the "bicycle" network) passed over in favor of creating a new operation and new image for radio.

So, National Public Radio (NPR) was created, and the interconnected public radio network was initiated in May 1971. National Public Radio differs from its television counterpart in at least two significant ways. First of all, NPR is a producer as well as a distributor; it produces much of its own programing, especially in the public affairs area. Secondly, NPR makes some specific grants directly to the public radio stations; it receives the funds from the corporation, then actually specifies them for individual station projects. The Public Broadcasting Service, on the other hand, does not make direct grants to individual PTV stations with CPB dollars; the stations receive television grants directly from the corporation.

(Recent activities of CPB, PBS, and NPR are discussed in Chapter 6, when we look at the current status of these Washington organizations as part of "The National Picture" of public broadcasting.)

For a couple of years after the creation of PBS and NPR, the NAEB continued with its professional station representation functions—with the ETS division representing television stations and the NER carrying on a similar function for radio. However, after a major realignment of national organizations in 1973 (Section 6.4), ETS merged with the original PBS and was absorbed into the new PBS corporate structure and image—with many of the ETS personnel being transferred to the new PBS. The original functions of the ETS division had been met by this time; ETS had performed its purpose well by initiating the study that resulted in the creation of the new public broadcasting hierarchy; and most of the charges to ETS had been superseded by either the CPB or PBS.

Also, as NPR became the dominant force in public radio activities, the NAEB decided that it no longer needed to support its NER division either. It would discontinue both its television and radio operations. Thus, at its 1973 convention, the NAEB again (adapting itself to shifting professional responsibilities and organizational relationships) approved a reorganization plan and a new constitution—reshaping the association strictly as a professional society composed of individual members. At that point, a new organization was formed—the Association of Public Radio Stations (APRS)—to handle some of the radio representation functions previously conducted by the NER (Section 6.5). No more station activities were to be assumed by the NAEB. After almost fifty years, the last vestiges of the old ACUBS were dropped by the professional society.

But what was to become of NET?

Day at NET: The Merger After the climactic decisions and turning points of 1967, NET was destined to play a different kind of role in national noncommercial television. Entering upon a new phase of its existence, NET soon ceased to function as the primary distribution agency for the 130 stations then on the air, and it turned even more to program production. This was a continuation of its deliberate campaign to upgrade the quality of the programing image of the fourth network. Following the 1963 reorganization of NET, the earlier concept of building up station production capabilities by "farming out" network production to affiliated stations was gradually abandoned (Section 4.2). By 1967, NET had established its own production crews, and was producing more than half of its material itself.[17]

As other agencies and systems were devised to handle the actual networking of public TV programs, John White felt that this would be a good time for him to step down as NET president. He had directed the growth and development of the Center for close to a decade; he had seen the Center assume a position of national preeminence; and he felt that it was time for new leadership. He resigned in 1968; and the NET board selected James Day, general manager of the innovative San Francisco community ETV

station, KQED, as the new NET president. Both Day and Hartford Gunn (new president of PBS) had long held positions of respect and leadership among their peers in the ranks of station managers.

Under Day, NET continued to play a key role as producer of quality noncommercial ETV programing. Then, in 1969, NET merged with WNDT, the New York City community PTV station (which subsequently changed its call letters to WNET) in order to command its own production studios. No longer the distribution center, no longer the focal point of all PTV activities, WNET does continue as one of the heaviest station producers of PBS materials.

New Faces, New Associations In the evolution of movements as well as institutions, it is a common phenomenon that the individuals who have led the fight for a new era often find themselves pushed aside once the era arrives. Examples can be found within political parties, civil rights movements, business corporations, and the like. So also with the educational radio pioneers who had struggled for more than two decades to establish their medium, only to find new progressive leadership taking over the expanded movement (at the 1948 NAEB convention) just as they were on the threshold of ETV. Similarly, agencies that had led the fight for the establishment and recognition of ETV throughout the country (specifically NET) found themselves relegated to the background just when they were on the verge of a breakthrough. Undoubtedly, jealousies and power struggles—as well as the need for fresh approaches—play an important role in such transitions (for example, many station managers felt that NET exerted too much centralized control, ignoring station input).

This is not to question the leadership and national reputation that NET had furnished the movement for more than fifteen years; but, once NET had succeeded in gaining nationwide recognition for the movement, perhaps it was inevitable that new faces, new associations, new sources of

[17]John F. White, "National Educational Television as the Fourth Network," *in* Koenig and Hill, *The Farther Vision*, p. 88.

financing, and a new image would take over for the next thrust forward.

No one can pretend to know what the future holds for educational telecommunications. The potential is unlimited; the stakes are high. In preparation for consideration of the present and the future, we have examined the past—for the purpose of underscoring the main events—in five dramatic acts: (1) The early history of educa-tional radio (1917–1948); (2) the FCC freeze (1948–1952); (3) the decade of growth and evolution (1953–1962); (4) regrouping and professional introspection (1962–1967); and (5) the new identities (from 1967). It has been an exciting and sometimes turbulent drama. It has been a fast-paced half-century. And the past has been but prologue.

two

**The
Public
Stations**

With this chapter we begin a look at the structure of public television and radio in the United States. Before continuing the story of some of the national organizations started in Part One, we need to look more carefully at the individual stations that are the most visible segment of the public broadcasting system. Exactly how are these "public stations" organized and operated? How are the stations structured? Who makes what kinds of decisions? Where does the money come from? How are state and regional interconnections of stations operated?

In subsequent chapters of Part Two, we will enlarge our view to consider the status of the national operations introduced earlier; we will look at the Public Broadcasting Service interconnection and at public TV and public radio programing; we will see what research tells us about the audiences for noncommercial TV and radio; and we will discuss some of the continuing basic issues in public broadcasting.

5.1 Twelve Components of a Public Broadcasting Station

Certain common elements are present in varying degrees in every open-circuit noncommercial TV or radio station—although some of these elements may be taken for granted in a given station. These twelve components could be used as a set of guidelines or criteria against which to evaluate the structure and performance of any local public station.

1. *Public Broadcasting Philosophy:* First of all, there must be some sort of basic philosophy or approach toward the use of media for noncommercial purposes. This may be a formal stated policy. It may be the result of a conscientious examination of some of the considerations reviewed in Chapter 1 of this book. It may be as simple as an unspoken commitment on the part of a handful of individuals to do something as constructive and positive as possible with the broadcast media. It may be dedication to a specific

social or educational purpose. But stated or not, there always is some sort of basic philosophy or policy—some sort of institutional directive—that forms the underpinning for every public TV or radio station.

2. *Community Needs:* As with any public or educational project, there must be specific felt needs before the project can be justified. There must be a problem before we can look for a solution. Some of the more common "needs" have centered around a lack of intelligent public affairs information; the need for basic adult education, classroom instruction, and children's out-of-schooling material; a lack of cultural outlets; specific ethnic or minority needs; particular social problems; or other specific or general categories or combinations of needs to which a public station could respond. Some needs are generally understood and unstated; others are spelled out in the charter of the station, in its articles of incorporation, in the enacting legislation that created the station, or through formal ascertainment procedures (Section 9.2).

3. *Governing Board:* Regardless of the type of station, there always is some form of governing and policy board responsible for the station. There may be a single board of directors or governors—such as a special board of civic leaders licensed to govern and set policy for a community station (Section 5.2). It may be a special state agency that holds the license and sets policy. It may be a board of education that has both responsibilities. Or governance and policy may be divided at the board level, as in the case of many university stations, with a board comprised of civic leaders advising the station on policy matters, and a board of trustees or regents having ultimate responsibility for the license. In any event, one or more groups are charged with responsibility for the overall affairs of the station, and these groups are in turn responsible to some sort of constituency—the voters in a district, the taxpayers in a state, or the citizens of a community. In every instance, ultimate power flows back to the people in some way through some type of representative board.

4. *Management and Staff:* The actual daily operation of the station is up to the full-time (and part-time) staff and volunteers who are responsible to the board. The staff consists of several different categories, such as administrative and management persons, production and engineering personnel, on-camera "talent" (either station staff or outside celebrities or specialists), community volunteers (who can perform a variety of tasks), and interns (many university-affiliated stations depend extensively upon student help) (Section 5.3).

5. *Financing:* One obvious component of any station is, of course, the money to run the operation. For noncommercial station operations, the funds come from a wide variety of sources: tax support (federal, state, and local), foundations, private memberships, gifts and donations, corporate underwriting, and auctions and other specific money-raising schemes. The sources of financial support vary greatly according to the kind of station operation (Section 5.4).

6. *Community Involvement:* In virtually every type of noncommercial broadcast operation, there is some degree of community involvement—either an extensive commitment or a rather minimal kind of effort. It may take many forms. The top management of the station may be active in civic affairs, prominent in community organizations; the station may have formal advisory councils and community committees to work with the station in specific areas. The station must be identified with the community—whether the "community" is a school district, a neighborhood, a section of a state, or a certain segment of the population. The members of the community must have ways to make themselves heard by the station.

7. *FCC License:* Turning to the material and legal aspects of a broadcast station, one obvious requirement is the Federal Communica-

tions Commission license. Like its commercial cousin, the public station is fully licensed by the commission (although under somewhat different operating requirements). Inherent in the license is the need to adhere to all of the applicable FCC rules and regulations and to operate the station in the public interest, convenience, and necessity.

8. *Physical Plant:* Another component common to all stations is the physical plant: generally, the building and facilities to house all administrative space, programing and traffic operations, and specifically the engineering and production facilities and equipment needed to originate local programing. Again, operations differ greatly from station to station: one station may rely heavily upon network productions, producing relatively few local productions; another station may go in heavily for local originations; yet a third may be a major production center for regional and national programing and may have extensive facilities.

9. *Nonlocal Programing:* In addition to the need to do some local productions (to meet local community requirements), a public TV station must also be able to procure programing from other-than-local sources. This need has been the principal focus for most of our investigation of the history of public broadcasting (from the Educational Television and Radio Center to the Corporation for Public Broadcasting). All stations share a common need for networking and interconnection arrangements that will give the local outlet sources for regional and national programing, and these interconnections help to bind and unite the individual stations into a greater whole. (Sections 5.5 and 5.6).

10. *School Relations:* Almost every public TV station (and many public radio stations) devotes a great portion of the daytime programing schedule to instructional School TV productions. Classroom instruction is the heart of many station operations, while for others it is only a peripheral service. In any event, in virtually every station organization, there is some sort of arrange-

ment (office, personnel, consultant) for handling both formal and informal relations with the local schools—whether or not the station actually produces any instructional programing (Chapters 11 and 16).

11. *Publicity and Promotion:* Like commercial stations, the public TV and radio stations must make their potential audiences aware of what is being scheduled. Public stations typically enjoy good relations with local newspaper columnists and critics and can often obtain much free publicity through the local press. Paid advertising is another obvious form of publicity; national commercial underwriters who have supported the production or distribution of some prestigious national PTV series (such as Mobil Oil underwriting *Masterpiece Theatre* or Polaroid supporting *Nova*) often will also pay for advertisements in the local press. Many public stations also print their own program guides, which are distributed both to viewers/listeners who become members of the station and to newspapers to publicize the program schedule.

12. *Audience Feedback:* Finally—just as no communication process is complete without some sort of feedback—no broadcast operation is complete without some type of audience response or feedback. In the commercial business, feedback consists of ratings and—ultimately—products and services sold in response to advertisements. In noncommercial broadcasting, the feedback is not quite so tangible and easy to obtain. Sometimes data from rating services apply—but generally PTV audiences are so small that ratings do not carry much significance. Stations often rely much more on telephone response and letters in reaction to a given program. Other forms of community feedback are also important: press reviews, feedback from organized clubs and agencies, decisive comments and actions from community leaders and opinion makers, and the like.

These, then, are twelve elements or components that are common—in one form or

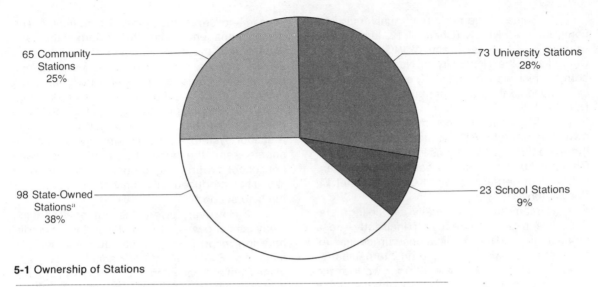

5-1 Ownership of Stations

ªIncludes stations owned and operated by state boards of education.

Source: Data in 1975 NAEB Directory of Public Telecommunications (Washington, D.C.: NAEB, 1975), pp. 106–120. The Directory lists 259 PTV stations including Samoa, Guam, Puerto Rico, and the Virgin Islands.

another—in virtually every PTV or public radio station. Analysis of any station in terms of these twelve essentials would result in a reasonably good profile of that particular station operation.

5.2 Four Categories of Station Ownership

For convenience in studying station structure and operation, it is helpful to look at specific categories of station ownership. Depending upon who owns and finances the stations—and for what purposes—stations have certain similarities and differences. Traditionally, there have been four principal categories of station ownership (figure 5-1).[1] The following discussion

is geared primarily to television station ownership distinctions, but the same types of categories and divisions also exist among public radio stations.

University Stations Almost 30 percent of all public TV stations in the country are licensed to universities and colleges, virtually all of which are public state-supported institutions of higher education. Many of these university licensees operate a single station for the immediate campus and for the greater metropolitan community surrounding the university. Typical of such operations are station KUHT, Channel 8, University of Houston, Texas—the first ETV station in the country, dating back to 1953; WKAR, Channel 23, East Lansing, Michigan—licensed to Michigan State University—originally on the air in 1954 on Channel 60; and WHA-TV, Channel 21—also on the air in 1954—the video counterpart to the country's first educational radio station at the University of Wisconsin.

[1]For a detailed and easy-to-read analysis of these four categories, see Frederick Breitenfeld, "The Four Faces of Educational Television," in Allen E. Koenig and Ruane B. Hill (eds.), The Farther Vision: Educational Television Today (Madison: University of Wisconsin Press, 1967), pp. 35–49.

Some of the early major university stations, such as KUON (Channel 12, Lincoln, the University of Nebraska) and WUNC (Channel 4, Chapel Hill, the University of North Carolina)—both of which were on the air by early 1955—have since become the nuclei of major state networks (Section 5.5).

Many two-year community colleges also own and operate PTV stations. For example, there are three such stations in California: KOCE, Channel 50, Huntington Beach; KVCR, Channel 24, San Bernardino; and KCSM, Channel 14, San Mateo.

Many of the university and college stations are related—directly or indirectly—to a broadcasting or communications curriculum. As such, students are often used on internships (or otherwise receive course credit) for work at the station; this practice aids considerably in cutting production expenses. Often the chief station executives also hold academic appointments, teaching perhaps one class a semester.

University stations sometimes, however, run into identity problems. As an integral part of the academic community, the station management is very much attuned to the "ivory tower" intellectual atmosphere—frequently drawing heavily upon university resources and faculty for erudite and cultural programing. At the same time, the station management is aware of the need to relate to the general community—trying to aim for a more general, less elite audience.

School-Owned Stations   If the university station has an identity problem, this situation is intensified among stations owned and operated by school districts. School-owned stations—which account for a little less than 10 percent of all the stations in the country—have real split-personality problems.

Licensed to a local board of education, the school station obviously has the primary mission of serving the schools in its district; yet there is also the need—inherent in the station's FCC license—to serve the general community. The station management usually is aware of the need to do more in this area, and the station staff would like to produce some community public affairs programing, some local cultural materials, and some good out-of-school children's programing. However, the station management reports to a board of education that is charged with running a school system—not a community cultural resource—and the board frequently must deny nonschool programing proposals that it would like to see produced but simply does not have the funds to support.

However, given this potential conflict, many school-owned stations have done a creditable job in producing some outstanding PTV materials. Station KRMA, Channel 6—licensed to the Denver Public Schools in 1956—pioneered in several general adult ETV series for NET. Other examples of school-owned ETV stations include: WTHS, Channel 2, Miami, Florida (Dade County School District); WETV, Channel 30, Atlanta, Georgia (Atlanta Board of Education); KNME, Channel 5, Albuquerque (licensed jointly to the University of New Mexico and the Albuquerque Public Schools); and WNYE, Channel 25, New York City (Board of Education).

State-Owned Stations   Like university stations and school-owned stations, public TV stations owned and operated by a special state authority or commission are supported principally and directly by taxes. This category of stations accounts for almost 40 percent of all the PTV stations on the air.

The South, generally, was the first region to see the potential of educational TV for alleviating some of the serious shortcomings of its public school systems. Alabama was the first state to set up a separate state commission to build and operate a series of ETV stations. Its first two stations—WBIQ, Channel 10 in Birmingham, and WCIQ, Channel 7 in Mount Cheaha—were on the air by 1955. Twenty years later, the

Alabama Educational Television Commission was operating nine stations that blanketed the state.[2]

Many other states eventually followed suit with separate state ETV commissions, state-wide operations run by a state university or state board of education, and even statewide closed-circuit systems. By 1975, fifteen states had separate, identifiable, specially established state commissions or authorities that owned and operated one or more PTV stations; most of these were operating an interconnected state network (Section 5.5).

Whereas school and university stations face programing problems in terms of defining their audiences, state-owned stations face a programing problem in defining their political role. State ETV commissions and authorities, by the very nature of their establishment and tax support, are responsible directly to the state government—usually reporting straight to the governor or to the state legislature. With this lack of insulation from political pressure, the threat of self-imposed political censorship is ever-present. The danger is not so much outright political control as it is the tendency toward bland, noncontroversial, apolitical programing.

Community Stations   This category of PTV station ownership is unique in that it has no direct local or state tax base; the community station exists and survives because of the determination and commitment of the lay citizens of a given community (Section 3.5). Typically, a community station is licensed to a nonprofit corporation established specifically for the purpose of owning and operating a noncommercial PTV station, rather than being licensed to a tax-supported institution.[3]

The fact that these community stations have no guaranteed source of tax support is both a drawback and an advantage. The drawback, obviously, is financial insecurity; there is no easy and reliable guarantee of where next year's budget is coming from. The advantage, on the other hand, is much greater flexibility and freedom in programing; no single financial kingpin can tell the community station how or what to program. Diversity of funding resources results in freedom of programing policies. This lack of guaranteed tax support does not necessarily mean that a station has to operate on a bare subsistence level—especially in the major metropolitan areas. Several of the larger community PTV stations—drawing on a wide variety of areas of support—have annual operating budgets in excess of $5 million, with a couple above the $10 million mark (Section 5.4).

About one fourth of all PTV stations fall into the community-station category—including those in a wide variety of smaller cities, such as Yakima, Washington (KYVE, Channel 47); Peoria, Illinois (WTVP, Channel 47); and Harrisonburg, Virginia (WVPT, Channel 51). However, the larger metropolitan community stations have always been the leaders in the PTV field. The first big stations were in the major centers where the FCC was able to reserve a noncommercial TV channel in the VHF spectrum. These were among the first major PTV stations on the air: WQED, Channel 13 in Pittsburgh; KQED, Channel 9 in San Francisco; WGBH, Channel 2 in Boston; WTTW, Channel 11 in Chicago—all on the air by the end of 1955.

---

[2]Ironically, in 1975 the FCC refused to renew the licenses for the Alabama Educational Television Commission—at least temporarily—for alleged racial infractions dating back several years. The Alabama commission was then permitted to refile for the licenses.

[3]In most respects, the community station is unique to noncommercial television—although, strictly speaking, viewer-supported noncommercial radio stations (such as the Pacifica stations) are community stations. Community public radio is growing rapidly, both independently from television (as in Minnesota where a network of stations is developing local support) and in conjunction with public television (many community corporations operate both a public TV and a public radio station, although incorporated primarily for public television).

In other major metropolitan centers, the choice VHF channels were already committed to commercial interests before the initiation of the FCC "freeze." Thus, development of these stations came along later: WETA, Channel 26 in Washington, D.C. (1961); KCET, Channel 28 in Los Angeles (1964). In New York City—where there was no room for a VHF noncommercial reservation—Channel 13 was purchased from commercial interests, and community PTV station WNET (originally WNDT) went on the air in 1962.

It must be realized, of course, that this classification of all PTV stations into four convenient categories is somewhat arbitrary and simplified. Not all stations neatly fit these compartments. For example, the ETV stations in Georgia are owned and operated by the state Department of Education. Should they be labeled as school stations or state-owned stations? And how about the statewide operations in Nebraska, Maine, and North Carolina, which are run by the state university system—into which pigeonhole do these stations belong? Where does one place WNYC, Channel 31 in New York City, which is licensed to the city itself?

Other divisions and categories are also used by some classifiers. Saettler, for example, divides all stations into *three* varieties: community stations, state network stations, and single-agency stations (including school, university, and WNYC).[4] The important principle behind any attempt at categorization is simply to point out that the purpose and licensee of a PTV station will dictate, to a great extent, its structure, financing, and programing policies.

5.3 Inside the Public Broadcasting Station: Structure and Organization

Turning to the individual station, a case study is the best way to describe how public tele-

vision and radio stations are internally organized. The KPBS stations at San Diego State University provide a good model, as these stations combine several of the characteristics of the different types of stations discussed above. Both the television station, KPBS-TV (Channel 15), and the FM radio station, KPBS-FM (89.5 mHz), are located on the campus of San Diego State University, and are basically *university stations*. The licenses are in the name of the Trustees of the California State University and Colleges "on behalf of San Diego State University." KPBS-TV receives a line-item appropriation from the state legislature (within the University's budget), so the station can be said to be owned by a *state agency*. The television station was established through a "joint powers agreement" with the schools in the area, so it has some of the characteristics of a *school PTV station*. The stations (TV and FM) also are advised by a community advisory board comprised of leaders in the area, which functions in some ways like the governing boards of *community PTV stations*; and the stations look to memberships, auctions, and corporate underwriting for some of their financing.

Organization of the KPBS Stations   Many universities have separate managers and organizations for their television and radio stations. At San Diego State a single manager is responsible for both; but where there are separate organizations, the TV and radio station managers usually report to an administrator who coordinates the work of the two managers in areas such as budgeting, facilities, engineering support, and community relations. This person could be a director of broadcasting, director of extension, coordinator of communication services, vice president of auxiliary services, director of public relations or development, audio-visual director, or a similar administrator.

The internal structure of a public broadcasting station is often similar to that of a commercial station—eliminating the sales department and substituting, perhaps, a development depart-

[4]Paul Saettler, *A History of Instructional Technology* (New York: McGraw-Hill Book Company, 1968), pp. 231–235.

ment (fund-raising unit). The KPBS stations are organized into five separate departments; at other public stations, there may be as few as two and as many as nine or ten separate divisions—depending upon how the tasks are assigned. Regardless of the number of departments, the activities are basically the same from station to station. The five KPBS departments—plus the administration activities and the involvement of the community advisory board—are shown in the accompanying chart (5-2).

TV Program Department The program director, as head of the program department, is responsible for the program schedule and for developing local programing, but he is not responsible for studio operations. In some stations, these three functions—programing/scheduling, program development, and production—are all under the program director. In other stations, these might be under three or more different administrators.

At KPBS, the TV program director is re-

5-2 Organization Chart of KPBS TV and FM

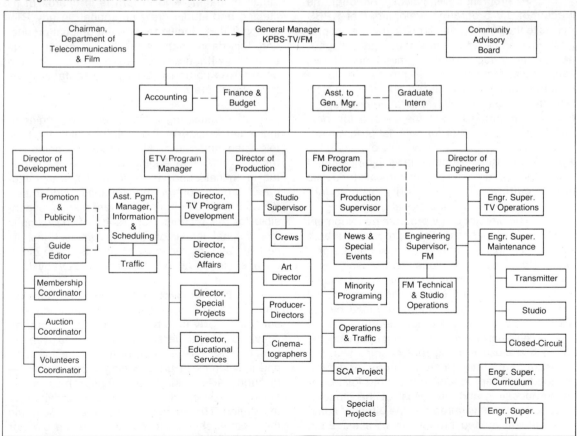

sponsible for all activities connected with scheduling and promoting network, syndicated, and (once they are produced) local productions. The TV program director is responsible for the "look" and "sound" of the station. He decides which nationally distributed programs the station should broadcast, and he participates in discussions on the local productions that should be undertaken.

Individuals who report to the program director have specific responsibility for program scheduling (traffic), program development, program information (primarily through the program guide), special projects, announcements (continuity), and viewer correspondence.

TV Program Development   Although the director of TV program development at KPBS reports to the TV program director, the area of program development functions almost as a department within a department. As mentioned above, many stations separate local program development from program scheduling. At WNET in New York, the station created two autonomous departments for program development—one for local programing and another for national program development. At KUON-TV, University of Nebraska, separate departments under the assistant general manager for programing are responsible for public affairs, cultural affairs, minority programing, special projects, and instructional programing.

The director of TV program development at KPBS-TV proposes the local programing, both series and single programs, which should be produced. Most local productions are discussion programs of current community activities and issues, but music, dance, and even some experimental dramatic programs have been produced by the KPBS-TV program development department.

As a university broadcasting operation, the station has close working relationships with the department of telecommunications and film, the academic department responsible for teaching broadcasting and film. Most of the student-produced programs aired by the station come out of the advanced production classes. Each

semester, the students produce several programs in the color studio, usually including an original drama or comedy, a variety show, a children's program, and two or three discussion or documentary programs.

Production Department   When a program is proposed by program development, the director of production usually takes part in the initial discussions, since it will be up to him to determine whether the resources of the station are adequate for the proposed program or series. Under the director of production are such people as the TV director/producers, graphic artists, cinematographers, photographers, and the TV studio manager, who in turn is in charge of all staging and studio crew personnel. In any PTV station with a significant amount of local production, this department is the hub of much of the activity—with many creative and production people who comprise a large percentage of the total station staff.

Engineering   The director of engineering (chief engineer) has responsibility for both television and radio. All engineering personnel report to the chief engineer, directly or through the technical supervisors for radio, transmitter operation, TV control, maintenance, and lab studio operation. The chief engineer also is responsible for the technical operation of the campus instructional television system, which transmits courses and instructional segments on an eight-channel closed-circuit television (CCTV) system.

Development and Membership Services The director of development is concerned with soliciting "private" money to support station activities. He is personally concerned with attempting to interest corporations, businesses, and individuals in underwriting programs and—through various kinds of memberships in the station—in supporting the general activities of the station. The auction, the volunteer office, and the membership office come under the director of development—who is also responsible for promotion, public relations, and the production of the program guide.

Development activities are becoming increasingly important to all types of stations; they have always been of paramount importance to the community station. The head of development in a large community station often assumes an importance second only to that of the general manager, and the development staff may be one of the largest at the station. Without success in this area, the station ceases to remain viable.

FM Program Director   Like the TV program director, the KPBS-FM program director is responsible for the "sound" of the radio station. He is also responsible for most other aspects of the FM station; in fact, he functions essentially as the manager of the station for day-to-day activities. Individuals under him are responsible for the usual station functions of traffic, programing, production, continuity, and the like. In larger independent FM operations, several of these areas assume the status of separate departments.

Accounting and Administration   Although not separate departments, there are individuals and activities under the general manager that are concerned with payroll, purchasing, and "business affairs." These are shown in the chart (5-2) as support operations under the general manager.

Instructional Services   Most stations have an instructional services or school programing department. The person in charge of this area may be variously titled director of educational services, director of instructional television, supervisor of school services, director of in-school TV, instructional TV coordinator, or the like. Often this position is the station's chief contact and liaison with the person designated by the school district or the metropolitan ITV agency (Section 11.4) to coordinate all ITV programing and activities for the local schools. Stations without a separate school services department may provide

housing and administrative support for a school district officer or for the metropolitan consortium representative. In San Diego, for instance, the county superintendent's office arranges for and schedules school programing during the daytime hours. In some of the largest stations, the instructional services department may undertake nationally funded projects to produce instructional telecommunications materials for schools.

Regardless of the specific divisions and lines of authority, the internal structure of most public TV and radio stations can be related to the model outlined above. The specific needs, purposes, funding, and programing patterns of an individual station dictate the individual organizational chart that it finds most functional.

**5.4   Financing Public Television and
            Radio Stations**

The basic sources of money for public broadcast stations have already been identified in broad terms—the Corporation for Public Broadcasting, Ford Foundation and other philanthropies, local and state governments, school systems, corporations, and viewer memberships. In this section we will look at these sources of money from the station's point of view—or, more accurately, from the station *manager's* point of view. If *you* were a station manager, where would you find the money to purchase new and replacement equipment, pay personnel, obtain programing, buy supplies and services, and pay the rent and utilities?

The pie-chart (5-3) illustrates the income for all public broadcasting in one fiscal year. The distribution of income for your station would depend upon current sources of income nationally (federal dollars have been increasing each year while foundation contributions are shrinking). The sources of income for your station also will depend upon the type of station you manage.

State Government   If you happened to manage a public university station, your largest

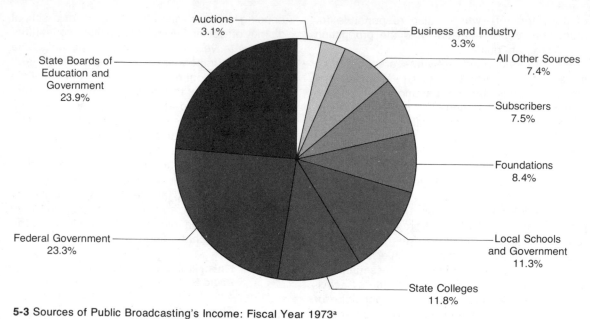

5-3 Sources of Public Broadcasting's Income: Fiscal Year 1973[a]

---

[a]*Based upon a total 1972–1973 income of $238,840,000.*

Source: Status Report on Public Broadcasting, 1973, Advance Edition *(Washington, D.C.: Corporation for Public Broadcasting, December 1974), p. 25.*

single source of income probably would come from your state legislature. Your funds might be a part of the total budget for your campus, or your station might be a separate budget item under the general heading of "education" or "higher education." As a part of that item, or in addition to it, you would receive aid that is "intangible"—that is, support that does not come to you in "hard dollars" that can be spent for goods and services. For example, a university station seldom is billed for rent, utilities, custodial, and similar services.

In total support for public broadcasting, state governments provide more than one third of all funds furnished by all sources of money. For individual stations located on college campuses, the percentage of support ranges from about one quarter of the total station support for the larger

urban university TV stations to virtually 100 percent for the smaller public radio stations—with all possible percentages in between.

If you were the manager of a state-owned station operated by a separate state commission or state department of education, your source of income would be essentially the same as for a university station. Probably, you would receive less indirect support—although state-owned stations are sometimes located on university campuses even though they are funded through special appropriations.

If yours is not a university or state-owned station, you probably still will have some sort of income from the state government as a part of your school TV income, as when the state matches local dollars for instructional programing.

School Systems   If you were managing a station that is owned by a school system, probably 80 to 90 percent of your total budget would be supplied by the local tax dollars that support your schools—with the remaining amounts coming from national sources such as the CPB.

As we have noted before, school systems are an important source of income to all other types of public TV stations (and, on a limited basis, to many public radio stations). Most PTV stations broadcast school programs during the day on a contractual basis. The contract usually provides for air time, and often includes provisions for studio time and production services "at cost" (which might be only half or two thirds what it would cost some other legitimate agency to use the facilities of the station).

In total, slightly more than 10 percent of all financial support for public broadcasting comes from local school systems. Interestingly, the figure is about the same for both public radio and public television—the difference comes in the way funds are spent. Public radio money is almost completely spent to support school-owned radio stations; school funds for television are spent both to own and operate stations and to contract for broadcast time and production services from stations owned by others.

Federal Government   Most of the federal support for public telecommunications is appropriated by Congress for the Corporation for Public Broadcasting (CPB)—which then disburses the funds to local stations, networks, production companies, and other agencies. Some federal dollars (about 20 percent) are provided directly to stations through the Educational Broadcasting Facilities Program by the Department of Health, Education, and Welfare (HEW), and by other governmental agencies for special programs. During the 1970's, the matching facilities grants shifted from a concern with new station construction to an emphasis on equipping existing stations with basic color capabilities.

Of greatest interest to you, as a local PTV station manager, would be the annual CPB community service grants (CSG). These federal funds are given by Congress to the CPB on a formula written into the five-year public broadcasting funding bill passed in late 1975, which provides two federal dollars for every five non-federal dollars raised locally and nationally by public broadcasting (Section 6.3). The CPB, working with PTV managers, develops a CSG formula for disbursing the money appropriated by Congress. This formula guarantees that a large proportion of the federal money will be passed on directly to stations (as required by the funding act), and it ensures that the money will be equitably granted—that comparable stations will receive comparable grants.

The CSG dollars are essentially unrestricted as to their use. In the early 1970's, the grants were used primarily for local public affairs programing, national syndicated series, promotion of both local and national programs, and staff positions. However, as the amounts provided to local stations increased, and as the Station Program Cooperative and the Station Acquisition Market were instituted (Section 7.3), an increasing percentage of the CSG now is used for purchasing national programing and services arranged for and distributed by the Public Broadcasting Service.

Other federal funds are channeled through the program and project grants made available to local stations by the CPB. Project grants also can be created directly by agencies of government. These are important primarily to the larger community PTV stations, which must support themselves entirely by grants, contracts, and donations. The seven largest community stations—sometimes referred to as the national production centers for PBS (Section 6.3)—devote much staff time and station energies to submitting program and project grant proposals to CPB, HEW, and a host of agencies concerned with the arts, public health, science, safety, and welfare.

With all forms of support combined, the federal government provides about one quarter of all public broadcasting dollars.

Philanthropic Foundations   Once the largest single source of national dollars for public TV, private philanthropic foundations now provide a decreasing percentage of general financial support each year. As the amount of dollars given by various government agencies (local, state, and federal) and other sources increases, the *proportion* of foundation dollars dwindles. The Ford Foundation and other private foundations now provide only about 7 or 8 percent of the total support of public broadcasting.

If you were the manager of a typical station, you could not expect too much in the way of a direct handout for your station operations from private philanthropies. Most of these foundation funds are earmarked for national programing or projects. Some money is contributed directly to stations for specific local programing projects of special significance or innovation.

Business and Industry   Underwriting for PTV programs by corporations became increasingly important after the mid-1960's, and the matching formula for federal dollars (in the 1975 five-year funding bill) promises to continue the importance of corporate underwriting. Each year, $10 million to $15 million is given by companies to support PTV programing. Most of this is given to national production centers to support specific programs, such as Mobil's support of *Masterpiece Theatre* or Exxon's support of *Dance in America*.

Stations have found that local businesses are often willing to pay for extending the broadcasting schedule and for supporting the local acquisition of nationally syndicated programs. Many PTV stations rebroadcast *Sesame Street* and *The Electric Company* on Saturday and Sunday mornings through local grants; and syndicated programs such as *Day at Night* (produced and hosted by former NET president James Day) are often preceded and followed by some announcement like "Broadcast of this program has been made possible in part by a grant from the Widget Manufacturing Company."

Auctions   The first TV auction to raise money for public broadcasting was held in San Francisco in the mid-1950's. The brainchild of James Day, it received national recognition, partly because it was successful and partly because of Kim Novak's bed sheets. The movie actress had donated the sheets to KQED's auction as a sort of "gag" item. They were purchased early in the week of the auction by a tie company, which then donated to the auction a large number of ties made from the sheets—they in turn became an "in" thing for viewers to bid on.

More than half of all PTV stations now hold auctions annually, not only because they provide income to the station but also because they involve large numbers of volunteers and businesses who develop a sense of involvement and participation in station activities through the auction. Gross income from auctions ranges from $10,000 for a small station to more than $500,000 for any of several of the largest community stations. All together, close to $10 million a year is raised through auctions—which is about 3 percent of the total support provided to public broadcasting.

Memberships/ Subscriptions   Most PTV stations now solicit memberships from their viewing audiences. Although donations in any amount are accepted, stations usually set a "minimum" membership fee of $15 to $25; this entitles the member to a monthly program guide and other information about the station and its programing. Higher amounts are, of course, encouraged—and most stations offer incentives (premiums) such as books, records, special publicity, and the like, to those who pledge higher levels of support ($50, $100, or other "plateau" amounts).

Memberships are solicited during regular programing in short spot announcements. Stations also have periodic "Pledge Nights," when volunteers man telephones on camera to receive membership pledges. Once each year, a nationally coordinated pledge drive is held, during which stations that have memberships solicit new members (and renewals) between programs. Special programs are offered during these annual "festivals," and nationally produced appeals by national figures and PTV stars are distributed to stations for use during the special week. Two or three additional national minidrives are usually scheduled with somewhat less fanfare.

Memberships (or subscriptions, as they are sometimes called) can provide anywhere from nothing to more than 90 percent (in the case of a few subscriber-financed radio stations such as the Pacifica Foundation FM chain) of a station's budget; 20 percent of total income is a typical amount for a medium-sized TV station. Nationally, individual contributions provide about 7 or 8 percent of all the financial support to public broadcasting. Although these are not large figures, memberships—along with auctions and local business underwriting—are important because they provide the discretionary dollars needed by stations to support activities that they otherwise might not be able to justify to a school board or to

a university administration. In some instances, local production is directly tied to those community funds, just as national programing is often tied to community service grants.

5.5   The Phenomena of State Networks

In each of twenty-four states, some designated state-level body owns and operates PTV stations. In some states, it is the state university system; in others, it is the state department of education; more often than not, it is a separate state agency established solely to operate a statewide ETV network. To some extent, stations in each state have either permanent or temporary interconnection facilities.

State Commissions and Authorities   As indicated in the table (5-4), by 1975 fifteen states had established a special state authority or com-

5-4 Categories of State Networks					
State	Type of net[a]	No. of stations	State	Type of net	No. of stations
Alabama	C	9	North Carolina	U	8
Arkansas	C	1	Ohio	C	3
Connecticut	C	4	Oklahoma	C	2
Georgia	D	8	Oregon	U	2
Hawaii	C	2	South Carolina	C	7
Iowa	C	6	South Dakota	C	5
Kentucky	D	13	Tennessee	D	3
Louisiana	C	1	Vermont	U	4
Maine	U	3	West Virginia	C	2
Maryland	C	3	Wisconsin	C	3
Mississippi	C	8	Samoa	D	(6)[b]
Nebraska	U	9	Guam	C	1
New Hampshire	U	5	Puerto Rico	D	2
New Jersey	C	4			

[a]C = State commission or authority; D = State department of education; U = State university network.
[b]Samoa has six channels assigned to one station.

Source: Data in 1975 NAEB Directory of Public Telecommunications (Washington, D.C.: NAEB, 1975), pp. 106–120.

mission to own and operate PTV stations and networks. The earliest, as mentioned above, was established in Alabama; a few others were created before 1967. The impact of the Carnegie Commission report and of the Public Broadcasting Act of 1967 resulted in activation of ten of the fifteen state commissions during the five-year period from 1966 to 1971.

Most state-commission activity has occurred east of the Mississippi River. This might be expected, since the western states are comparatively expansive and contain enough "empty" territory that interconnected facilities are not so practical. Purposes and emphases among the state commissions vary geographically—the southern states have been more concerned with adapting ETV for in-school uses, while the New England states have tended more to use state facilities for public broadcasting purposes.

In addition to the fifteen states that have state authorities to own and operate PTV stations, several other states also have established statewide networks to tie together stations that are not state-owned. In Pennsylvania, for example, community stations (seven of them) and one university station were developed fairly early in the ETV movement; there was no need to create a separate state commission to put stations on the air and operate them. However, in order to provide a statewide interconnected network, the state did create the Pennsylvania Public Television Network in 1971. Neighboring Ohio had several established ETV stations (dating back to the 1950's) but established a statewide agency (the Ohio Educational Television Network Commission) which put three additional stations on the air and arranged interconnection network facilities for the other nine community, university, and school stations. Many different kinds of patterns have evolved throughout the various states.

State University Networks   Six states have evolved statewide network operations built around the state university system: Maine, Nebraska, New Hampshire, North Carolina, Oregon,

and Vermont. Again, the pattern has varied from state to state; there was no set formula for all to follow.

Both the Maine and Vermont networks were conceived and funded from the start as statewide operations—with the university being designated as the operating agency. In both instances, all stations for the state operation were built and operating within a two-year period.

In Oregon, the State Board of Higher Education built two stations to better serve their campus communities. Although the two stations are tied together by microwave, the combined coverage is far short of a complete statewide network.

Each of the other three state networks (Nebraska, North Carolina, and New Hampshire) was started when the state university independently established an ETV station on its main campus in the 1950's. In each state, the network came into being when the state legislature funded the statewide operation and built the entire network approximately a decade after the first station had gone on the air.

The Nebraska ETV operation is probably the best example of a totally integrated statewide operation. Station KUON, Channel 12 in Lincoln, was inaugurated by the University of Nebraska in 1954, under the leadership of Jack McBride. McBride was later instrumental in the study that led to the establishment of the National Instructional Television Library, and, specifically, the Great Plains National Instructional Television Library in Lincoln (Section 4.3). In the early 1960's, the Nebraska Educational Television Commission was established (with McBride as general manager), followed by the establishment of eight additional stations between 1965 and 1968. Finally, the State University of Nebraska (S-U-N) was created in the early 1970's. Inspired by the Open University of Great Britain, S-U-N combines television and other media with a large variety of off-campus instructional arrangements to enable citizens throughout the state to obtain university education without being physically present on the

5-5 Nebraska ETC Center, Housing Station KUON-TV, Headquarters for the Statewide ETV Network, the Great Plains National Instructional TV Library, and the State University of Nebraska (Courtesy: KUON-TV, University of Nebraska)

Lincoln campus. All these ETC agencies are housed in one modern telecommunications center (illustration 5-5).

Two other statewide telecommunications networks are not strictly public broadcasting operations. Both Texas and Indiana have established a system of microwave links to tie together all major institutions of higher education in their respective states. The Texas Educational Microwave Program (TEMP) and the Indiana Higher Education Telecommunication System (IHETS) involve complex arrays of microwave channels, computerized scheduling, and small but efficient centralized operations to beam a wide variety of specialized materials from campus to campus. Technical demonstrations, medical information, special guest lectures, and shared course content can all be transmitted among different institutions using the TEMP and IHETS facilities. Several other states have less elaborate arrangements for sharing higher education resources through telecommunications and print media.

State Department of Education Networks

Three southern and border states have determined still another pattern for a statewide ETV hook-up. Georgia was the first state to establish a chain of stations through its State Department of Education (DOE). The first DOE station was put on the air in 1961. Others followed within the next few years. By 1965, the state DOE stations were interconnected, and by 1968, the nine-station network was completed. Cooperating with the University of Georgia station (WGTV in Athens), which began telecasting in 1960, the statewide network furnishes a full schedule of both instructional and public TV programing.

Both Tennessee (in 1967) and Kentucky (in 1968) followed with a statewide operation licensed to their respective state boards of education. Kentucky now operates the largest state network in terms of the number of individual stations—thirteen. Kentucky's inauguration of its network (which is actually operated by the Kentucky Authority for Educational Television) was a unique start in that it put its first eight stations on the air simultaneously (September 23, 1968)—as opposed to virtually every other state operation, which put stations on the air one at a time.

In addition to the various statewide networks mentioned above, it should also be pointed out that several states that do not have operating

state authorities do have some stations tied together with microwave links: California, Florida,
Michigan, Minnesota, New York, North Dakota,
and Washington all have at least two stations tied
together for occasional interconnected programing.

5.6   Regional Public Networks

One other type of station association/
network/interconnection arrangement has been
created in order to meet certain needs that were
more than local yet less than national. The state
networks and associations were formed largely
as a result of political-geographical-financial convenience, that is, coinciding with state boundaries. What was still needed was a means to enable
stations to pool resources, ideas, and programing
on a larger-than-state-level basis.[5]

Reasons for the Regional Networks   In a
detailed 1970 study of the six regional networks,
James Robertson (former NETRC vice president
for network affairs and NER executive director)
defined several specific reasons why the various
regionals were formed, needs they hoped to
answer, purposes they hoped to fulfill.[6] Before
looking at the interconnections themselves, it
may be helpful to examine the rationale behind
their existence.

Videotape Exchange   One of the most
obvious reasons for the creation of regional net-

works was for the potential exchange of both
instructional and public TV programing among
stations with common interests. This was a common need felt in the 1960's, although interconnection facilities have now eliminated much of
this concern.

Regional Acquisition of Programing   It was
foreseen that several stations, banding together,
could jointly lease or purchase properties that
were too expensive for individual stations to acquire. Such joint ventures have worked in some
instances and still hold promise for the future.

Cooperative Regional Production   It was
hoped that resources could be pumped into productions that would have regional interest for
viewers in a multistate area. However, as it developed, "regional programing as something
exclusively of interest to audiences within a region
is largely a myth."[7] Nevertheless, the regional
networks have played a vital role in stimulating
station productions of more-than-local interest.

Production Workshops and Seminars   Regional workshops could facilitate in-service training sessions and professional seminars, without
the expense of attending national conferences
and training programs.

Interconnected Facilities   Provision of
interconnected facilities has become an increasingly important function of the regional operations. Although PBS now furnishes virtually all
of the needed interconnection service, some of
the regionals—especially in the Rocky Mountain
and Western states—furnish the centers for
program delay operations (to accommodate different time zones) and regional "splits" (special
programing fed only to a single region).

Station Activation   Some of the regionals—especially in the Eastern, Southern, and

---

[5]The terms "network" and "interconnection" are often used
synonymously, although a fine distinction should be made. As
used by the Carnegie Commission (*Public Television: A Program for Action,* New York: Bantam Books, 1967, pp. 53–54),
"interconnection" refers to the physical electronic hook-up
between a number of stations and one or more central distribution points (we often use the term "network" in this sense).
The interconnection (or network) can then be used for one of
two purposes: *networking,* which refers to the immediate
dissemination and actual simultaneous transmission of a
common signal by all stations interconnected; and *distribution,*
which refers to the delivery of a program from a central point
to all interconnected stations, each of which can transmit it
immediately or record it for later use.

[6]James Robertson, *The Future Role of Regionals* (Washington,
D.C.: Corporation for Public Broadcasting, 1970), pp. 2–5.

[7]*Ibid.,* p. 27.

Upper Midwestern areas—have been active in encouraging the activation of new stations. This vital function had been undertaken at various times by the Fund for Adult Education, NCCET, ETRC, and JCET; but after 1963, when NET pulled back on related network activities (including station activation), in some areas the regional network stepped in to help new stations get on the air.

Professional Communication   From the demise of NET's expanded activities in 1963 until PBS was fired up in 1970, there was a lack of national network focus. The Educational Television Stations division of the NAEB picked up much of the slack, but station managers still had a need to communicate with one another about a wide variety of professional concerns, interconnection, station affiliation, and lobbying matters. The regionals enabled station personnel to get together and consolidate their positions—on a regional basis, among close associates and neighbors—before coming together at various national forums to argue their positions.

The Six Regional Public Networks   The six regional PTV networks that have developed cover most of the United States and give almost every station an opportunity for affiliation—although not every station has decided to participate.

Eastern Educational Network (EEN)   The first regional (established in 1960), EEN served as the prototype for many of the other regional networks. EEN grew out of discussions among WGBH-TV, Boston, and neighboring New England stations—it rapidly expanded, and today EEN includes more than forty stations in ten states, from Maine to Virginia and inland to Pennsylvania. It also has affiliate memberships—other major producing stations with which it exchanges programing—such as Chicago, Los Angeles, and San Francisco. More than a dozen full-time professional staff members are employed to operate the network and oversee the quarter-million-

dollar annual budget. The network has been active not only in PTV programing but also in instructional materials. It has operated the Northeastern Regional Instructional Television Library Project (Section 4.3), as well as several other special projects. Today, EEN stations reach about half of the U. S. population who can receive noncommercial television.[8]

Midwestern Educational Television (MET)   The second regional network in the country was established in 1961 on a somewhat different basis; MET was created around the community station, KTCA, St. Paul-Minneapolis, initially as a vehicle for extending the coverage of the Twin Cities station. It then became a powerful force in encouraging new stations throughout the Upper Midwestern region. Several stations in the Minnesota–North Dakota–South Dakota area were substantially aided in their early station activation procedures by MET. Today MET has about ten affiliates in the three-state area.

Southern Educational Communications Association (SECA)   Incorporated in 1967, SECA was formed to undertake a variety of cooperative endeavors in the field of educational communication. However, the PTV regional network has become its most visible activity. Initially supported by a grant from the South Carolina Educational Television Commission, SECA is still housed in Columbia, South Carolina. More than fifty stations in more than a dozen states are represented in SECA, whose territory reaches north to Maryland and West Virginia and west to Texas and Arkansas. With an annual budget of close to $100,000, SECA has been active in the programing area—although it has not been able to accomplish as much in School TV as some of its members would like. The association also holds a major annual conference that serves both public

---

[8]For a detailed discussion of the early history of EEN, see Donald R. Quayle, "The Regional Network," *in* Koenig and Hill, *The Farther Vision,* pp. 107–129.

and school telecommunications professionals in the South.

Central Educational Network (CEN)   Founded in 1967, CEN was designed to be a counterpart to EEN. More than thirty stations in a dozen states are included in the Chicago-based network. The $125,000 annual budget supports a three-person staff. The triangular territory of CEN extends from North Dakota on the north to Kansas on the south and Ohio on the east. This region obviously overlaps with MET, and several stations are affiliated with both regional networks. Central Educational Network is more a mutual network than is MET; many of its stations contribute to its programing projects. It has also been very active in handling regional workshops and seminars—significantly aided by the hospitality and background of the Midwestern universities that played such a prominent role in the earlier days of educational broadcasting and broadcast education.

Western Educational Network (WEN)   More loosely organized than the other regionals, WEN has existed since 1968. It is a confederation of more than twenty stations, from Alaska and Hawaii inland to New Mexico, Utah, and Idaho. Although the stations have in common the interconnection feed from KCET, Los Angeles, the vast geographical area has made difficult the organization of programing activities. Professional services and workshops (such as those sponsored by SECA and CEN in their respective regions) are carried out in the western states by a separate professional association, the Western Educational Society for Telecommunications (WEST); therefore, to a certain extent, the existence of WEST has reduced the pressure to create a network organization in the western states. As of 1976, WEN was still struggling with plans to establish a position for a full-time executive director to organize and coordinate network activities.

Rocky Mountain Public Broadcasting Network (RMPBN)   The sixth regional network was established in 1969 as a major activity of the Rocky Mountain Corporation for Public Broadcasting. The corporation itself was set up by the Federation of Rocky Mountain States, an agency of eight states incorporated for various educational and social programs. The RMPBN has about ten member stations in five states: Colorado, New Mexico, Arizona, Utah, and Idaho. Originally set up in Denver, the RMPBN is now operated with a small staff out of Albuquerque. One advantage of the network is that the parent corporation, being spawned from the larger federation, has direct access to the governors and legislatures in the several states. As a result, it has more political clout than the mutual networks of the other regionals.

Like the state associations and networks, the regional networks have evolved a number of different patterns, each meeting the individual needs of the stations in its region, each carving out its own unique niche in the complex structure of public broadcasting entities.

The regional network fits between the national and the state network but has direct relationships with and responsibilities to the individual stations. It can be a most significant element in the total ETV complex of the country and must be nurtured and strengthened if it is to play, as it can, a major role in establishing noncommercial broadcasting as a vital cultural institution in this country.[9]

---

[9]*Ibid.*, p. 129.

The National Picture

Part One of this book presented the common heritage for instructional telecommunications and public broadcasting—the common path followed by both up to the passage of the Public Broadcasting Act of 1967. In this chapter, we will continue the story of some of the institutions of public broadcasting in order to better understand its present structure.

6.1 Public Broadcasting 1969–1970

The Public Broadcasting Act of 1967—creating the Corporation for Public Broadcasting—was passed in November of that year. By the summer of 1969—barely a year and a half after the signing of the act—Frank Pace (as chairman of the CPB board), John Macy (as president of the corporation), and a few board members, consultants, and working "task forces" of interested station managers had already accomplished the following:

Begun a regular national interconnection network service under an "Interim Interconnect Group," which provided programs such as *Mister Rogers' Neighborhood, The French Chef* with Julia Child, *World Press,* and *Washington Week in Review*.

Awarded $2 million to individual TV stations.

Completed a comprehensive study of public radio after observing first-hand some 200 stations.

Established twelve CPB "career fellowships" to speed the development of young talent in public broadcasting, and four "foreign broadcasting fellowships" to study other public broadcasting systems.

Strengthened the six regional networks, distributing $10,000 to each to help with administrative expenses.

Granted $125,000 to the ETS program service of the NAEB to enable it to increase its acquisition of programs from 250 to 500 productions yearly.

Granted $450,000 to NET to help establish a special projects unit to provide in-depth, network coverage of major national and international news.

Awarded $65,000 to the NAEB National Educational Radio (NER) Network to be used for improved public radio programing.

Established the National Center for Experiments in Television at KQED, San Francisco, and the National Center for Audio Experimentation at WHA, University of Wisconsin.[1]

---

[1]*CPB Newsletter,* March 1969 (first issue), p. 3. The newsletters and regular "memos" of the CPB, entitled *CPB Newsletter* (March 1969 to July 1972), *CPB Memo* (June 1971 through August 1974), and *CPB Report* (from September 1974), were basic references throughout this chapter.

6-1 The French Chef, Julia Child, Visits *Mister Rogers' Neighborhood* (Courtesy: Public Broadcasting Service)

Creation of PBS   During the first year of CPB, John Macy and his staff had been able to arrange an interim TV network service, which was operating smoothly by the fall of 1969. In addition to the programs already mentioned, the interconnection would now carry programs which demonstrated what could be done when larger amounts of money were available for national programing. *Sesame Street, The Advocates,* and *The Forsyte Saga,* all programs produced or obtained partially with CPB funds, were introduced in the fall of 1969.

If CPB was able to run the interconnection with few problems, then why was a separate organization, the Public Broadcasting Service, created to do the same thing? For one thing, Congress had specified in the Public Broadcasting Act that CPB was not to "own or operate any television or radio broadcast station, system or network . . . or interconnection or program production facility."[2] It was one thing to make initial arrangements for interconnections with the telephone company on an experimental basis, and quite a different matter to make plans for years ahead—as was now required.

There were other reasons why Macy was eager to establish the Public Broadcasting Service. Both Pace and Macy understood well the need to differentiate clearly between the administration of federal funds and the administration of programs and services. If public broadcasting was to succeed as a free voice, there needed to be an insulation wall between the giver of federal funds (Congress) and the users of the funds (the production centers, stations, and their network). This insulating function should always be a principal role of the CPB. Macy decided that CPB could go on making grants for *programs* as it had been doing, but there was need for an organization to which CPB could grant money for the *operation of the network.*

---

[2]Public Broadcasting Act of 1967, Public Law 90–129, Section 396(g)(3).

John Macy announced the creation of PBS in October 1969. The organization would be primarily station-run; policy would be set by a board of directors consisting of five PTV station managers, one NET representative, one representative from CPB, and two from the public at large. In February 1970, the PBS board announced that it had selected Hartford Gunn, manager of WGBH in Boston, as its president.

Establishment of NPR   There was one more operational unit needed to make public broadcasting complete—a radio network system. Because of the different structural heritage of noncommercial radio, Macy and the CPB board decided that responsibility for *programs* should be assigned to the same organization that was responsible for *network operations*. In the spring of 1970, National Public Radio was incorporated as the "primary national program service for public radio" (Section 4.7), and its network operations started a year later.

Status of Public Broadcasting in Summer 1970   The beginning of the 1970–71 season was another of those high moments in educational telecommunications—activity, excitement, and general optimism pervaded the national scene. Congress was supporting CPB through annual appropriations—although Macy, Pace, and others wanted a guaranteed long-range basis for obtaining federal funds (and everyone agreed PTV needed more money than what was being authorized each year).

Live interconnected television service was a reality; PBS was off to a strong start under its new president, Hartford Gunn, who had selected Washington as the home base for PBS rather than New York "to move public broadcasting out of the shadow cast by commercial broadcasting."[3] National Public Radio (NPR) was rapidly becoming established in Washington under the direction of Donald Quayle—who left CPB in August to be the new president of NPR. Quayle had been in charge of the TV interconnection for CPB prior to the establishment of PBS, and before that he had achieved respect in the educational telecommunications field as the head of the Eastern Educational Network. Public broadcasting was beginning to make a name for itself. *Sesame Street* and *The Forsyte Saga* were more popular than anyone had predicted, enjoying large audiences and many awards in their first year on the air. The 1970–71 season would add such programs as *Civilisation* with Sir Kenneth Clark, *The Vanishing Wilderness,* and *Kukla, Fran and Ollie* for both children and adults.

Even the White House seemed to be supporting public broadcasting, although President Nixon had inherited public TV from President Johnson (along with Vietnam and a restless public). In his "Message on Education" in the spring of 1970, the President "singled out CPB for special attention, and recommended it receive a three-year authorization for funds—seen by many as a significant step toward permanent financing."[4] Nixon turned over the task of working out a long-range funding bill to his recently established White House staff group, the Office of Telecommunications Policy (OTP). (See below.)

Whispering in the Wings   Amid all this optimism there were some signs of conflict, however. During the 1970–71 season President Nixon and his OTP director, Clay Thomas Whitehead, began to include public broadcasting in their statements about lack of "fairness" in the criticisms of Nixon made by the media (about his handling of the Vietnam conflict and the domestic civil disorders).

In Congress, some legislators were criticizing specific PBS programs that seemed to them to lack objectivity. In PBS there was increasing discomfort about the lack of control it had over programs funded directly by CPB. Stations were wondering if PBS was really controlled by them.

[3]*CPB Newsletter,* April 1970, p. 2.

[4]*Ibid.,* p. 1.

Officials of NET and NAEB were still unhappy about being bypassed in setting up the TV and radio networks. A further blow was dealt to NET in the fall of 1971 when CPB created yet another organization, the Washington-based National Public Affairs Center for Television (NPACT), to produce essentially the same kinds of documentaries and "actualities" that NET had been doing through its "Washington Bureau."

6.2   Struggles for Control: 1971–1973

The opening shots in the battle for control of public broadcasting were fired by the federal government late in 1971. Almost everyone in public broadcasting would become involved in the struggles, but it would take vigorous new leadership from the lay public to arrange the peace a year and a half later.

Whitehead at the 1971 NAEB Convention Those who attended the NAEB convention in November 1971 expected to hear good news about the present and rosy predictions about the future. Much had been happening in Washington, and conferees looked forward to briefings on the activities of CPB, PBS, NPR, and OTP by those in charge of the organizations. There was special anticipation for briefings on the new Children's Television Workshop reading series, *The Electric Company,* and on the newly created National Public Affairs Center for Television—especially since immediately before the convention NPACT announced it had been able to woo two respected newsmen away from commercial broadcasting, Sander Vanocur and Robert MacNeil.

There were no surprises at the conference until Clay Whitehead, director of the OTP, began to speak. Instead of describing progress on the long-range financing bill (which everyone thought would soon be delivered to Congress), Whitehead delivered a stinging denunciation of public broadcasting—declaring that the administration's plans to improve CPB financing "lay in the distant future" if public broadcasting con-

tinued to be "structured as it is today and moving in the directions it seems to be headed."[5]

The Basic Issues and the Big Club   To understand this turnaround by Nixon and Whitehead, one should consider all that was happening at the time. The Vietnam war was going badly; students were rioting; crime was increasing; minorities were militant; and the media, especially commercial television, were becoming increasingly critical of the administration. Nixon—who had not enjoyed good relations with the press for several years—and Vice-President Agnew were fighting back with charges of "elitist gossiping" and "East Coast liberal bias" on the part of news programs, which, they said, should be "objective" and "balanced."

Nixon had created the Office of Telecommunications Policy to establish policy on domestic and governmental broadcasting and telecommunications, and to provide guidance on international issues in such matters as satellite communications. Although Whitehead was only thirty-one at the time he was appointed OTP director, he had already demonstrated considerable skill in both politics and bureaucratic management with the RAND Corporation and as a White House consultant, talents needed in his new assignment.[6]

Whitehead initially concerned himself only with commercial TV criticism of the administration. However, Nixon and Whitehead became increasingly disenchanted with public broadcasting—as segments of programs such as *The Great American Dream Machine* and special documentaries focused attention on the ills of society and criticized various administration policies.

The ultimate exacerbation of their discontent came with the establishment of NPACT

[5]John P. Witherspoon, memorandum to public TV station managers, dated November 5, 1971, quoting Clay T. Whitehead's speech at the NAEB conference, October 20, 1971. Witherspoon's memo was reprinted in the *CPB Memo,* November 8, 1971, special attachment.

[6]Fred C. Esplin, "Looking Back: Clay Whitehead's OTP," *Public Telecommunications Review,* March/April 1975, pp. 17–22.

and the hiring of Vanocur and MacNeil in the fall of 1971. Not only were these men noted for their forthright opinions ("courage to challenge the status quo" or "liberal bias," depending on one's own personal views), but also they were given salaries of $85,000 and $65,000 per year—considerably higher than those of congressmen and most civil servants.[7]

When it became obvious that OTP would not be submitting to Congress a long-range funding bill, the chairman of the House Subcommittee on Communications and Power, Torbert Macdonald, introduced a bill providing for five-year funding of public broadcasting at levels approaching the recommendations of the Carnegie Commission. Whitehead used the hearings on the bill to voice Nixon's concerns about public broadcasting.

Administration criticism of the PTV establishment over the next year or so was centered around three broad issues: (1) *Centralization:* Whitehead's main thrust in the 1971 NAEB speech was that public broadcasting was trying to become just one more centralized New York-Washington based network—as evidenced by the establishment of a "fixed schedule, real-time network contrary to the intent of the 1967 Act."[8] This was contrary to the spirit of the Carnegie report and the Public Broadcasting Act, both of which emphasized the need for localism and the strengthening of local station operations; more funds should be given directly to stations through station support grants. (2) *Public Affairs:* OTP questioned the whole concept that educational broadcasting should even be involved with public affairs—since federal funds were involved, it would be wise to stay away from current controversial programing and stick with cultural and educational materials. (3) *Salaries:* the salaries paid

to Vanocur and MacNeil (and to the highest CPB officers) were partially supported with federal dollars, and therefore they certainly should not be higher than salaries paid to our elected representatives in Congress.

Internecine Quarrels Begin   As often happens when a group is faced with a serious threat, communications begin to break down within the group—in this case, the public broadcasting community. Most of the issues underlying the Macdonald bill were ephemeral and matters of definition (such as "balance" and "objectivity"); but attention began to concentrate on the one substantive issue in the hearings—whether CPB should be required to give individual PTV stations a fixed percentage of its funds (to counteract the "centralization" charge), and what that percentage should be. Some said that no formula should be applied; the Macdonald bill specified 30 percent. The then-president of the NAEB, William Harley, said the percentage should increase to 50 percent during the five-year period covered by the bill. The chairman of the NAEB board and the vice-chairman of the ETS board each testified that the percentage should rise to 70 percent for the individual stations by the end of the period. Macdonald and his committee indicated that those representing public broadcasting should decide such details outside the hearing room, and that such divided testimony was "harmful to the chances for passage of the bill."[9]

Other issues that distressed public broadcasting groups were more difficult to discuss openly. Some have been mentioned, such as the NAEB and NET pique at being overlooked in the establishment of the PTV and radio network services. In addition, PBS was becoming annoyed at being caught in the middle on the matter of programing. PBS had essentially nothing to say about what programs would receive federal or foundation dollars, since CPB and the Ford Foun-

[7]*CPB Newsletter,* February 1972, p. 1. Both Vanocur and MacNeil took one-third salary cuts to come to NPACT, but their salaries were still considerably higher than the $36,000 base pay of congressmen at that time.

[8]"CPB Responds to Whitehead Charges," *CPB Newsletter,* February 1972, p. 7.

[9]"House Holds Hearings on Financing; Macdonald Five-Year Plan Supported," *CPB Newsletter,* February 1972, p. 6.

dation granted money directly to stations for programs and series; yet PBS received the criticism, since it was "the network" as far as viewers and most politicians were concerned. After it got in trouble with influential politicians for carrying a program critical of congressmen who had financial dealings with banks, PBS began flexing its muscles over the matter of programing. Finally, in February 1972, while the hearings were being conducted on the Macdonald bill, PBS refused to carry "The Politics—and Humor—of Woody Allen," which included a half-hour satire on President Nixon. This assertion of control bothered CPB, which thought *it* should be the final arbiter of what should be distributed to PTV stations. Even within the CPB there was not complete harmony; Pace and Macy, who had been appointed by President Johnson, were beginning to have trouble working out policy with a board that was by this time half comprised of Nixon appointees.

The Nixon Veto   After a series of meetings, the organizations involved did become united in their support for the Macdonald bill; and in the summer of 1972 both the House and the Senate voted for it by overwhelming margins (254 to 69 and 82 to 1). Whitehead assured both broadcasters and congressmen that Nixon would sign the bill. Without warning, Nixon vetoed the bill—repeating his concern that "localism" was being undermined by both CPB and PBS. A newspaper article reported that "to say that the President's action was a surprise would be a Titanic understatement."[10] Pace, Macy, and the director of television activities for CPB, John Witherspoon, resigned after many public statements condemning Nixon for continuing interference and pressure, culminating in the veto of a bill everyone favored.[11]

The CPB–PBS Feud   In September, Tom Curtis, vice-president of Encyclopaedia Britannica and former Republican congressman from Missouri, was elected chairman of the CPB board of directors. Immediately afterward, Henry Loomis, deputy director of the U.S. Information Agency and former head of the Voice of America, was appointed president of CPB. Curtis and Loomis were selected in part because they were relatively conservative in their views. They were both initially seen as strong Nixon supporters. They believed that public broadcasting should not bite the hand that feeds it, and that CPB should be more cautious in making grants in the areas of public affairs and social commentary.

During the fall of 1972, Curtis, Loomis, and the CPB board debated the question of who should control the interconnection. Gunn was adamant that final decisions about what could and could not be transmitted on the PBS network had to remain vested in PBS. CPB felt that it should have that right. On January 10, 1973, at its regular meeting, the CPB board passed a formal resolution calling for "a new relationship with the interconnection organization of public broadcasting, the Public Broadcasting Service."[12] The CPB board specifically felt that it, and not PBS, should have the "ultimate responsibility for decisions on program production support or acquisition [and] the pre-broadcast acceptance and post-broadcast review of programs to determine strict adherence to objectivity and balance in all programs or series of programs of a controversial nature."[13]

Sides were uncertainly drawn. Gunn and the PBS board said CPB was trying to centralize power and thus would make public broadcasting all the more vulnerable to White House and congressional pressure. Even though ETS was com-

[10]Bernie Harrison, "The Cookie Monster's Last Stand?" *The Evening Star* (Washington, D.C.), July 5, 1972.

[11]John Witherspoon, who as director of television activities for CPB was a key individual in mapping CPB strategy during the 1971–72 battle with the Nixon administration, writes: "It would not be correct to tie the resignations directly to the veto,

although that was of course a critical event. Control and White House pressure, in which the veto was a specific tactic, were the fundamental points." (Personal communication.)
[12]*CPB Memo,* January 15, 1973, special insert.
[13]*Ibid.*

prised of the same public TV stations that supposedly controlled PBS, it seemed unsure in which direction it should go; PBS did not know how to deal with the internecine squabbling; and OTP sat back and smiled. As a former PBS staff member said later:

Up to and throughout the period of the White House assault a constant stream of bickering emanated from ETS (accompanied by bitter dissension on a host of additional matters by NET and other organizations) that was not always well handled by PBS. These notes of disharmony added to the directionless image of public broadcasting and thereby played directly into the hands of OTP and CPB.[14]

The picture was complicated by the fact that the PBS board consisted primarily of PTV station managers—professionals in the field—and the CPB board consisted of representatives of the lay public. The CPB board followed the classic democratic pattern of a representative group of lay citizens (trustees, directors, or whatever they may be called) who actually direct the corporation or company or school district or ETV station; the lay board members set the policy, make the basic decisions, and hire the professionals (the station managers) to carry out the day-to-day operations and execute the board's policies. Public Broadcasting Service, however, did not fit this pattern; it was an organization directed and run by the professionals themselves—serving as their own board of directors.

Another complication was that the corporation board saw itself as the higher authority that actually created the structure of PBS in the first place. Moreover, as a lay board appointed by the President, the CPB board felt it had a mandate to watch over the station-run interconnection.

Thus, without a separate policy board of lay citizens—which could give PBS autonomy and independent status—the PBS leadership was bargaining from a weak position. The CPB board felt, with some justification, that—if PBS wanted to negotiate on an equal plane—it would have to become something more than an operational group representing the professional station managers.

The National Coordinating Committee of Governing Board Chairmen   Public broadcasting desperately needed a lay knight in shining armor to give it a new sense of direction; and it found one in the form of Ralph Rogers, a self-made Texas millionaire. Rogers was chairman of the board of Texas Instruments Corporation when he was first asked to serve on the board of PTV station KERA-TV, Dallas. Soon he was chairman of the KERA board of governors and found himself traveling to Washington to lobby for the Macdonald long-range funding bill.

These lobbying efforts made Rogers realize that PTV had a powerful untapped resource in its lay governing boards. With the help of a Ford Foundation grant and through organizational work by the ETS division of the NAEB, the National Coordinating Committee of Governing Board Chairmen (of PTV stations) was formed, with Rogers as its chairman and prime mover. The coordinating committee could, thought Rogers and the NAEB, share ideas on local fund raising and, hopefully, organize support for federal legislation favoring public broadcasting.

When negotiations between CPB and PBS broke down at the beginning of 1973, Rogers represented the ideal arbitrator. He could identify with both the lay board of CPB and the station managers who largely made up the PBS board. His allegiance, however, naturally was with the stations; and so he began to work with PBS President Hartford Gunn to devise a way to strengthen the bargaining position of PBS.

Rogers' strategy was clear, clean, and simple. The PTV stations needed to restructure their national organizations. There was need to

---

[14]Williard D. Rowland, "'Public Involvement': Anatomy of a Myth," *Public Telecommunications Review,* May/June 1975, p. 14.

create a new entity, combining the representative functions of both PBS and ETS, that would control the PBS staff—who, in turn, would run the network. The new organization should be run by lay representatives of the governing boards of the stations, who after all were ultimately responsible to the communities served by the PTV stations. Creation of the new organization not only would present a unified front to the CPB but also would clarify relationships among stations, the PBS staff, and the CPB. So, the original Public Broadcasting Service was disbanded, and a new structure was created (Section 6.4). For the sake of continuity and public image, the "PBS" name was retained; but the new entity was definitely the voice of the lay public—community leaders and business executives—as well as station managers. As Rowland said, "No longer could the White House and the CPB Board pretend that the network was out of the hands of the stations. Nor could they continue to claim they did not know with whom they were dealing."[15]

The CPB did not back down easily from its position that it should control the interconnection. After arranging a compromise with Rogers, Curtis took it to the CPB board in good faith—and it was rejected. Immediately, the story circulated that the White House had contacted conservative board members and pressured them to vote against the compromise. Curtis was so angry at this apparent betrayal that he resigned the chairmanship of the board. Again from Rowland:

Apparently embarassed by the allegations about the way it was continuing to be manipulated by the administration and unprepared for the subsequent harsh reactions of the newly reorganized station organization and its broadly based political supporters, CPB began to reconsider its precipitous actions. In May the board elected Dr. James Killian, former chairman of the Carnegie Commission, as chairman of the CPB. In

elevating Dr. Killian, one of the leaders among the moderates, CPB was signaling both the White House and the stations that it had had enough of the direct confrontation.[16]

The Compromise   The feud officially ended in May 1973, when a joint resolution, accepted by the CPB and PBS boards, left the decision on CPB-funded programs to the CPB program department but provided a way for PBS to dissent from CPB decisions. A monitoring committee of three CPB representatives and three PBS board members would decide on non-CPB-funded programs that either CPB or PBS thought were not balanced or objective. Scheduling of the interconnection would be by another similarly balanced group, with a seventh nonaffiliated person acting as chairman of the scheduling committee.

The significance of the May joint resolution is best summed up by a statement by James Killian shortly afterward:

When the history of the Corporation for Public Broadcasting is written, 1973 will be remembered for its new focus and the renewed strength that has emerged.

The new focus of the Corporation comes out of a long dispute that ended in a landmark partnership agreement with the Public Broadcasting Service representing the local non-commercial television stations. Under this agreement . . . an increased portion of CPB's funding will flow to the individual stations.

These partnerships have several major objectives. They will increase the strength, independence, and responsibility of the individual stations. The decisions about which established programs to support will shift more and more from CPB to the local stations, thus eventually freeing CPB to concentrate on developing new programs

---

[15]*Ibid.*, p. 15.

[16]*Ibid.*

through research, experimentation and diversity, and the establishment of new standards of excellence for public broadcasting.

New institutions do not respect their own strength until it has been tested. Public broadcasting is no exception. During the last year, it has settled its internal squabbles, forging a new unity for the public broadcasting industry as a whole, and thus strengthening its vital independence. This year of achievements was made possible by men and women of purpose and goodwill responding to the voice of the American people who made it apparent that they care about the future of public broadcasting and about its unity and independence as a precondition of that future.

The Corporation is proud of the role it has played in forging that new unity and reaffirming that independence.[17]

Rogers, representing the lay leadership of the community PTV stations, had involved the *public* in public broadcasting as it never before had been involved, and had brought a new balance to the public broadcasting scene.[18] The battle with CPB ended, and soon Whitehead would be supporting a long-range funding bill in cooperation with PBS and CPB.

---

[17]James R. Killian, Jr., "The New Focus," *Corporation for Public Broadcasting Annual Report, 1973,* p. 1.

[18]It has often been pointed out, however, that the lay boards of community public TV stations do not necessarily reflect the actual viewing public (Section 8.5). This lay leadership is generally drawn from a social and economic elite that represents the same kind of cultural viewpoint as that of the professional educational broadcaster. James Fellows, president of the NAEB, writes: "There are some who feel that the lay representatives, important though they have been, do not in fact constitute 'the public.' This is probably correct and it would be wise to correct any impression that they *are* the public.

"The value of such lay persons is beyond dispute. They have been of value at very critical times locally and nationally; their value in the negotiations with CPB was that they represented a separate but equal echelon from the socio-economic levels which constituted the CPB Board. They were not really more 'representative' of the public than the professional managers." (Personal communication.)

6.3 The Corporation for Public Broadcasting

The Corporation for Public Broadcasting exists to promote the growth and development of the nation's noncommercial television and radio systems. "The goal of CPB is to create the conditions and provide the broad initiatives that will enable this service to develop."[19]

According to the Public Broadcasting Act of 1967, the Corporation for Public Broadcasting is authorized to:

1. Facilitate the full development of educational broadcasting in which programs of high quality, obtained from diverse sources, will be made available to noncommercial educational television or radio broadcast stations, with strict adherence to objectivity and balance in all programs or series of programs of a controversial nature.

2. Assist in the establishment and development of one or more systems of interconnection to be used for the distribution of educational television or radio programs so that all noncommercial educational television or radio broadcast stations that wish to may broadcast the programs at times chosen by the stations.

3. Assist in the establishment and development of one or more systems of noncommercial educational television or radio broadcast stations throughout the United States.

4. Carry out its purposes and functions and engage in its activities in ways that will most effectively assure the maximum freedom of the noncommercial educational television or radio broadcast systems and local stations from interference with or control of program content or other activities.[20]

---

[19]*Corporation for Public Broadcasting Annual Report, 1973,* p. 7.

[20]Public Broadcasting Act of 1967, Public Law 90-129, Section 396(g)(1).

The Corporation for Public Broadcasting receives funds from Congress (through appropriations) and from private foundations (by special grants). It distributes money to the Public Broadcasting Service and National Public Radio for the operation of the networks; to producing agencies (which usually are stations) for national programs; to local stations for national programing, local programing, and general support; to individuals for special training opportunities; and to task forces and consultants who develop plans and projects that will further the goals of the organization.

Generally, the CPB budget has increased over the years as the corporation has received larger amounts from Congress. The chart (6-2) shows the CPB funding by Congress from the beginning of the corporation, projected through the end of the 1970's. The sharp increase in funding beginning in fiscal year 1976 results from the passage, in December 1975, of the long-awaited Public Broadcasting Financing Act. This legislation enabled the CPB, for the first time, to estimate the funding it would receive in future years— necessary for long-range program planning.

The Structure of the CPB   As we have seen, broad policies for CPB are established by its board of directors, which is appointed by the President and confirmed by Congress. The Public Broadcasting Act of 1967 requires that the fifteen-member board should have no more than eight members from the same political party. Individ-

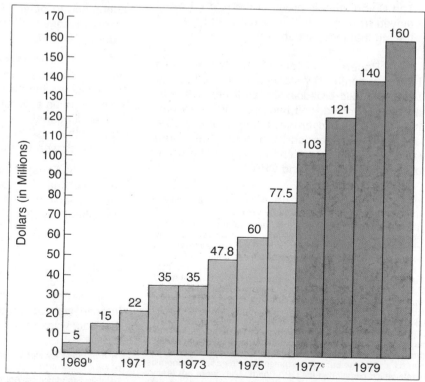

6-2 Federal Funding of Public Broadcasting[a] (Amounts in Millions of Dollars)

[a]Not including facilities.

[b]"1969" indicates fiscal year 1968–69, etc.

[c]Figures for 1977 and succeeding years are estimates, as provided in the Public Broadcasting Financing Act.

Source: CPB Annual Reports, and the Public Broadcasting Financing Act.

ual members are to be "eminent in such fields as education, cultural and civic affairs, or the arts . . . a broad representation of various regions of the country, various professions and occupations, and various kinds of talent and experience. . . ."[21] Each board member serves for six years; the board elects its own chairman.

Titles and lines of responsibility in organizations tend to change to meet different management styles and different individual capabilities and personalities. The president of CPB, Henry Loomis, established an administration consisting of an executive vice-president who assists in broad planning and coordination of all activities, a senior vice-president for broadcasting who maintains principal contact with PBS and who coordinates the radio and TV activities, a vice-president for finance, a general counsel, and eight directors—one each for TV activities, radio activities, public affairs, educational activities, engineering, research and development, communications research, and finance.

CPB Activities   Since the principal function of CPB is to pass along money, a review of its grants will cover most of its activities.[22]

Program production grants are meant to provide support for high-quality programs of sufficient general interest to merit national distribution and to support specific programs of regional or local interest. National TV programs funded by CPB are usually produced by one of the seven PBS stations designated "national producers"—WNET (New York City), WETA/NPACT (Washington, D.C.), KCET (Los Angeles), KQED (San Francisco), WGBH (Boston), WQED (Pittsburgh), and WTTW (Chicago)—and by the Children's Television Workshop (New York City). However, any station can apply for program grants; and other organizations with producing capabilities (such as Southern Educational Communications

Association) and independent producers (such as Family Communications Incorporated, producers of *Mister Rogers' Neighborhood*) are also eligible for grants.

A special kind of program grant is the "step-up" grant. These funds are given to a PBS station when an exceptional local program can be modified to make it suitable for national distribution.

Program grants normally are made during the first half of the calendar year, but proposals can be submitted at any time. Step-up grants are made at any time, subject to the availability of funds.

Program development grants are given to any qualified station or institution that needs funds to plan, research, and/or produce *pilot* programs for a proposed series. Program development proposals can be submitted and awarded at any time.

Community Service Grants are given annually to public broadcasting stations to be used for purposes of their own choosing. Most grants are used to produce local programing and to obtain national programs, but the funds are also used for personnel, promotional activities, development of needed community support, and so forth. Grants are made to "eligible stations" defined in criteria of minimum staff and power standards. This encompasses virtually all PTV stations; but many low-powered and school FM stations are not eligible for radio CSG funds (Section 6.5). The amount a station gets is determined by a formula that specifies a minimum amount any station will receive (such as $50,000 for a PTV station) and an additional amount based upon the annual unduplicated nonfederal budget.

Station Program Cooperative Grants have been given during the first several years of the Station Program Cooperative (Section 7.3) to increase the number of programs that stations can "purchase" through the cooperative. The grants are made to PBS rather than to stations and are decreasing each year.

---

[21]The Public Broadcasting Act of 1967, Public Law 90–129, Section 396(c)(2).

[22]CPB *Annual Report, 1973, op. cit.*, pp. 15–17.

PBS interconnection grants are made annually to PBS to pay for all out-of-pocket costs of the interconnection. The amount is based on the current contract with the Bell Telephone System.

NPR interconnection and programing grants are given each year to operate the radio interconnect, the network program service, and the "options" tape library (Section 7.6).

Minority training grants are given to stations to encourage them to bring in new minority persons for training in management or creative professional areas. The grants can also be used to train and upgrade individuals already employed by a station. On application and approval, the CPB will pay up to one half of the salary of a minority candidate for a two-year period.

Research grants are usually given to answer some specific questions about noncommercial broadcasting audiences, employment practices, funding, program development techniques, technical matters, and so forth. Experimentation in programing has also come under the research office—particularly support of the National Center for Experiments in Television located at KQED, San Francisco, which was organized in 1969 to explore new creative directions in TV production.

Cooperative Grants for the Arts are offered in cooperation with the National Endowment for the Arts to support arts programs involving public media.

Task force and project funding are special grants for long-range policy and information. The CPB's Information Systems Project maintains data on all aspects of public broadcasting, obtained primarily through an annual Television and Radio Survey. Task force activities include the Task Force on Long-Range Financing of Public Broadcasting, which issued its report in 1973, and the ACNO/Education Task Forces, which issued a joint report in 1975 (Section 6.6).

6-3 CPB Funding Allocations (by Categories of Grants) (Amounts in Millions of Dollars)

Year	1971	1972	1973	1974	1975
Community Service Grants	2.9	4.7	5.0	15.2	24.0
TV National Programing Grants	8.7	11.4	13.3	12.4	11.2
PBS Interconnection Grants	6.8	9.8	9.3	8.1	9.8
Public Radio (total)	2.3	4.5	5.1	6.8	9.0
All Others	1.3	4.6	2.3	5.3	6.0
Totals	22.0	35.0	35.0	47.8	60.0

Source: Robert K. Avery, "PBS's Station Program Cooperative: A Political Experiment," paper presented at the International Communication Association Convention, April 1975, and CPB Annual Reports.

The breakdown of how the CPB allocated its funds among major grant areas for the period 1971 to 1975 is indicated in the table (6-3). The Community Service Grants to local stations rose sharply beginning in 1976 since the Public Broadcasting Financing Act provides that from 40 to 50 percent of all funds given to CPB must be distributed directly to the stations.

6.4   The "New" Public Broadcasting Service

In March 1973, public television station representatives approved the formation of the new membership organization—retaining the old name of "PBS." The new organization resulted from a merger of three former ones: The National Coordinating Committee of Governing Board Chairmen, the ETS division of the NAEB, and the original PBS (Section 6.2).

The creation of the new PBS paved the way for the May 1973 partnership agreement between PBS and CPB. As a result of that agreement, PBS's direct reliance on CPB funding diminished; the stations themselves pay for the

PBS membership and professional services and representation of the licensees' interests before CPB, Congress, the Executive Branch, and the general public. In addition, under a contract with CPB, PBS would "manage the distribution of programs over a delivery system furnished to CPB pursuant to an order and tariff filed with the FCC."[23]

Three Organizations in One   To understand PBS, one must understand that it really is three organizations in one. The PBS staff at times reflects one face and at times another. Actions by PBS can seem unpredictable if this is not understood.

Face One: A Representative Organization The "new" PBS is in all ways an organization of, for, and by the PTV stations that are members. From the beginning, PBS was designed to be controlled by stations, but the new charter clarifies and formalizes this in unequivocal terms. If PTV stations do not like the way the interconnection is being run, they can only blame themselves if no action is taken.

Face Two: The Voice of the Public   Certainly, PBS is run by PTV stations; however, this does not mean that it is run by station *managers* alone, as was true under the old charter. "PBS is governed by an elected board of governors made up of laymen from the boards of local television stations and a PBS board of managers comprised of and elected by television station managers."[24] Thus, PBS is governed by these two boards: the lay board of governors is the senior policy board for PBS; the board of managers recommends policy on professional matters. The chairman of PBS, initially Ralph Rogers, is a lay citizen representing the public. The involvement of lay public in the governance and in the actual operation of the network is not advisory or peripheral—it is direct and authoritative.

---

[23]"Public Broadcasting Service Factsheet," undated mimeo, circa July 1974.
[24]*Ibid.*

Face Three: A Staff and Organization   No matter how much people would like to believe that an organization is controlled by its members or directors, it always has a corporate mind and staff personality of its own. Public Broadcasting Service is a large organization involved with activities and problems that the member PTV stations are not even aware of; and the PBS staff operates the interconnection on contract to CPB, with funds directly given to PBS by CPB. It is not surprising that at times PBS is perceived as "them" instead of "us" by local stations.

The PBS Staff Organization   At the top of the PBS structure are the two board leaders who are primarily concerned with policy and liaison with other agencies of government and broadcasting. The chairman of the board (representing the board of governors) is unpaid, and after 1975, Ralph Rogers, initial chairman, began gradually reducing his direct personal involvement in public broadcasting. In January 1976, Hartford Gunn was officially appointed to the newly created position of vice-chairman of the PBS board, the top-ranking salaried PBS officer.

Reporting to Rogers and the PBS board, the new (in 1976) president of PBS, Lawrence Grossman, is directly responsible for the operation of the interconnection and all staff activities coordinating and supporting the national service.

Offices within PBS are headed by vice-presidents, directors, or coordinators. Since these titles and relationships change from time to time, we simply will identify different PBS offices and services—without giving the exact titles of their administrative heads: (1) *planning:* activities relating to corporate planning and future directions; (2) *programing:* obtaining and scheduling programs on the interconnect; (3) *public information:* contacts with the press and public; (4) *development:* obtaining grants, underwriting, and other "outside" money; (5) *member services:* providing professional assistance to individual PTV stations; (6) *national affairs:* lobbying, primarily before and with legislators; (7) *international:*

maintaining contacts with public systems in other countries and making contacts for obtaining programs from other countries; (8) *educational services:* developing instructional programs and helping schools use prime-time programs; (9) *library:* maintaining, promoting, and developing the Public Television Library (see Section 7.4); and (10) *legal:* the office of the general counsel.[25]

PBS Activities   As outlined above, the activities of the PBS staff can be divided into those that CPB and others (principally the Ford Foundation) support, and those that annual dues from PTV stations support. The costs for runninng the interconnection—including the operation of the technical center in Washington and the delay centers in Denver and Los Angeles—are given directly to PBS by CPB. CPB and Ford also pay a portion of the cost of the Station Program Cooperative (Section 7.3).

The *dues-supported services* can be divided into eight categories. The following outline of services was adopted after PBS reorganized in 1973. The percentages indicate the estimated distribution of dues income among the different activities.[26]

Programing (35%)
Determining station needs through surveys and panels
Coordination with CPB: Program funding, priorities, general plans
Evaluating SPC program proposals and administering the SPC[27]
Program scheduling

Evaluating program offerings
Standards and practices
Coordinating the Public Television Library

Program and Public Information (20%)
National advertising
Program promotion: national
Program promotion: stations (program descriptions, schedule information, promotional material, previews, on-air promotion, program guide support)

Planning and Research (10%)
System financing
Station and program finance requirements
Operational research

Member Services (5%)
Industry/management newsletter
Station assistance
Management services and information
Training and professional development

Station Relations (5%)
Assessment of station needs
New station development and planning
Representation to operational management
Station–PBS information flow
Membership and other station meetings

Development (5%)
National programing underwriting
Underwriting for other services
Station Independence Plan

Representation (10%)
Government liaison: Executive Branch; Congress; FCC
National organizations (ACNO, NEA)
Meetings, elections, advisory committees, governing board chairmen

Administration (10%)
General administration

---

[25]*Ibid.*

[26]From notes taken by one of the authors during a presentation on the "new PBS" by Ralph Rogers, Hartford Gunn, and PBS staff members, to public TV station managers, closed-circuit over the PBS interconnection in the spring of 1974. Some activities, such as administering the SPC, have been added to the activities described in the telecast.

[27]The Station Program Cooperative (SPC), Public Television Library (PTL), and Station Independence Plan (SIP) are self-supporting activities, but overall supervision is provided by the PBS departments indicated.

Legal

Business and finance

Occupancy

Thus, approximately half of all dues money is spent on program acquisition, scheduling, and promotion; and half is spent on management, representation, development, and similar administration activities. This might seem like an unusually large allotment for "indirect" and overhead expenses such as research, management, and representation; however, since the dues income represents only a fraction of the total budget for PBS, management and other overhead costs become a much smaller proportion of total gross income than the above percentages indicate. One should also remember that PBS took on all station representation and professional services when the old PBS organization merged with the ETS division of the NAEB.

A picture of the day-to-day operations at PBS can be found in the current directory and guide for PBS services, such as the following alphabetical list of activities and services excerpted from a recent directory.[28] This gives some idea of the scope of interests and tasks of PBS:

Advertising (to promote PTV programs)

Billings/invoices: PBS, PTL

Broadcast operations

Cable TV: engineering, materials, policy on PBS carriage, PTL programing

Captioning for Deaf Project

Cassettes: future applications, experiments with network, PTL programing

Clearances

Clippings

Communications problems (nontechnical)

Development: national program support, National PTV Week (pledge nights)

Dial Access Communications System (DACS)— the dataphone Teletype system on which messages, computer print-outs, billings, promo information, and the like are sent to stations

FCC affairs

Interconnection: chronic problems, quality and reliability of day-to-day signal, network originations, immediate technical problems, network outages, preemptions, refeeds, system planning and installation

International broadcasting

Legal affairs

Liaison: PBS staff, other PTV agencies, ACNO, NAB, OTP, NCTA

Long-range financing

Mail from viewers

Mailing lists: maintenance, changes to lists

Meetings: boards of directors, national membership, regional meetings, staff participation in meetings of others

Membership: dues, new stations, existing members

Monthly program listings

Music libraries

Music: use on PBS

National affairs

National publicity

New station activation

Newsletter

On-air promos

Operations log

Press previews

Program content coordination: cultural affairs, educational programing, public affairs

Program submissions/proposals: completed programs, proposed productions for PBS net, PTL submissions, engineering problems/considerations for proposed productions, production problems/ considerations for proposed productions

Program surveys/evaluations

Program timings

---

[28]*PBS Guide to Staff and Services* (Washington, D.C.: Public Broadcasting Service, 1974).

Promo slides: PBS, PTL

Public information

Publicity photos

Purchasing

Reference library

Rights

Satellite communications: engineering aspects, international use, policy, background, FCC activity

Scheduling information

Screenings for national media

Spot copy

Staff appearances

Station Program Cooperative

Tape distribution: noninterconnected stations, PTL

Technical evaluation: film, PTL, videotape

Transcripts of PBS programs

Underwriting

Visiting PBS

The scope and variety of activities and concerns of a large network operation are only hinted at in this directory listing. The major operation of PBS, of course, is the interconnection. The map (6-4) shows the national hook-up.

Even as the land-based interconnection was being refined and modified, however, plans were under way to replace the system with a satellite hook-up. PBS and CPB had been laying the groundwork for the transfer of the networking operations to a satellite system, and at a meeting

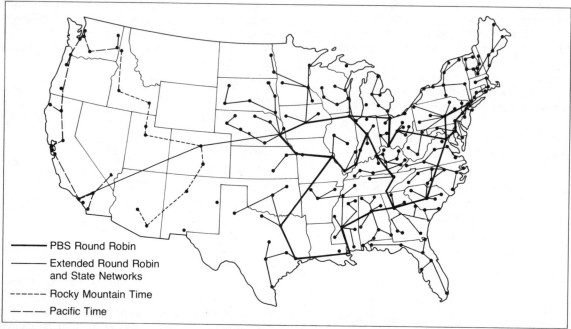

PBS Round Robin

Extended Round Robin and State Networks

Rocky Mountain Time

Pacific Time

6-4 PBS Interconnection System: 1974

Source: Status Report on Public Broadcasting 1973, Advance Edition *(Washington, D.C.: Corporation for Public Broadcasting, December 1974), p. 58. (Courtesy: Public Broadcasting Service)*

of all stations early in 1976, the PBS affiliates voted to endorse the plan—although it would cost the stations additional money at the outset. Plans for the satellite interconnection—which initially called for implementation by 1978—would ultimately result in a substantial savings to the entire public broadcasting movement, as well as providing additional channels with more flexibility.

6.5   National Public Radio

National Public Radio, like PBS, is several things in one, and the situation is compounded somewhat by the name itself. With capital letters, National Public Radio (NPR) "is the network of more than 170 public radio stations from Maine to California and from Puerto Rico to Alaska . . . America's only interconnected public radio network . . . a membership organization funded by the Corporation for Public Broadcasting (CPB)."[29] We also speak of national public radio in a more generic sense to mean anything and everything that involves noncommercial educational radio at the national level.

One distinguishing characteristic about NPR is that it is a *producing agency* as well as a network. Although it is a membership organization, representation of public radio stations on national issues is found in a different organization—the Association of Public Radio Stations (APRS)—which represents public radio stations in dealing with NPR and other organizations and agencies (see below).

The differences between NPR and PBS are all a part of the system of checks and balances that have evolved—partially as a result of historical precedents (depending upon the functions and activities of previous and current organizations) and partially in response to existing needs (wherever there is a vacuum some agency will be formed to fill that need and take on the appropriate tasks).

NPR Activities   National Public Radio operates as a two-way network, so it is easy to confuse network and production activities. Programs are fed to the stations *from* NPR headquarters in Washington, D.C.; and individual stations send programs and reports *to* NPR for national distribution. This is exemplified by *All Things Considered,* the daily ninety-minute news and public affairs program. Approximately one third of every program comes from member stations—even for this regular NPR-produced network program. Many other programs are produced in their entirety by member stations, and 85 percent of the catalog of program "options"—the taped programs available to any noncommercial station—is comprised of member station contributions. Some programs, such as the live broadcasts of congressional hearings, are produced entirely by NPR.

NPR and member stations also produce short feature pieces that can be inserted into local station programing. Additionally, the NPR staff seeks good foreign radio broadcasts; features from the BBC, the CBC, and Radio Nederland are common on the network.

NPR has developed an inexpensive way to provide inserts, features, and news stories to stations on an automated basis. The patented system, called "netcue," uses tones on the leased network line to automatically start and stop a tape recorder permanently connected to the network feed. When an insert or program is about to be sent, the correct tones are sent first—activating the recorder—and then the program material is transmitted.

NPR Program Services   The programs produced and acquired by NPR are made available to potential users through five distribution services: the Interconnected Service, the Network Tapes Service, the Scheduled Tapes Service, the Program Library, and the Audio Archive.[30]

---

[29]"National Public Radio Is . . . ," a factsheet for National Public Radio, undated, circa March 1975.

[30]"Program Standards and Practices of National Public Radio," NPR Staff Base Document V, mimeographed, undated, p. v-3.

6-5 NPR Network Studio Audio Console (Courtesy: National Public Radio)

The *Interconnected Service* is the primary means of distributing topical programs, free of charge, to NPR member stations. It is made possible by a dedicated, 24-hour-per-day, leased-line interconnection provided by American Telephone and Telegraph and associated companies. By 1976, an average of approximately fifty hours of programing per week was transmitted by means of the interconnection.

The *Network Tapes Service* is an alternative means of distributing programs, free of charge, to NPR member stations. This service is used when higher fidelity than is possible on the interconnection is required—and to allow delivery of programs in stereo. Presently, the service is used primarily for distributing tapes of musical performances.

The *Scheduled Tapes Service* is used to make programs available to *all* noncommercial radio stations at nominal cost. Charges are determined by the number of hours of material ordered. Programs in the service consist of material originally released in the Interconnected Service and Network Tapes. Programs ordered are distributed on tape for broadcast according to an established schedule.

The *Program Library* contains some program material that is in limited demand (and, therefore, not distributed through the other NPR services) and some programs that may be obtained for broadcast only through special arrangements with producers and suppliers. Charges are determined by the lengths of the material requested and by any special arrangements with producers.

The *Audio Archive* includes all the programs and audio material distributed through the other NPR program services, the library of programs acquired from the NAEB, and miscellaneous audio, such as recordings of actuality that have not been used in programs. Audio Archive materials can be used for broadcast except where specifically prohibited. Materials are distributed on tape to requesting users.

NPR Staff   NPR is organized into two divisions—a corporate division, which provides leadership and direction for the organization as a whole, and a programing division, which includes the different departments responsible for the acquisition, production, scheduling, and transmission of the network programs.

The corporate division includes the president, the director of business affairs, the director of station relations, and the general counsel (legal affairs). The programing division is under the leadership and direction of the vice-president of NPR; this area includes the departments of program administration and operations, cultural programs, national news and information, specialized audience programs, system program resources, development, public information, and engineering.[31]

Qualifications for Membership in NPR Many noncommercial radio stations do not qualify for CPB Community Service Grants, and as a consequence do not qualify for NPR or APRS membership (see below).[32] To be a "CPB-qualified" (or full-service) noncommercial "NPR member" station, the public radio broadcaster must be able to show that the station has at least five full-time professional management-level employees, that the station broadcasts 18 hours per day 365 days of the year, and that the station has an unduplicated nonfederal annual budget of $75,000. These may not seem difficult criteria to meet, but many low-powered FM stations that are operated by high schools and colleges either for training students or for transmitting instructional materials into classrooms find it difficult to qualify for CPB funds. In 1976, there were more than 180 full-service NPR stations, and another 650 noncommercial radio stations that did not qualify for CPB funds.

Stations that are not CPB-qualified can still use NPR network programs. If they can receive clearly a station that is on the network, they can retransmit NPR programs. Any taped program listed in the Scheduled Tapes Service catalog can be requested by a non-CPB-qualified station. In practice, those stations that do not qualify for CPB funds tend to be in areas with one or more full-service stations, and they therefore tend to develop their own individualized program service with little use of NPR programing.

The total number of educational radio stations and those that qualify for CPB grants, are indicated in the table (6-6).

Association of Public Radio Stations Although the APRS is not a part of NPR, it is usually referred to in the same breath—since the stations eligible to belong to NPR are eligible for membership in APRS, which was formed in 1973 shortly after PBS merged with the ETS division of the NAEB. APRS took over the functions of another NAEB division, National Educational Radio (NER), for representing public radio before federal agencies and Congress. It parallels almost exactly the functions and activities of the national affairs division of PBS. Both work to inform the general public, legislators, and governmental agencies such as the FCC, about public broadcasting—APRS focusing on radio and the national affairs division of PBS concentrating on TV.

APRS was formed to fill a definite need— a vacuum created when the NAEB dissolved its NER division—since NPR did not assume all the NER functions (the way that PBS assumed the ETS activities). Thus, APRS represents, to some extent, the professional services counterpart to the network services of NPR. It is interesting that the name of the organization, the *"Association of Public Radio Stations"* even parallels the title of the NAEB predecessor of almost fifty years earlier, the *"Association of* College and University Broadcasting *Stations"* (Section 2.3). The need is still there; the vacuum must be filled; and we come full circle a half-century later. The organization *will* be created to do the job.

6.6 Other National Organizations

Several national production organizations, such as NPACT and CTW, that provide

---

[31]"Manual of Policies and Procedures of National Public Radio," mimeographed, undated, circa January 1973, amended March 1975 and May 1975.

[32]It is possible for a station to be CPB-qualified and still not belong to NPR. The Pacifica stations are the only prominent examples.

6-6 Number of Educational Radio Stations: 1921–1975 (Selected Years)

Year	AM stations[a]	FM stations	Total stations on the air[b]	CPB-qualified stations	Year	AM stations[a]	FM stations	Total stations on the air[b]	CPB-qualified stations
1921	1	–	1	–	1958	37	151	188	–
1922	67	–	67	–	1960	37	175	212	–
1923	88	–	88	–	1963	38	237	275	–
1924	102	–	102	–	1965	38	268	306	–
1925	90	–	90	–	1966	37	296	333	–
1928	71	–	71	–	1967	36	326	362	–
1930	50	–	50	–	1968	36	362	398	–
1935	39	–	39	–	1969	35	403	438	73
1938	38	1	39	–	1970	35	462	497	96
1940	38	4	42	–	1971	35	501	536	109
1945	38	9	47	–	1972	35	563	598	132
1948	37	27	64	–	1973	35	642	677	147
1950	37	73	110	–	1974	35	717	752	158
1953	37	112	149	–	1975	35	804	839	170
1955	37	123	160	–					

[a]*Data for the number of AM "educational" stations is difficult to obtain, as there is no official classification for these stations. The figures from 1940 on are approximations of the number of AM licenses held by educational institutions, being used for other-than-commercial purposes.*

[b]*Figures indicate the total number of stations operating as of December 31 each year.*

Sources: Federal Communications Commission Annual Reports, as quoted in David Eshelman, "The Emergence of Educational FM Broadcasting," NAEB Journal, March–April 1967, p. 60. Federal Communications Commission "Educational Television," Information Bulletin, April 1971, p. 5, quoted in Status Report on Public Broadcasting 1973, Advance Edition, (Washington, D.C.: Corporation for Public Broadcasting, December 1974), p. 8. S. E. Frost, Jr., Education's Own Stations (Chicago: The University of Chicago Press, 1937), p. 4. "Summary of Broadcasting," Broadcasting, selected issues.

PTV programs have already been mentioned. Additionally, there are other nonproduction organizations centered in Washington, D. C., serving individuals or groups, that are a vital part of the national picture today. Each was formed to fill a particular need; each continues to play its unique role in the educational telecommunications scene; and all of them can trace their spiritual heritage—if not their direct lineal ancestry—back to the early days of educational radio and ETV.

NAEB   The National Association of Educational Broadcasters, which has played such a prominent role in the story of educational tele-

communications, celebrated its fiftieth anniversary in 1975. Because it now serves only individuals, its activities are limited to what membership dues, publications subscriptions, special grants, and workshop or conference registrations will finance.

Although the NAEB identifies itself as "a professional association of individuals who are utilizing communications technology for educational and social purposes,"[33] it has always emphasized broadcasting, and it continues to do so.

---

[33]*1975 NAEB Directory of Public Telecommunications* (Washington, D.C.: National Association of Educational Broadcasters, 1975), p. 2.

It thus tends rather to be more concerned with *broadcast* programs than with the larger view of television as a medium comparable with print that can exist in an unlimited number of forms for an unlimited number of educational and instructional purposes. It also tends to cater to public broadcasting professionals, in part because there is no other organization for *individuals* in public broadcasting.

The NAEB leadership has made efforts to correct this imbalance, however, and to place more emphasis on nonbroadcasting telecommunications concerns. As early as 1971, former NAEB president William Harley announced

> the formation of a policy to support the expansion of educational broadcasting stations into comprehensive public telecommunication centers, incorporating new technological delivery systems (cable, cartridge, ITFS, audio cassettes, etc.) as a way of undertaking an increasing number of essential communications responsibilities.[34]

[34]*NAEB Memo on Instruction,* May 15, 1971.

The most important activities of the NAEB are the annual NAEB conference and its publications, some of which are shown in the illustration (6-7). The NAEB publishes the *Public Telecommunications Review (PTR)* bimonthly and the *NAEB Directory of Public Telecommunications* annually. Both publications are considered "must" reading for professionals in public broadcasting. Many school media specialists (particularly those who use broadcast instructional materials in their schools) and industrial training specialists also belong to the NAEB and rely on its conferences and publications to keep them informed of issues and practices in the field.

Other activities of the NAEB include ITV conferences, educational broadcasting institutes and other professional training programs, special projects (often funded by the government or foundations), reports and directories, and similar nonperiodical publications. The membership is encouraged to become active in "professional councils," which publish monthly "letters," participate in the planning of the annual convention, serve on special projects, and in many other ways

6-7 Representative NAEB Professional Publications in the 1970's

attempt to serve council members' special professional interests. Twelve professional councils existed by 1975, with others under consideration: broadcast education, business officers, development, engineers, graphics and design, instruction, music programing, producers, radio programing, research, state administrators, and television programing.[35]

In 1975, William Harley retired after serving as NAEB president since 1960 (when the position of the full-time paid chief executive was first established). He was succeeded by James A. Fellows, who had served the NAEB in a number of posts, including corporate secretary and, later, executive director.

AECT   The Association for Educational Communications and Technology is the professional association for the broader audio-visual field; it is the principal organization for school and industry educational media specialists, such as audio-visual directors and TV utilization coordinators (Section 2.3). Its Division of Telecommunications (DOT) is organized for those AECT members who are especially interested in the electronic media, principally television. There is considerable membership overlap between the AECT and the NAEB.

JCET   The Joint Council on Educational Telecommunications, which was crucial in establishing noncommercial television channels in the early 1950's (Section 3.4), is now concerned with policy issues in such fledgling activities as cable communications for educational purposes, satellite communications, instructional TV fixed service (ITFS), and television disc recordings. Working with a small full-time staff, under Frank Norwood as executive secretary, the JCET provides a Washington voice and a coordinating forum for its many constituent groups.

---

[35] *1976 NAEB Directory of Public Telecommunications* (Washington, D.C.: National Association of Educational Broadcasters, 1976), pp. 54–63.

National Friends of Public Broadcasting   The National Friends is now the principal nonaffiliated organization of lay persons who are interested in and concerned about public broadcasting, particularly PTV. The organization represents volunteers in public broadcasting, benefactors, and business people who believe that public broadcasting will succeed in the end only if the *public* gets involved. The National Friends assists local volunteer groups and "friends" organizations with suggestions and information on fund-raising activities, auctions and pledge drives, obtaining volunteers, and the like. Since the merger of the National Coordinating Committee of Governing Board Chairmen with PBS in 1973, the National Friends has become important as the single lay group that is not affiliated with any other (professional) agency.

ACNO   The Advisory Council of National Organizations is both an advisory group and an independent organization in its own right. Although it was organized by CPB—as a consultant council to the corporation board—and has always been funded by the corporation, ACNO operates autonomously to a large extent. Formed in the early days of CPB primarily to organize support for funding bills in Congress, ACNO is an association comprised of representatives from many diverse organizations, from the AFL–CIO and the American Association of School Administrators at one end of the alphabet to the United Auto Workers and Veterans of Foreign Wars at the other. In all, about fifty organizations belong to ACNO.

ACNO has been growing in importance in the past few years as the noncommercial broadcasting community has become increasingly aware of the importance of involving the public in public broadcasting. ACNO is only partially representative of the lay public, however, since the individuals representing the membership organizations are themselves often professional telecommunicators. Within each ACNO organization the individual delegate is likely to have the title of "media specialist" or "communications director" and to have a great deal of sophistication in the media. For example, the representative for the American Association of School Administrators during the first several years of ACNO has been Martha Gable, who helped give birth to school television in Philadelphia in the late 1940's (Section 2.5).

This, then, is a thumbnail summary of the various organizations—professional associations and lay groups—that complete the national picture (as centered in Washington, D.C.) in the late 1970's.

**Public
Broadcasting
Programing**

Turning specifically to programing for public TV and radio stations, we want to examine different programing philosophies, ways of categorizing programing, and sources of programs. Although we will be primarily concerned with PTV programing, most of these considerations and comments will apply equally to public radio programing.

7.1   A Perspective on Public TV and Radio Programing

Imagine for the moment that public broadcasting does not exist. Only commercial stations are on the air, with the existing gamut of disc jockeys, game shows, soap operas, news, sports, comedy, variety programs, and dramas. Then suppose that *you* were given the responsibility of designing a noncommercial system of broadcasting in the United States—a system that would not have to rely upon advertising for its support.

What kinds of programing would you specify for this new system of broadcasting? Would you decide to offer the same kinds of programs that commercial stations are offering— only give producers more freedom to be creative and give audiences a chance to watch something without being interrupted by commercials? That could be quite an improvement. Or would you make the new system of broadcasting something like our art galleries and museums—something to be enjoyed by the elite and others who are willing to put forth a little effort to improve themselves—and so offer symphonies, operas, serious drama, discussions of art and literature, and the like? Or would you offer programs for ethnic minority groups—programs in Spanish and Vietnamese and programs designed by and for blacks, Chicanos, native Americans, Chinese, and other cultural minorities? Or would you feature how-to-do-it programs in cooking, playing the guitar, training dogs, or whatever? Maybe you would concentrate on adult education, maybe literacy programs. Would you build studios and let anyone with something to say or do come into a studio to "do their thing"? Or would you reserve most of your broadcast time for credit courses? Perhaps you would try to do many of these things. In this case, how would you establish priorities for the limited amount of money and desirable broadcast hours available?

Noncommercial Broadcasting as an Alternative Service   The above programing alternatives have one thing in common—they all are examples of things that commercial stations might conceivably do but really cannot afford to do on a money-making basis. The federal government decided to support public broadcasting stations "with the proviso that they provide quality programming which commercial stations, in their drive for ratings and profits, either could not or did not provide enough of."[1]

In previous chapters we have compared television with print media; at this point, let us compare the role of newspapers and magazines in our daily lives to develop a perspective on public TV and radio programing. Almost all households subscribe to a daily newspaper, and most adults buy magazines regularly, from the racks or by subscription. In this comparison, the daily newspaper is roughly analogous to commercial TV. The cost of the paper is paid for by advertising (the amount paid by the subscriber for most daily papers is roughly equal to the cost of delivering it, just as the cost of the TV set must be paid to receive otherwise "free" commercial TV programs). The daily newspaper basically contains popular information—current events, human interest stories, household hints, features, comics, fashion news—something for everyone.

Because different people want different amounts of information or features that are touched on superficially or not at all in a newspaper, many subscribe to magazines—news-in-depth types like *Time* and *Newsweek,* how-to-do-it magazines such as *Popular Mechanics* or *Family Handy-*

man, partisan political periodicals, or perhaps magazines on art, music, or community affairs, written for the elite or for the special interest audience. Like the specialized magazines, public broadcasting can offer what the advertising-supported media cannot satisfy because a particular interest is too specialized for a mass-appeal medium. Public television and radio can offer *alternatives* to commercial broadcast stations.

In another analogy to print, John White (former NET president) used to compare ETV to a library. You would not go into a library, he pointed out, and expect to read every book on the shelves—although commercial broadcasters hope you would do that with their media (watch and/or listen to everything that is offered). To the contrary, you go into a library and select only those particular materials that are of interest or value to you; that is the kind of service that noncommercial television and radio offer.

Programing Decision Makers   If we now agree that public broadcasting should be an alternative service, and that it should be daring, innovative, stimulating, and enlightening, we should consider for a moment *who* is going to decide *what* kinds of programs will be provided to American viewers and listeners. How are the desires, needs, and priorities of the audience to be determined and met? With magazines, the problem is simple—either people buy the magazine or else the magazine cannot pay its bills and it folds; but how do we determine if "enough" people are viewing a particular PTV program?

Later, we will discuss ways viewers and listeners can indicate preferences, and ways that public broadcasters can ascertain community needs. Even after soliciting audience needs and desires, decision makers for programing at both local and national levels must ultimately use the educated-guess approach in selecting programs, with an intuitive feeling for the process. The program director at each local PTV station has the most to say about which programs are or are not to be aired, since this individual selects the PBS

---

[1] Arthur Barron, "PBS: Public Television Is More Than Meets the Eye," *American Airlines Magazine,* June 1975, p. 26.

programs, other nationally available programs, and the local programs that will be carried by the station.

At the national level, the programing committees of the Corporation for Public Broadcasting and the Public Broadcasting Service are the most influential groups affecting what is ultimately seen at home. All these individuals and groups at local and national levels make an earnest, honest effort to serve all the viewing publics, but they are among the first to admit that means for determining precise needs and exact priorities are lacking.

Although the systems for selecting programs are less than perfect, public television is organized as a grassroots system. The construction of the national program schedule involves the local station directly through the Station Program Cooperative (Section 7.3). The network is an interconnection distribution service, not a live network service in the way commercial networks are centrally constructed and operated. The lay public—the viewing audience—is involved at every level of public broadcasting, from the local station to the direction of CPB. These basic structural characteristics of public broadcasting help to keep programing balanced and to keep priorities in proper perspective.

7.2   Categories of PTV Programing

One way to describe PTV programing—and at the same time continue to examine programing philosophies, approaches, and priorities—is to look at types of programs that have been discussed in the past and that now are used to describe program services and priorities.

Evolution of NET Programing Patterns

From the inception of the Educational Television and Radio Center, officials of the Center recognized that, if their programing plans were to reflect any type of organized approach to an educational scheme or any kind of balanced program fare, the programing operation obviously had to be based

upon some rational plans for acquiring programs in specific areas. In a far-reaching and farsighted memorandum of late 1952, Robert Hudson suggested thirteen possible subject areas for ETV programing.[2] This was the first mention of any attempt at national ETV program organization: (1) problems of the Free World, (2) economic interdependence, (3) cause of industrial peace, (4) the education of a child, (5) art, (6) natural history, (7) literature: plays, (8) American foreign policy, (9) music, (10) Washington (current events), (11) the twentieth century, (12) opera workshop, and (13) ways of mankind (social anthropology). Hudson did not intend this rambling list to be a totally inclusive catalog; it was expected to be no more than an indication of "certain areas of knowledge and certain specific subjects which may suggest types of program series which the Center's Board and staff will wish to give consideration to."[3] It was the first record of anyone's giving any thought to the categorization of a national ETV programing service, and it provided the seeds for the program categories that would be adopted later.

By 1955, the ETRC was thinking in terms of four broad areas of programing concentration comprising a breakdown of the major academic disciplines plus children's programing: *social sciences* (including public affairs), *humanities and fine arts, physical and natural sciences,* and *children's programs.* This academic breakdown was apparently inspired by ETRC President Harry Newburn's university background and strong pedagogical orientation (Section 3.8).

These four general areas were then divided into ten specific program categories for the purposes of administration and program procurement: (1) history and civilization, (2) the

[2]This twenty-four page memorandum from Robert Hudson to C. Scott Fletcher dated November 16, 1952, dealing with "The Educational Radio and Television Program Development and Exchange Center," was to become the working document that the Fund for Adult Education followed in setting up the ETRC.
[3]*Ibid.*

individual and society, (3) public affairs, (4) literature and philosophy, (5) music, (6) the arts, (7) natural and physical sciences, (8) child interests, (9) youth interests, and (10) special.

The concept behind this categorization was that the ETRC would distribute, each week, one half-hour program (in a series) devoted to each of these specific themes—for a total of five hours of programing a week.

In practice, this approach to programing did not work out as planned. Occasionally, the Center wanted to do more than one concurrent series in a given area—while there might not be much priority for a series in another area.

Eventually, this system of rigid quotas was scrapped, and there evolved a system of six primary program areas: (1) humanities, (2) fine arts, (3) science, (4) social science, (5) public affairs, and (6) children's programs. One program officer was placed in charge of developing programs and series in each of these six categories—which still reflected a commitment to the traditional academic disciplines (the first four areas), with a growing awareness and use, however, of programing categories and thinking borrowed from commercial broadcasting (the last two areas).

Under John White the NETRC system of program categories continued to loosen up, and the number of definite areas was cut down further—no vestige of the traditional academic divisions remained. By the mid-1960's, the Center was working in terms of three broad categories: *public affairs, cultural programing,* and *programs for children.*[4]

Under PBS and the Station Program Cooperative (Section 7.3), program planning and acquisition have become even more flexible and less systematized. To some extent, of course, flexibility is to be desired; but, on the other hand,

it can be argued that public broadcasting now suffers from a lack of planning and commitment to a balanced programing approach encompassing all subject areas and aspects of the full range of man's activities and concerns.

Demise of the Sequential Series One other interesting trend in national noncommercial television programing can be traced over its first two decades. When the ETRC was outlining its original programing policies, one basic concept was that programs should normally be presented in sequential series. An early Center brochure stated: "An attempt has also been made wherever possible to obtain programs in series in the belief that the cumulative impact of such recurring presentations will be more effective."[5]

This philosophy can be traced back to Robert Hudson's 1951 views on the potential effectiveness of the TV-Radio Workshop series *Omnibus;* he pointed out that a weekly series of nonrelated cultural attractions was not likely "to persuade anyone to engage in consecutive thinking" (Section 3.3).[6] The underlying premise, of course, was that if a series of programs is constructed sequentially—with each program building upon the preceding one and, in turn, being requisite for the one that follows it—then the overall series can get into much greater detail than would otherwise be possible. This would be the only way to probe any subject matter with any depth; and the Midwestern universities that pioneered in adult educational television had tended to prepare programs in this fashion, probably because both credit and noncredit offerings to the community tended to be series of lectures.

Early ETRC programing followed this approach as much as possible. Much—if not most—of the early Center programing was organ-

[4]See John F. White, "National Educational Television as the Fourth Network," *in* Allen E. Koenig and Ruane B. Hill (eds.), *The Farther Vision: Educational Television Today* (Madison: University of Wisconsin Press, 1967).

[5]*Presenting National Educational Television* (Ann Arbor: Educational Television and Radio Center, 1955), p. 3.

[6]John Walker Powell, *Channels of Learning: The Story of Educational Television* (Washington, D.C.; Public Affairs Press, 1962), p. 62.

ized into specific series dealing with discrete subject areas. A program series might include any number of individual programs—four, thirteen, twenty episodes—whatever it would take to cover the intended content. For example, in a given time slot a twenty-five program series on child psychology might be followed by four programs on urban renewal, which might be followed by a sixteen-part series on American Indians; but each program in each separate series would build upon the information presented in the preceding program or programs.

This type of sequential series programing obviously demanded a certain dedication and planning on the part of the audience members. An individual viewer could not easily start watching the fourth program if he missed too much basic requisite information in the first three programs. Conversely, a viewer might not start viewing a series if he knew that his schedule would not allow him to watch the two concluding programs three months hence. This programing concept was rather limiting—from the standpoint of attracting the largest possible audience; and, while noncommercial TV stations have never formally entered the ratings race with their commercial counterparts, every ETV station manager wants to attract an audience large enough to justify his station's existence.

Therefore, the sequential series concept has gradually been given secondary status. Under White's leadership, NET began to program more and more single programs, short series, individual specials, one-shot cultural or public affairs programs, and continuing *nonsequential* series. Nonsequential programs (such as *Wall Street Week, Nova, Black Journal*) are continuing series of individual programs dealing with a common general topic or taking a specific perspective wherein each separate program stands alone as a distinct entity; it is not necessary to view the programs in any given sequence; nor is any preceding program essential if a viewer is to understand the current offering.

The shift toward nonsequential programing had become so pronounced by 1975 that an examination of a random week of the program schedule of one large community PTV station revealed only five series (excluding daytime instructional programing) that could, by any definition, be considered sequential.[7] Two of these series were concerned with yoga instruction and practice; one was a continuing BBC serialized drama (sometimes referred to as a "high-class soap opera"); one was a college-credit course in psychology; and one was a true noncredit sequential series dealing with mental health. Even during these programs, frequent references to "those viewing for the first time today" or "those joining recently," imply that one is not expected to view every program in sequence (except, of course, for those enrolled in the college course for academic credit).

Note (Section 7.3) the number of nonsequential series selected by the stations in the 1975–76 SPC II. Clearly, by the mid-1970's, the average viewer could watch an evening of public television offerings without missing much because earlier programs in a series had not been viewed.

Programing Types Today: the CPB Categories   Public television programing has become so diverse that any set of categories is meaningful only for a given purpose, such as categorizing the SPC, PTL, or PBS Flex catalogs (Sections 7.3 and 7.4). The Corporation for Public Broadcasting has said that there are four fundamental ways of looking at public broadcasting programing: by program *content, audience* designation, audience *age,* and program *format;*[8] a single program would have a description that would include specifications for each aspect. The CPB has used these categories to list some of their priorities for the 1970's:

---

[7]Station KCET, Los Angeles, for the week of May 11–17, 1975.
[8]*CPB Report,* October 14, 1974.

Program Content: Programs are needed on dance, visual arts, science, public health and safety, history, contemporary life, international issues, and national issues.

Audience Designation: Programs are needed for women, rural, blue-collar, and ethnic minority audiences.

Audience Age: Programs are specifically needed for children (eight to twelve), for teenagers (thirteen to nineteen), and for seniors (sixty and older).

Program Format: Programs are needed that make more use of documentary, magazine, animation, "docu-drama," and variety formats.

As broad as these guidelines are, they represent good criteria to follow as we look at specific programing activities and program sources.

7.3   The Station Program Cooperative

The Station Program Cooperative (SPC) is described by PBS as "a unique system of television program selection and financing through which the nation's public television stations may participate in the funding of those nationally distributed programs they wish to broadcast."[9] The SPC, originally proposed by Hartford Gunn in a journal article in 1972,[10] had been discussed before the CPB–PBS feud (Section 6.2); but the 1973 shake-up and restructuring of PBS made clearly evident the need for a way to involve the local stations more formally and more directly in the selection of nationally distributed programs.

The SPC Process   There are four steps in the implementation of the SPC:

1. Determination of national program needs of the stations, which includes station surveys, audience research data, data on which programs are repeated and requested, and consultation with advisory panels within and outside of the public broadcasting community.

2. Solicitation of program proposals from potential producers (stations) based on the results of the determination of national program needs.

3. Preparation and distribution of a catalog of program proposals for the use of participating licensees in the selection process.

4. The selection by the licensees of those programs they wish to carry. This process consists of: (a) a number of bidding rounds in which stations express interest in specific programs but do not commit themselves to them; (b) elimination rounds in which programs begin to be purchased or are dropped from the selection process due to insufficient interest by a number of stations; and (c) purchase rounds in which final purchase selections are made by licensees.[11]

The stations which in the end "purchase" each program or series are in effect a consortium created by the SPC to share in the production costs of the series. Each program/series in the SPC catalog has a slightly different list of stations that are sharing in the costs of that offering.

A Case Study: "College for Canines" and SPC II   To illustrate how the SPC operates, let us follow a program proposal from the inception of the idea to the airing of the first program.[12]

Spring 1974: A member of the program development department of station KPBS-TV, San Diego, meets the owner and trainer of a San

[9]*The People's Business: A Review of Public Television, 1974* (Washington, D.C.: Public Broadcasting Service, July 1974), p. 30c.

[10]Hartford N. Gunn, Jr., "Public Television Program Financing," *Educational Broadcasting Review,* October 1972, pp. 283–308.

[11]*The People's Business . . . 1974,* p. 30c.

[12]*KPBS Guide,* August 1975, pp. 4–5.

Diego dog obedience school called Canine College. Impressed with the personality of the owner, she mentions to the director of program development that a dog obedience series might prove popular on public television.

October 1974: The director of program development submits a proposal to PBS.

November 1974: PBS distributes sheets on all programs proposed for the second Station Program Cooperative (SPC II) to program directors, and asks that all proposals be ranked on a five-point scale.

January 1975: PBS selects the canine proposal for inclusion in the SPC II catalog on the basis of ratings received.

February 1975: PBS distributes the SPC II catalog along with information on non-SPC programs that will likely be available to the interconnection.

March 1975: The first noncommitting bidding round begins. Twenty-seven stations are seriously interested in the series—enough to keep it in the bidding.

June 1975: After a total of twelve bidding rounds, *College for Canines* has been selected by 103 stations from Hawaii to New York.

7-1 The PTV Series *College for Canines* during Production (Courtesy: KPBS-TV, San Diego State University)

7-2 Two Successful PBS Cultural/Historical Series: *Ascent of Man* (left) and *The Forsyte Saga* (right) (Courtesy: BBC-TV and Time-Life Films)

July 1975: Production on the series begins.

September 1975: The series is completed; all of the programs have been recorded.

April 1976: The series is first aired on the PBS interconnection.

Other SPC Selections Some of the other representative series selected along with *College for Canines* for the 1975–76 season included the following:

Hollywood Television Theater: 5 dramatic productions (KCET)

Nova: 21 science adventures for curious grownups (WGBH)

Evening Edition with Martin Agronsky: 261 half-hour news analyses (WETA/NPACT)

Erica: 13 half-hour programs on needlework (WGBH)

Wall Street Week: 49 half-hour programs on investments (Maryland Center for Public Broadcasting)

The Romagnolis' Table: 13 half-hour programs on Italian cooking (WGBH)

Bill Moyers' Journal: 26 hour-long programs on current affairs (WNET)

Black Perspective on the News: 52 half-hour programs featuring black journalists from the national press with current newsmakers (WHYY, Philadelphia)

Consumer Survival Kit: 26 half-hour programs on consumer concerns (Maryland Center)

World Press: 52 half-hour programs discussing the treatment of issues by foreign newspapers (KQED)

Soundstage: 15 one-hour programs with superstars from the musical recording world (WTTW)

Washington Week in Review: 52 half-hour programs reviewing past week's headlines (WETA/NPACT)

Sesame Street: 130 one-hour programs (Children's Television Workshop)

The Electric Company: 130 half-hour programs (CTW)

Lilias, Yoga and You: 52 half-hour programs on yoga and exercise (WCET, Cincinnati)

Theater in America: 8 programs of various lengths showcasing regional theater companies (WNET)

Firing Line: 46 one-hour programs featuring William F. Buckley and guests (SECA)

Book Beat: 45 half-hour programs featuring authors and reviews of good books (WTTW)

In all, thirty-eight series—making up a little less than half of the 1975–76 PBS national schedule—were selected and "purchased" by most or all of the stations (only a station including a program in its final bidding round may finally air the program).[13] These final selections were culled from a total of 135 proposals that were listed in the SPC II catalog. These proposals were divided into the following six categories (the figures in parentheses refer to the number of proposals in each category):

---

[13]*Public Television National Programming from the Station Program Cooperative 1975–1976* (Washington, D.C.: Public Broadcasting Service, August 1975).

7-3 PBS Public Affairs Series: *Black Journal* (top) and *World Press* (bottom) (Courtesy: Public Broadcasting Service)

Arts and Humanities (37)

Contemporary Life, including cooking, pets, how-to-do-it, sports, and physical fitness programs (15)

Public Affairs (23)

Science and Medicine (6)

Variety (7)

Target Audience, including ethnic and racial series, women's, children's, and instructional programing (47).

Later station program cooperatives have had similar distributions of programs, although reduced CPB and Ford Foundation support, combined with the emergence of the Station Acquisition Market (see page 125), have reduced the number of programs selected through the SPC.

Reactions to the Station Program Cooperative The amazing thing about the SPC is that—as cumbersome as it is—it works. The bidding and the eliminating of programs become possible only because a computer can be connected directly to the Dial Access Communication System (DACS), and, during a bidding "session," each of the 150 or so licensees (many of whom own more than one station) communicates directly with a computer at PBS in Washington, D. C.

The SPC has not been without its critics, however. Although the SPC is the epitome of the democratic, nationwide, electronic "town meeting" grassroots process, it also has the drawbacks of a 150-member committee decision-making process. Former NPACT President James Karayn has been one of the most vocal critics from within public broadcasting's ranks.

Concerned with what he considers a lack of serious investigative journalism in the programs selected, Karayn suggests that the Cooperative may serve to perpetuate programs that appeal to the largest possible audience, and indirectly, the greatest number of potential contributors. Large audiences and fund raising appeal should not be the sole criterion upon which programming selections are made.[14]

Arthur Barron, a film maker and critic, is another who was troubled about the small number of documentaries and public affairs programing in the early SPC's.

Does this mean that programs which tackle controversy and explore issues of topical concern will be played down? Given the limited funds available, will small and less expensive projects drive out more elaborate and ambitious projects? (From the SPC I catalogue, "Lilias, Yoga and You," which costs around $1,700 per program, was selected over WNET "Opera Theatre," which would have cost approximately $45,000 per program.) Also, will there be enough talent at local stations to produce national shows of the highest professional and artistic quality? Will voting majorities mean "group-think," the end of innovation, experiment, creative audacity?[15]

The Advisory Council of National Organizations (ACNO) has at least twice repeated its concern that the SPC does not, and will not, serve the minority audiences most needing to be served—again because stations will bid on the programs that will bring them a large audience.[16] ACNO advised the CPB that it should underwrite programs in the following areas that would be inadequately served by the SPC: (1) women's programs; (2) programs for children from ages

[14]Robert K. Avery, "PBS's Station Program Cooperative: A Political Experiment," a paper presented at the twenty-fifth annual convention of the International Communication Association, April 23, 1975. Many of Avery's comments regarding Karayn were based on an article by Michael J. Connor, "Public TV Is Viewing Its Long-Term Future With More Optimism," *The Wall Street Journal,* August 23, 1974.

[15]Barron, "PBS," pp. 27–28.

[16]*CPB Report,* May 10, 1974.

10 through 17; (3) programs for working Americans; (4) programs that deal with government as it relates to the American public; and (5) minority programs.[17]

Hartford Gunn has maintained that the group of programs selected by the SPC could likely have been chosen by staff members or a committee, but the process makes the choices those of the stations. "That's the first and the most important argument for the cooperative."[18]

It must be emphasized that the SPC furnishes less than half of the total PBS national schedule—and, obviously, even less of the total program schedule of each station. The remainder of the national schedule is made up of repeats from previous seasons; programs supported by CPB, foundations, and government agencies; and—increasingly—programs underwritten by corporations (see below).

Funding the SPC   Basically, the local station pays for the programs that are bid on during the SPC. Each station is charged for the programs it selects in the final round of bidding. It can use part of its Community Service Grants, its regular operating funds, membership dues, grants, or any other source of income that it has. Not all stations pay the same amount for a given program. A "pricing factor" is worked out at the beginning of the bidding process that takes into account both the operating budget and the potential audience of each station. In SPC II, for example, station WNET (New York City) had a pricing factor fifteen times that of KAVT in Austin, Texas, and therefore had to pay fifteen times as much for each program.

On the other hand, stations do not have to supply all the dollars necessary to pay for every program. Part of the costs of many programs are underwritten by corporations (see below), and some programs are partially supported by

the stations or networks that propose them—especially nonformal educational programs that meet a regional or local need but also have national significance. In addition, the CPB and Ford Foundation have made special grants to PBS to help pay for SPC programs; these funds have been used to match dollars provided by individual stations (SPC I and II) or given outright to stations to purchase programs in the Station Program Cooperative (SPC III).

Corporate Underwriting and the SPC   We mentioned corporate underwriting when we discussed funding from the point of view of the local station manager (Section 5.4); but the subject deserves detailed attention at this point, for this form of quasi-sponsorship is an important source of funds for public TV programing—and it is becoming increasingly important each year.

The FCC formally recognized corporate underwriting, and the use of identifying credit lines, in 1970. Before that time gifts from corporations had been obtained, but no one was sure what should be done about credit lines. FCC regulations on the one hand prohibited commercial references on noncommercial TV stations and on the other hand required that anyone providing financial or other tangible assistance to a program had to be identified by name. So the practice has evolved of giving the sponsoring corporation visual and/or aural credit for the donation—without any reference to the corporation's business or products.

Corporations were quick to recognize the value of corporate underwriting. As early as 1964 Jacobi wrote:

> Now that educational television programs attract a meaningful and measurable audience, [corporate underwriting grants] are highly effective instruments for enhancing a company's public relations. . . . It serves public good, the underwriter's name is associated with the program or series throughout its life on the air, and most ETV stations

---

[17]*CPB Report,* September 16, 1974.

[18]Hartford N. Gunn, Jr., "Inside the Program Cooperative," *Public Telecommunications Review,* August 1974, p. 18.

average two or three repeats of the same show within a short period of time.[19]

Corporate underwriting exists at several levels of public broadcasting, often coupled with the SPC. At the highest level of commitment, corporations pay for an entire series of programs. Mobil Oil has been one of the most active companies in PTV underwriting, often spending more than $1 million a year to underwrite such programs as *Masterpiece Theatre*. Xerox *(Civilisation),* Gulf Oil *(National Geographic Specials)* Exxon *(Theatre in America),* and Sears Roebuck *(Mister Rogers' Neighborhood)* are other representative corporations that have long and actively been involved with corporate underwriting.[20]

The motivations of companies in supporting PTV programing are both altruistic and self-serving:

Corporate social concern undoubtedly plays a role in this largess, but public relations benefits are also a strong factor. It is no accident that oil companies, which have had a serious image problem, have been in the forefront of funders. Although corporations are not permitted to advertise directly on the shows they underwrite, they are given prominent screen credit.[21]

Corporations prefer to give most or all of the costs for a particular series so that they can identify themselves with the series, and promote the series in print advertisements. Mobil often buys entire pages in newspapers and magazines to announce at the beginning of a new season the PTV programs it is underwriting. In addition to this series underwriting, however, corporate funds are also solicited on a more general basis—both at national and local levels. Within days of the announcement of the fall schedule in July or August, PBS publishes a description of new programs not underwritten by corporations (that is, those being paid for by the SPC) for use by its development staff and local stations to obtain corporate underwriting for the costs of broadcasting the programs. Some stations have even experimented with what might be called "run-of-schedule" underwriting, promising X number of credits throughout the week at different times for Y number of dollars. Most stations find this uncomfortably close to selling time and have not picked up the idea. Some have tried it and found it did not work well. These and other problems in connection with corporate underwriting are further explored in Chapter 9.

Station Acquisition Market "SAM," the Station Acquisition Market, grew out of the Station Program Cooperative and operates as an adjunct to it. SAM originated as a result of the increasing popularity of syndicated programs and series among PTV stations, especially the regional networks. Noting the better per-station rates the networks were able to negotiate, PBS proposed in 1975 that a procedure be developed for acquiring special programing for the interconnection through negotiations at the national level.

A SAM is like a regular SPC purchase for the individual PTV station; the programing will be scheduled on the interconnection if enough stations buy it, and only those stations purchasing the program may broadcast it when it is fed by PBS to the system. SAM is different from SPC in that (1) a SAM can occur any time that the PBS Programming Department feels that a program or series warrants the creation of a SAM; (2) because the PBS Programming Department decides whether a SAM will be offered, and negotiates the terms of the SAM, there is more centralization

[19]Fredrick Jacobi, "Educational Television: Growing Public Relations Opportunity," *Public Relations Journal,* March 1966, pp. 10–11.

[20]One of the *National Geographic* specials, "The Incredible Machine," was the first prime-time public television program to receive higher ratings than its commercial competitors in major markets. The success of this program is analyzed in a special issue of *Public Telecommunications Review* subtitled "The Incredible Machine: Lessons of a Winner," November/December 1975.

[21]Barron, "PBS," p. 29.

of activity and authority with SAM compared to the SPC; (3) a SAM generally involves either programing that already exists (such as the classic films from the Janus Film Collection) or public affairs programing (such as *The MacNeil/Lehrer Report*), while SPC proposals generally are series which stations wish to produce if others are willing to share in the costs of production; and (4) a fixed price for each station is negotiated before a SAM even is announced, allowing the station to decide quickly on a yes-or-no basis whether it wishes to participate in the SAM. If a very large number of stations agree to purchase a SAM offering, the price per station will begin to drop after a certain point is reached.

7.4   A PTV Program Director's View of Program Sources

As has been mentioned (Section 5.3), the PTV program manager is the person most directly concerned with meeting the programing needs of the different groups in the area served by the station. To meet these needs and interests, the program director considers all the programs available from the PBS interconnection, as well as other nationally available programs, regional network programs, and local programs that are needed and can be afforded.

Programs Available from the PBS Interconnection   In considering the programs available from PBS (to many of which the station has committed itself as they were "bought" through the SPC), the program manager can accept or reject programs without concern for commitments by others. This is in sharp contrast to commercial broadcasting where a network can commit stations to carry a particular program. Even corporate or agency underwriting for national programing cannot commit stations to broadcast a given program—it is possible (though hardly conceivable) that a corporation could pay $1 million to create a series and then find that no station wanted to air it.

In addition to those SPC series to which the station has committed itself, the station may later wish to broadcast other SPC programs; arrangements to carry these programs must be made after the bidding on programs has been completed, and—after an announced cut-off date—these programs must be paid for without the benefit of any matching funds. There are programs on the interconnection that have been paid for outside of the SPC by corporate underwriting, the CPB, or foundations, or very often by a combination of these sources of funds. Stations also occasionally offer local programs to PBS for distribution on the interconnection; often these are documentaries on local situations that the station and PBS feel have national interest. Finally, many programs from past seasons are repeated at least once on the interconnection. In most instances, a program has clearances from performers and copyright owners for a three-year use of the program (with a maximum of four plays or "releases" per station per program).[22] Programs not repeated on the interconnection are therefore usually offered to local stations through the PBS-Flex catalog at least for the time remaining under the original rights agreements.

PBS-Flex   The PBS Flexible Program Service is actually a library activity operated as a part of the Public Television Library (see below). A loose-leaf catalog lists available programs and the conditions for their use. Stations can then order whatever programs they want. The primary differences between PBS-Flex and PTL programs is the source of the programs (national funding for PBS-Flex as opposed to local or regional funding for PTL) and the fact that stations usually have already used PBS-Flex programs (and likely have promotional material on hand and are familiar with the programs). Another possible difference is that some series in PBS-Flex restrict the number of "releases" permitted on any given PTV

---

[22]A "release" is a one-week play of the program, which could be a single showing or repeated showings in a seven-day period.

station; thus it is necessary to see whether the series was run at all (if it was not, and if it was an SPC program, extra fees must be paid), and if it *was* broadcast when transmitted on the interconnection, how many times it was broadcast by the station.

Public Television Library   The PTL is the cooperative program exchange that was begun in 1965 by the Educational Television Stations (ETS) division of the NAEB (Section 4.4) and was known for several years as the ETS Program Service. In 1973, when ETS and PBS merged, the service became a department in PBS, and the entire operation moved from Bloomington, Indiana, to PBS headquarters in Washington, D. C.

There are few restrictions on what stations can offer to the PTL other than the fact that clearances must have been obtained for use by other stations and that the program or series have enduring and general interest. A committee of station representatives rate each program offered to PTL and decide if a program should be included in the PTL catalog. The ratings are included in the description of programs and series to guide PTV program directors. Hundreds of individual programs and series (some of which contain as many as 26 programs) are now available from PTL. As a program exchange/library service, member stations can order any program(s) from PTL for a fee that covers duplication, handling, and other costs.

The offerings are described in thick looseleaf notebooks divided according to program types. Different notebooks contain cross-reference sheets filed by program titles. Each sheet in the notebooks describes a program or series and any special conditions of its use: program numbers and titles, exact timings, the ratings of the screening committee and PTL engineers on the general program quality and technical quality, the date first made available to PTL, the date that rights to the program expire or other considerations require that the program be removed from the catalog, the storage formats (film or videotape) that can be requested (also whether the program

is in black-and-white or color), and whether promotional pictures are available.

A brief look at the categories will give an idea of what the PTL catalog contains. Many of the programs within each category are of the "nonformal education" type, in many cases because the program was underwritten by a governmental or nonprofit agency that wanted to get across a particular message.

Public/Community Affairs: Includes many programs on public health issues, such as mental retardation, autistic children, alcoholism, and drug abuse. Other programs include ecology, penology, education, aging, community relations, and the like.

Performance, Cultural Programs, and Arts: Includes many series of programs that were once offered as credit or noncredit courses, such as a series on black music, or on a particular art technique. Individual programs and some series offer a "showcase" for talent that often goes unrecognized.

How-to-Do-It Programs: Mostly in series, on everything from needlepoint to auto mechanics.

Children's Programing: Relatively few offerings; often produced in cooperation with a museum, zoo, or library. Some series have been produced by stations or regional networks to help alleviate a particular community problem, such as race relations.

Sports: Very few offerings; usually instructional series on a sport, or a documentary-type program on an athlete or sports person.

Miscellaneous: Any program or series that does not seem to fit any of the above categories.

Many PTL and PBS-Flex programs are now being made available to schools and libraries in video-cassette format, extending the usefulness of the PTL programs. This will be an increasingly important activity for PTL as new PBS programs are cleared for nonbroadcast use, and as distribution methods for STV programs improve (Section 14.2).

NET-Flex   The NET Flexible Program Service should be included in any discussion of program sources, more for its historical significance than for its importance today. The service, located in Ann Arbor, Michigan, but operated under WNET direction, once was the supplement to the NET "bicycle" network (1954 to 1968) just as PBS-Flex supports the national PBS service. NET programs that stand the test of time can still be ordered from NET-Flex, but it is not considered an important source of supplementary programing today. The technical center in Ann Arbor provides duplicating services to PTV in the Midwest, and back-up duplicating services for WNET.

Free-Loan Films   For years, companies and agencies have been producing 16-mm films to promote their products and activities. Many films produced for use in schools contained very little mention of the company producing the film. In the early days of commercial TV, and again when noncommercial TV began, these free-loan films became a staple of every station's programing. They still are sometimes used to fill empty slots in the program schedule.

There are two major "clearing houses" for free-loan films, both of which publish special catalogs describing films cleared for TV, and offer advice and special services to TV stations. Modern TV, a division of Modern Talking Picture Service, has six libraries across the country. Association-Sterling Films has five libraries throughout the United States. Films of every description are found in the catalogs, but both distributors list more travel films than any other single type of program. Association-Sterling also distributes some films from federal agencies and from UNICEF. Both distributors offer the films free-of-charge, even paying postage to the station. The only requirements are that the stations

pay return postage, and that a card is completed showing the day, time, and estimated audience for the film.

Although many of these films are suitable for broadcast purposes, many of them, unfortunately, are not. Many are strongly biased to the point of view of the producer or contain so many references to the sponsor of the film (often visual) that they simply are too blatantly propagandistic to put on the air.

Free Government Films   Films, and occasionally videotapes, can be requested from several different governmental agencies. The U. S. Navy, for example, has a catalog listing more than eighty films cleared for television (one on sharks became a favorite of stations around the country in 1975 when the motion picture *Jaws* began setting records). Many other state and federal departments and bureaus have extensive catalogs of available programs.

Syndication and Sales   There are always a number of individuals and companies that have produced, or are producing, programs that can be purchased, leased, or rented for a fee.[23] Sometimes the fee is established and inflexible, but often it is negotiated. The offerings are of three basic types:

Feature Films: Many of the feature theatrical films available to commercial stations are also available to PTV stations at a reduced price. Stations and regional networks have been most interested in classic films and foreign films, such as *The Silent Years* (a series of landmark silent films). In some instances, university stations have added introductions and follow-up programs to offer film-study courses for credit.

---

[23]Actually any syndicated program available to commercial television stations can be purchased by a public TV station. The *Los Angeles Times* (January 5, 1976) reported that *Mary Hartman, Mary Hartman,* the Norman Lear soap opera parody, was being carried by a PTV station in one market where no commercial station would purchase the syndicated program, and other PTV stations also have aired the series since it began.

Documentary Films and Special Series: The best examples of this kind of syndicated programing were the Time-Life documentaries and the Monty Python series distributed in the 1970's. The cost of programs of this sort is often so high that stations usually can afford such a series only when they are able to arrange corporate underwriting.

Topical Discussion Programs: Just as discussion programs have been successful in commercial TV as syndicated series, so have they proven successful in public TV. Both *Firing Line* with William Buckley and *Day at Night* with James Day began as syndicated programs. The Buckley series is now a regular PBS offering through the Station Program Cooperative.

7.5   Local Programing

Locally produced programs vary considerably from station to station. (In fact, local programing is one of the few ways one can deduce what category of PTV station one has tuned in during a trip to different cities.) Some stations do very little local programing; few do a great deal. The following types of local programs, therefore, should be considered very rough generalizations.

Community Affairs Programs: Probably the most important type of local productions are those programs designed to inform viewers about events and issues in the immediate community. Often, these are discussion or debate programs; but larger community stations schedule regular community affairs documentaries that resemble (in format) news features on commercial stations—except that they frequently examine issues in much greater depth. As more and more stations obtain portable videotape equipment, these community affairs programs increasingly take viewers out into the community—rather than having community news makers come into the studio to talk.

Community Talent Programs: Two basic types of community talent programs are seen on PTV stations—the young talent variety program and the "special." The talent variety program often resembles the local talent show that once was frequently seen on commercial stations. It is more likely to be offered by the smaller PTV stations to provide encouragement and a showcase for local talent. The "special" is offered on an occasional basis and highlights one or more recognized community groups in depth, sometimes (when there are performances involved) in programs of an hour or more.

Special Events: Many PTV stations have remote equipment that permits on-the-spot coverage of important community events. Often the PTV station will offer uninterrupted coverage of a live event (public debate, concert, legislative session) that would be covered by commercial stations in a very few minutes of time in a news broadcast.

Sports: The coverage of local sports is a type of special event that deserves separate mention. This is an important area of local programing for most university PTV stations, but even smaller community PTV stations have some sort of sports programing—even if it is only an in-studio review of high school sports with 8-mm film inserts shot by high school students.

Follow-up/Feedback Programs: Many PBS and PTL programs lend themselves to local follow-up programs to bring the issue discussed on the national program into community focus. Often the national producer recommends that local experts on the subject sit in the studio next to phones to allow viewers to call in on the air to ask questions or to make comments. A series that used this technique to good advantage was *The Killers,* a health series broadcast in the mid-1970's. A program on alcoholism, "Drink, Drank, Drunk," featuring Carol Burnett, was followed in many cities by local programs in which viewers living with alcoholics were told how to obtain help from community groups—and in some cases were able to call studio experts to get immediate help for a particular situation involving an alcoholic.

Student/Volunteer/Community Access Programs: In Chapter 5, we mentioned that PTV stations licensed to universities with broadcast curricula often schedule programs produced entirely by the students, recorded perhaps in advanced production workshops. Other stations may set up workshops for volunteers, high school students, college classes, or simply for anyone interested—out of which may come programs broadcast on a special basis. Some stations set aside time for anyone with something to say or do to come before the cameras, similar to (and often involving the same people as) the community-access programing on larger cable systems.

Instructional Programing: For many stations—especially school-owned and state-supported stations—instructional programing may be the crux and raison d'être for the entire operation. The locally produced in-school materials provide the justification for virtually all of the tax support. A college station, also, often does a great deal of both formal and nonformal instructional programing and may justify the physical plant (studios, technical equipment, and the like) primarily on the basis of local instructional programing. Even community stations often find that contracts with local school districts or with metropolitan instructional TV associations (Section 11.4) for local school productions are a substantial source of income.

The local PTV program director not only must choose from among all these program sources—and from these possible types of local productions to be undertaken—but also must consider what other stations in the area are offering at different times. As the director of programing at the Los Angeles community station put it:

> Another factor in scheduling programs is the type of show being considered and how it relates to commercial TV schedules. At KCET we make every effort not to slot outstanding programs opposite the same type of fare on one of the commercial stations.

In other words, we try to provide an *alternative* service for the viewer. Because of the large number of television stations in the Southland and the wide variety of programming available, we are not very successful in every circumstance, but we do try.

Considering these factors, putting together a local schedule is much like making a patchwork quilt. While it is possible to place most of the patches anywhere on the quilt, there are many instances where certain patches work better in specific locations. So it is with programs and the program schedule. As other stations change their schedules, and as new programs begin on our own station, every effort is made to present them in the most effective manner possible.[24]

7.6   What about Radio?

As we have been examining programing specifics, we have been talking only about public *television*—or have we? Most of the categories, the issues, and the problems facing TV program directors also apply to some degree to radio; but the public radio program director has both an easier job and a harder job—for the same reason; in most locations there are many radio stations, most of them programing for specialized audiences. Therefore, there is less pressure on the public radio program director to examine and serve specialized listener interests, and the public radio station can be more of a service for the elite without apologies. Nevertheless, the number and different types of radio stations also require that the program director keep aware of what all the local stations are offering and to be creative and conscientious about serving those whose interests are not otherwise being answered. He is still trying to provide an alternative service.

---

[24]"KCET's 'New Season,'" *Gambit* (KCET members' magazine), January 1975, p. 16.

Only a couple of programing types should
be further commented on in relation to radio. One
obvious category is music. Most public radio stations program uninterrupted classical music at
various times during the day; and many have
special programs on jazz, ragtime, ethnic, big-
band, or other music—often with commentary.
National Public Radio programs in these latter
nonclassical areas often feature performers and
guest critics who comment on the music and
the times.

In the area of community affairs and
special events, public radio stations frequently
tap into the microphone system in city and county
buildings in order to broadcast live the meetings
of local governing groups such as the city council.
Sometimes the city or county group itself will
pay the costs of broadcasting the proceedings.
Otherwise, with the portability of (relatively) inexpensive audio recording equipment, the more
ambitious local public radio stations find they can
do a very creditable job of covering local community affairs.

In summary, the local public radio station
program director finds himself in a situation similar to that of his PTV counterpart: same categories
of programing, similar sources of national programs, comparable local production options, and
the same opportunity to serve the public.

Issues in Public Broadcasting: Funding and Controls

In Chapters 8 and 9, we will review some of the current issues facing public broadcasting, look at the history behind these issues, and outline some of the future problems facing public broadcasting. Steve Millard, former editor of *Public Telecommunications Review* and director of publications for the Corporation for Public Broadcasting, once described the problems of public broadcasting this way:

> If public broadcasting draws large audiences, it is attacked for seeking the masses; if it programs for small, select groups, it is damned as an insufferable snob. If it tackles a tough issue, it is trendy, left-wing, unrepresentative, and misusing the taxpayers' money; if it presents fine drama and stimulating discussion, it is aloof and uninvolved. If it moves toward centralization, the specter of autocracy is raised; if it does not, there is the accusation that it is frittering away its public money without seizing the chance to make an impact on the national consciousness. Anything it does, in any realm, is sure to be attacked by someone as contrary to the spirit of the Carnegie Commission report or the Public Broadcasting Act— both of which mean many different things to many different people. Meanwhile, the struggle for federal funding limps along; federal appropriations increase, but freedom from annual accounting does not.
>
> In television and radio, there *is* no way to be both a Good Guy and a tough, visible force. . . . Now it is learning to live on a height that is also a precipice![1]

These are some of the dilemmas facing public broadcasting. As we discuss these and other questions, we should remember that, as Millard says later in his article, "The noncommer-

---

[1] Steve Millard, "The Story of Public Broadcasting," *Broadcasting,* November 8, 1971, p. 30.

cial media have traveled a long road to confront problems and opportunities of such magnitude."[2]

8.1   Federal Funding of Public Broadcasting

Federal funding has been a struggle since the passage of the Public Broadcasting Act in 1967. Even today, some say the federal government should not be giving *any* money to support public broadcasting, and the debate as to *how much* money should be given, and *how* the money should be given, goes on and on.

The Existence of Federal Support for Public Broadcasting   The funding pattern of the mid-1970's might best be called a partnership between the federal government and the private sector. The federal government supports some specific activities, but the bulk of the money is given on a matching basis—$1.00 of federal funds for each $2.50 collected from nonfederal sources (local and state government and private sources).

Some say that the government should be supporting public broadcasting totally, as in Japan and Great Britain. Several countries have evolved a dual arrangement whereby a government-supported operation and a commercially supported system exist side by side. At the other extreme, there still are some who argue that public radio and television should be completely self-supporting, perhaps with help from private philanthropies such as the Ford Foundation. Few, if any, would argue against the existence of non-commercial broadcasting, but the question raised by some is whether federal money should support it.

Kevin Phillips, for example, after citing examples of liberal public affairs programs "that are often contemptuous of everyday America," writes in *TV Guide:*

Alas, even though "public" television has become the plaything of the Manhattan and Washington Mercedes crowd, a lot of it is still being paid for by the middle Americans who drive two-year-old Chevrolets and live in the Maine-to-California equivalents of [middle class] Rego Park. Benjamin Stein, in a superb Wall Street Journal article, recently queried the fairness of having tax-payers pick up 25 per cent of the national PTV cost when the programming is so elitist-oriented. Said Stein: "'Public' television represents a subsidy by the lower-middle-class people, who pay the bulk of Federal taxes, to the upper-income groups, and a subsidy of the many to the few." It is income transfer from the poor to the rich, to over-simplify a bit.[3]

Stein himself referred to Milton Friedman, University of Chicago expert on government and citizens rights, who called the use of tax money to support public broadcasting "a disgrace":

He points out that under the present system of partially tax-supported public television, the average taxpayer, who is a great deal poorer than the average viewer of evening PBS fare, is subsidizing his financial betters. "It just shows," he says, "that the rich know how to manipulate the political process to get what they want out of it." Friedman says that if people want to watch television, they should pay for it. He recommends that all public television be pay television. That way only those who want to see something will have to pay for it.[4]

Public Television as a "Merit Good"   Of course, one could make the same argument for most tax-supported institutions. The elite tend to use the libraries and museums. (On the other hand, they tend not to use welfare.) Except for

---

[2]*Ibid.*

[3]Kevin Phillips, "Should Federal Money Support Public Television?" *TV Guide,* July 12, 1975, p. A–3.

[4]Benjamin Stein, "PBS Under Fire: Its Reach Is Short," second of a three-part series, *TV Guide,* December 20, 1975, p. 27.

such things as streets and schools, *most* tax-supported activities are focused on the needs of particular segments of the American population.

The question, as three economists put it, is whether public broadcasting can justify its existence as a "merit good," the way libraries are classified in economic terms. These economists—Owen, Beebe, and Manning—note that, from an economics point of view, all alternatives must be explored, and they conclude that federal money could be given to community groups to buy time on commercial stations.

> One can imagine a public-broadcast institution, still local, which simply bought time on local commercial stations (or combined resources to buy network time) for programs that fit our selection criteria. This would have at least three advantages. First, the heavy capital and overhead cost of maintaining 225 local stations would be saved. Second, public-broadcast programs would not be confined to the ghetto of a channel that many people have come to associate with "boring" programs. Third, a number of useful VHF television assignments would be opened up for commercial use—either as independents or as affiliates of a new fourth network, options that viewers *would* value highly.[5]

This argument, however, misses the central point—that public television and radio exist to provide an *alternative* service (as well as educational programs). Further, the first quarter-century of radio proved repeatedly that a *system* of noncommercial broadcasting was necessary; the interests and economic needs of commercial broadcasting are so fundamentally antithetical to specialized, minority audience needs—and there are so many different minority interests—

that limited times on commercial channels simply do not satisfy the basic need. Also, the Ford Foundation's *TV-Radio Workshop* (Section 3.3) proved the ineffectiveness of this type of thinking. It would be about as reasonable to suggest that society should do away with libraries and give community groups federal money to rent space for books in commercial bookstores. At least until such time as we are a "wired nation" and entertainment programing comes into all homes on a practically unlimited number of cable channels, most critics and observers agree that public broadcasting *is* a "merit good," and should be supported as such by public money. Stein summed up this position by saying,

> Public support of symphonies and art galleries is largely taken for granted as a way of raising the aesthetic level of consciousness and therefore the intelligence of the whole Nation. Only a fraction of the citizens of America ever hear those subsidized operas, but more people could hear them, and the fact that they are available is reassuring, even if not everyone uses them.[6]

Nature of the Federal Support   The next question, then, is how much support is needed—and how should it be collected? The Carnegie Commission was precise in its recommendations. It said that public television ultimately needed $270 million annually (in 1967 dollars), and that this amount could and should be raised by an excise tax on the sale of new receivers. The commission arrived at these recommendations only after considerable discussion.

> The Commission members wrestled with this issue throughout their deliberations. They were concerned over the potential influence of federal money on independent

[5]Bruce M. Owen, Jack H. Beebe, and Willard G. Manning, Jr., *Television Economics* (Lexington, Massachusetts: Lexington Books, 1974), p. 167.

[6]Benjamin Stein, "PBS Under Fire: 'We Need It Now More Than Ever,'" last of a three-part series, *TV Guide*, December 27, 1975, p. 32.

programing, particularly in the sensitive area of video journalism. The political characteristics of the annual budget and appropriations process raised the specter of congressional intervention against the funding of programs without universal appeal. The dependence of the system on the uncertainties inherent in the annual cycle would handicap planning. . . . The possible options were discussed but not deeply researched. The recommendation for a dedicated excise tax on the sale of sets was the preferred course, although there remained doubt about its political feasibility. At least it would constitute independent funding, outside of the annual appropriation [process], on a regular and assured basis.[7]

Congress passed the Public Broadcasting Act of 1967 without specifying how CPB would receive federal dollars. The idea of a dedicated tax on receivers continued to be pushed, but so did alternative plans that better suited the personal interests of one group or another. Even before the Public Broadcasting Act was signed, the Ford Foundation submitted its proposal for a broadcasters' nonprofit satellite system to provide funds and a free interconnection for educational telecommunications (Section 4.6). In 1969, the National Citizens Committee for Broadcasting proposed a 4 percent tax on gross receipts of commercial broadcasting, which would yield $120 million a year and more.[8]

In 1972, when various funding plans and formulas were being debated before the Macdonald subcommittee (Section 6.2), the CPB initiated an industry-wide study to develop "a common plan which all elements of the system could endorse and to negotiate with executive and congressional leaders on the basis of such terms. With the backing of all system organizations, a task force was formed of fourteen representatives of all elements under the chairmanship of Joseph Hughes, a recently appointed [CPB] board member."[9]

The committee evaluated all of the possible financing modes, individually and in combination, including a tax on commercial radio and television advertising revenue; a tax on community antenna TV subscription revenues; charge for access to the broadcast spectrum (payments for frequency space); using a portion of income taxes paid by commercial radio and TV stations and CATV operators; a dedicated excise tax on residential electric or telephone bills; general tax revenues; a "user charge" to be paid by each family owning radio or TV sets (similar to the annual British Broadcasting Corporation receiver license fee); and a public broadcasting development bank financed by bond sales.[10] Other ideas, which were not seriously considered, included a federal lottery and a billion-dollar endowment that would permit public broadcasting to exist on the interest earned.

Task Force Criteria   After establishing a detailed set of assumptions and projections, the task force estimated that total operating costs of the "ideal" national system of public broadcasting would come to $475 million annually. If the federal government supplied 30 to 50 percent of the needed support (which had been agreed upon as the ideal federal portion by the task force members), federal support should amount to about $200 million per year. Although the amount was derived by a different process, this calculation was remarkably close to the figure recommended by the Carnegie Commission—adjusted for the five-year inflation rate.[11]

[7]John Macy, Jr., *To Irrigate a Wasteland: The Struggle to Shape a Public Television System in the United States* (Berkeley: University of California Press, 1974), pp. 24–25.

[8]Dick Netzer, *Long-Range Financing of Public Broadcasting* (New York: National Citizens Committee for Broadcasting, April 1969).

[9]Macy, *To Irrigate a Wasteland*, p. 109.

[10]Corporation for Public Broadcasting, *Report of the Task Force on the Long-Range Financing of Public Broadcasting*, September 1973, pp. 49–53.

[11]Macy, *To Irrigate a Wasteland*, p. 110.

The task force then enunciated a list of basic principles, assumptions, and recommendations, which became the basic guidelines or criteria for all discussions of public broadcasting financing in ensuing years:

1. The principal share of the operating expenses of public broadcasting should continue to come from nonfederal sources. But it is entirely appropriate and necessary that federal funds be a part of a total financing plan.

2. The federal contribution should be designed so that it provides incentives for increasing nonfederal financing.

3. Financing public broadcasting should not impose unreasonable burdens upon any segment of the economy, but rather should be designed so that those who benefit also pay.

4. If federal funds are appropriated in whole or part by a matching system, a portion of those funds should be returned to the stations on an equitable basis which reflects local effort.

5. If federal funds are appropriated, there will need to be accountability to the Congress in the use of these funds.

6. The need for insulation against undue pressures from whatever source is particularly important with respect to the financing of programing.

7. Long-range planning, which is based on a reasonably assured level of future funding, is essential to a viable public broadcasting industry capable of producing high quality services and programs, locally and nationally.

8. The financing of facilities is as urgent as the financing of operating expenses, and the funding level must be increased to meet system needs.

9. It is both appropriate and vital that private underwriting of local and national program costs continues as an important method of financing.

10. The development of a plan for the system's growth, the strengthening of local planning and management capabilities, and the setting of priorities which can be translated into specific local and national objectives are all essential to the achievement of long-range financing.

11. The development of a strong and effective public broadcasting industry requires that the Corporation for Public Broadcasting continue to play a leadership role as envisioned in the Public Broadcasting Act of 1967.

12. Any long-range financing plan should be flexible enough to permit participation in cable, satellites, and other new technologies.[12]

The Five-Year Funding Act   The task force decided that, although a dedicated tax on receiver sales would be the best way to finance public broadcasting, any funding plan that would be acceptable to Congress would have to be based on a relatively simple matching formula. To provide predictability in financing, it was proposed that a five-year funding bill be enacted that would authorize two federal dollars for every five dollars the system could raise from other sources—providing $88 million the first year, increasing each year to $160 million annually at the end of the five years. To provide a measure of insulation from political control, the task force recommended that legislation authorizing these funds also *appropriate* the matching funds for the five-year period. (Normally a congressional authorization bill is followed at a later date by a separate bill to appropriate the necessary funds.)

In July 1974, the Office of Telecommunications Policy prepared a proposal for long-term insulated funding that followed all the recommendations of the task force except for the recommended amounts. The authorization bill that President Ford submitted to Congress in mid-1974 provided for an initial $70 million, increasing to a final $100 million by 1981. After a year and

[12]*Ibid.*, pp. 111–112.

a half of debate (and the resubmission of the bill to a new Congress), the Public Broadcasting Financing Act of 1975 was signed into law on the last day of 1975. Although the public broadcasting leadership was able to convince Congress and the President that the higher amounts of funding were necessary, the battle for the five-year appropriation was lost; CPB would continue to have to go to Congress each year with outstretched hands to receive the funds.

The 1975 financing act, of course, met with mixed reactions. Obviously, the prospects of receiving a total of $634 million from 1976 to 1981 was exciting and encouraging, but implicit in the failure to secure the automatic appropriations was the reality that the administration or Congress could reduce or withhold funding whenever CPB-PBS programing might be considered too liberal, too conservative, too controversial, too bland, or whatever. Public broadcasting must continue to fight for a source of *insulated* funding for its federal support.

8.2   Corporate Underwriting of Public Broadcasting

Even while the public broadcasting establishment wrestles with the problems of federal funding, the sources of nonfederal dollars are fraught with similar problems and controversies. Specifically, the securing of financial underwriting from commercial firms raises several issues.

The Basic Problem   We have already discussed the nature of corporate underwriting and the advantages both to the underwriter and to public broadcasting (Section 5.4). As discussed above, the form of federal financial support set by the 1975 financing act virtually guaranteed that this corporate kind of "development" would be aggressively pursued. However, as Anne Branscomb, a New York attorney, has pointed out, corporations seldom support *public affairs* programing, and their support of *entertainment* (cultural) programing raises both programing and ethical questions:

Since the purpose of a commercial underwriter in public television must necessarily be somewhat similar to that of an advertiser in commercial broadcasting—to obtain a large audience and a positive corporate image—there is a substantial question whether it is in the interest of public broadcasting to expand from this source or to permit greater commercialization. Although PBS allegedly maintains ultimate editorial control over the programming content, the availability of funding from corporate interests helps determine the direction of programming choices. This means more entertainment programming and fewer informational programs of a controversial nature.

Moreover, the fact that other media are used without restriction to advertise such commercially supported ventures raises the issue of whether the reserved channels are actually being used "commercially." For example, Atlantic Richfield uses the ARCO logo in its newspaper ads promoting PBS programs. The Mobil Oil Corporation, in its full page ads advertising *Masterpiece Theatre* and *Mobil Showcase Theatre,* actually refers to the program credits as "commercial interruptions."[13]

Some noncommercial broadcasters have also been outspoken in their contention that substantial underwriting by giant corporations has already undermined the basis for a "noncommercial" service. One spokesman for the listener-supported Pacifica radio stations stated, "Public TV has indeed become part of the commercial medium. . . . The gentility of its role should in no way be allowed to disguise the reality of it. Its boast of nonconformity, of swim-

---

[13]Anne W. Branscomb, "A Crisis of Identity: Public Broadcasting and the Law," *Public Telecommunications Review,* February 1975, p. 14.

ming against the tides of polluted commercialism, is empty pretension."[14]

On the other hand, the corporations point out that they are not using noncommercial television because it is an efficient advertising medium. Herb Schmertz, Mobil Oil's vice president for public relations, indicates their support of PTV is not for commercial gain.

> We could do just about as well on cost with syndicated, non-network shows. We'd probably reach a wider audience and get six minutes of commercial time per hour for our troubles. We're not involved in public television because it's a bargain.
>
> We saw public television as an emerging institution needing and deserving the support of a corporation with the responsibility to put back into society some of what it had earned; and we also believed that there was an audience for the kind of quality programming that the commercial networks weren't supplying.[15]

The Problem of de facto Censorship Unfortunately, there are also indications that, while corporations seem not to interfere with program content in the programs they underwrite, public broadcasting professionals *are* occasionally affected by underwriting. Probably the most blatant example of de facto censorship occurred in 1969, at a time when the newly created PBS was beginning to become defensive about government and press accusations that it carried biased programing. A program from the NET series, *The Nader Report*—which was dealing with examples of how advertising can be misleading—was to focus on a series of questionable Mobil Oil ads claiming its gasolines made auto engines cleaner, thus contributing to cleaner air.[16] At that moment PBS,

through WGBH, was negotiating with Mobil Oil for the largest single, private, nonfoundation grant ever given to public broadcasting—$1.1 million. The NET program was first aired without mentioning Mobil. Then, after the announcement of the Mobil grant and news of the alteration of *The Nader Report* both appeared in the trade press, one right after the other, the original program was aired—but only after Mobil executives were given the chance to comment on the specific examples. Av Westin, one-time producer of the *Public Broadcast Laboratory,* maintains that in these subtle ways underwriters for PTV programs can and do exert more pressure on content than do sponsors on commercial networks.[17]

Educational broadcasting pioneer Harry Skornia is more disturbed, and more specific in his examples, than any other writer to date on the matter of buying a kind of "protection" through underwriting. He lists a number of national problem areas that he thinks public broadcasting must attack; and then, for each, cites underwriting that might cause public broadcasters "to lose their nerve": energy (oil companies); nutrition (food companies); violence (commercial networks); drugs (pharmaceutical houses and commercial broadcasters); pollution (steel companies).[18]

Like others, Skornia believes the problems will not be solved until there is greater percapita funding of public broadcasting by the federal government, allowing public broadcasters to forget about their need for money.

8.3 Centralized Control of Programing and Production

Of all the issues in public broadcasting, the question of whether or not PBS should be a centralized network and the related question of programing control by CPB are undoubtedly the

---

[14]*Los Angeles Times,* January 2, 1976, Part IV, p. 1.

[15]*Ibid.,* p. 24.

[16]Les Brown, *Television: The Business Behind the Box* (New York: Harcourt Brace Jovanovich, 1971), p. 338. *See also* John J. O'Connor, "Can Public Television Be Bought?" *New York Times,* October 13, 1974, reprinted in Ted C. Smythe and

George A. Mastroianni, *Issues in Broadcasting* (Palo Alto, California: Mayfield Publishing Company, 1975), pp. 352–354.

[17]Av Westin, "Fourth Network," *Nation,* January 17, 1972, p. 69.

[18]Harry J. Skornia, "Has Public Broadcasting Lost Its Nerve?" *Public Telecommunications Review,* October 1974, pp. 34–38.

thorniest and the most frustrating. The issues are fundamental and philosophical, with no simple and correct answers.

Background of the Problem Even beford the Public Broadcasting Act of 1967 was signed, a thoughtful article by Stephen White reviewed the issue of centralization of public broadcasting: Should public television be a locally controlled service with some national programs or a national service with some local programing? The issue was constantly debated during the Carnegie Commission study, in part because it was a central question for such a committee and in part because the commission (particularly its chairman, James Killian) believed strongly in local control, while the Ford Foundation (particularly its president, McGeorge Bundy, and its consultant on TV, Fred Friendly) believed strongly that public television should be all that NET had been striving to become—a true fourth network:

> To begin with, the Carnegie design is persistently erected on the base of the local station. It anticipates complete local control of program scheduling, substantial local production, and a broad reliance on local talent. It explicitly opposes the concept of a fourth network. It enthusiastically supports electronic interconnection among the stations, but looks upon it as a distribution device, which will give every station access to programs produced anywhere within the system, without prescribing for any station the manner in which those programs are to be used—the station is in fact freer not to use a program than to use it, since it will have access to far more programs than it can possibly air.

> What underlies that design is the belief that a fourth network would fail to serve needs that are not now being met by commercial television. Any system structured as commercial television is structured will ultimately obey the same imperatives as commercial

television: maximization of audience, centralization of decision, and the appeal to a least common denominator. It may be a higher least-common-denominator than that served by commercial television, but there is no guarantee that it would be high enough to serve any real public needs.[19]

The Question of a Fourth Network Looking back on the early days of the ETRC immediately following the creation of noncommercial broadcasting in 1952, one of the first questions to be decided was "whether the Center should be simply a service agency to fulfill with programs the expressed needs of the stations as they developed, or whether it should have a directing philosophy of what constituted good educational broadcasting."[20] Should the Center furnish the stations with whatever type of programing they requested—whether it be entertainment, sports, original French drama, or how-to-do-it programs—or should it provide them with what the Center considered good educational fare?

The ETRC board took the position that if "the Center was to develop and serve national educational television . . . this required a definite concept of good education."[21] Thus, the Center would consider the stations' desires but would provide them with what constituted "good education"—by the Center's definition. As early as 1952, it was determined that the ETV movement would need a strong centralized programing policy—a national focus. Without this sense of direction and leadership, there could never develop a truly *national* noncommercial television programing service. This centralized leadership, of course, would have to be balanced by a careful responsiveness to the needs of the individual

[19]Stephen White, "Carnegie, Ford, and Public Television," *The Public Interest,* Fall 1967, pp. 15–16.

[20]I. Keith Tyler, "The Educational Television and Radio Center," *in* William Elliott (ed.), *Television's Impact on American Culture* (East Lansing: Michigan State University Press, 1956), p. 229.

[21]*Ibid.*

local stations, but the concept of a national programing policy was established.

It was also necessary, at the outset of the ETRC operations, to decide if the Center was to function primarily as a film exchange library—from which stations might draw programs as they needed them—or whether it should function as a "film network which would supply regularly scheduled programs to affiliated stations."[22]

Again, thinking in terms of what the Center should do in order to serve most effectively the idea of national educational television, the 1952 ETRC board elected the alternative that placed the Center in a strong centralized position. From an educational standpoint, a regularly planned, weekly service could best present stations (and audiences) with an educationally balanced program schedule—although it might leave the stations with less freedom of choice. Thus,

> It was also decided early in the history of the organization that a continuing and regular program service was essential. Theoretically, it would have been possible to concentrate on a few programs of outstanding quality to be sent out at irregular intervals. Practically speaking, however, this was not the case.[23]

Although actual interconnection was to be years away, the Center would act as a scheduled "network" service rather than as a lending library (Section 3.8). By 1966, almost half of the programing of a typical ETV station was being furnished by NET.[24] Not all this material was on a regularly scheduled basis, of course. At its peak NET was furnishing ten hours of scheduled programing per week to its affiliates. Other programing was distributed through its Flexible Service and other library arrangements. By the mid-1970's,

the PTV system was served by a network interconnection that delivered seventy hours per week of network programs to interconnected stations and by a Public Television Library with more than a thousand different programs available to stations.

In the late 1970's the system is actually a compromise between the Carnegie Commission/Killian approach and the Ford Foundation/Bundy/Friendly approach. Although local stations bid on (and are supplied) nationally produced programs (Section 7.3), the ideas for programs tend to originate at national production centers; and the development and (less often) pilot production of many of the programs on the interconnection are supported by CPB grants—providing a measure of national control and standardization. The general feeling appears to be that the compromise is reasonable and workable. Strangely enough, it was the challenge to the concept of a national service—by Whitehead and Nixon (Section 6.2)—that made many individuals realize that television is too expensive to hope for a simple interstation exchange of the best local programs. As we will see (Section 14.1), this parochial approach for many years kept school TV from achieving any real impact as a learning resource in our schools. There must be some degree of national consolidation and control in order to effect "economies of scale" that can provide large enough budgets to ensure quality productions.

The Question of Centralized Production Facilities   Another issue that has been repeatedly debated throughout the history of noncommercial television is the question of whether or not there should be national ETV production facilities. Some early educational broadcasters felt, for example, that the newly created ETRC should maintain its own production staff and studios. C. Scott Fletcher, however, felt that decentralization of the Center's production "would protect the Center against any temptation to build an empire of its own comparable to those of the

---

[22]*Ibid.*

[23]"ETRC Report to the FAE, October 1, 1955," p. 4.

[24]Carnegie Commission, *Public Television: A Program for Action* (New York: Bantam Books, Inc., 1967), p. 25.

141

commercial networks."[25] Fletcher's wishes were respected, and his policy is represented in Article XIII of the original bylaws of the ETRC:

The Corporation shall not engage in the technical production of motion pictures, radio or television programs, but shall have the power to employ others to do so on its behalf.

There is the tendency, of course, within every professional staff to build its own empire. The NETRC was no exception. As the Center, under John White's leadership, became increasingly concerned with the quality of its programing, it was only natural that the Center began to exercise more and more direct production control. By the early 1960's, NET was hiring its own film crews to develop certain programs. By 1967, White could state that "fully half of the 260 hours of new . . . programming that go into the weekly service each year is produced by NET itself"[26] (Section 4.7).

The feeling that the central national ETV organization should not, in itself, be a production agency was echoed fifteen years after the formation of the ETRC. In recommending the establishment of the "Corporation for Public Television," the Carnegie Commission stated: "Except for its obligations in providing interconnection and dissemination of information, and perhaps for the establishment of archives, it should not act as an operating agency."[27]

This restriction was spelled out even more explicitly in the language of the Public Broadcasting Act itself: "The Corporation may not own or operate any television or radio broadcast station, system, or network, community antenna television system, or interconnection or program production facility"[28] (Section 4.6).

This is a question—recycled after fifteen years—still not fully resolved, as public broadcasting debated in the late 1970's what the role of CPB should be in funding specific program series and overseeing the PBS interconnected network service.

The Question of Stations as Producers One other related issue remains. If there is not to be a network center for producing PTV programs, who should do the producing? Should the nationally distributed programing be produced primarily by the individual ETV stations themselves or by outside independent producers? From the beginning of the ETRC, C. Scott Fletcher and the Fund for Adult Education felt that the Center's programing money should be invested in ETV stations whenever possible; this would allow more dollars to be pumped back into the ETC stations themselves—endowing them with more resources.

It should be pointed out that in all of the early grants to the Center [from the Ford Foundation] it was intended that the money help the whole movement not only in delivering good programs but also by helping producing stations to build "muscle" in both facilities and staff.[29]

By 1958, almost three quarters of the ETRC national package was being produced by affiliated stations. However, more than two thirds of the station-produced programing was being turned out by only five of the stations (Boston, San Francisco, Chicago, Pittsburgh, and St. Louis).[30]

[25]John Walker Powell, *Channels of Learning: The Story of Educational Television* (Washington, D.C.: Public Affairs Press, 1962), p. 78.
[26]John F. White, "National Educational Television as the Fourth Network," *in* Allen E. Koenig and Ruane B. Hill (eds.), *The Farther Vision: Educational Television Today* (Madison: University of Wisconsin Press, 1967), p. 88.
[27]Carnegie Commission, *Public Television*, p. 40.
[28]Public Broadcasting Act of 1967, Public Law 90–129, Section 396(g)(3).
[29]Internal NETRC memorandum from Robert Hudson, July 6, 1962.
[30]Data collected from NETRC, "Series History Sheets."

Undoubtedly, contracts to produce for the Center contributed to the survival of a few stations during the early years . . . . The Center, on the other hand, felt its money was being unduly siphoned off to general station support at the expense of product quality, considered its control over production far too limited, and often deplored the technical level of the programs delivered.[31]

Because so few of the stations were benefiting from the "muscle-building" policy, John White slowly revised this approach. He argued that, with an increasing number of stations coming on the air, a relatively smaller percentage each year could benefit as production centers. What was needed—to help the greatest number of stations the most—was the best quality national package, regardless of who was the producer.

Thus, by 1966, the number of station-produced programs that NET acquired had dropped to roughly one third of its total.[32] There was no longer any argument that this would help put muscle on the stations' staffs in order to ensure station survival. It was merely a matter of contracting with the best production center in order to get the most value for dollars spent. As might be expected, there were still only a small number of producing stations that received most of the contracts; the "Big Seven" in the 1970's were New York (WNET), Los Angeles (KCET), Washington (WETA), Boston (WGBH), San Francisco (KQED), Chicago (WTTW), and Pittsburgh (WQED).

Interestingly, as PBS turned to its Station Program Cooperative (Section 7.3), the percentage of station-acquired productions started to increase—and many more stations were vying for lucrative CPB production contracts. The days of looking upon national production contracts

in order to put muscle on one's production staff are far from over. And the question of "How much centralization?" is far from resolved.

8.4   Distribution of Public Broadcasting Programs

The earliest financial support to public broadcasting was for equipment, but invariably programing concerns tend to push distribution problems to the background.

Yet *distribution*—the array of transmission lines, antennas, transmitters, and miscellaneous hardware—is the core of electronic media. Without adequate signal strength and technical quality, without the ability to compete in the pattern of commercial competition, educational broadcasting is doomed to be a voice in the wilderness, unable to meet any of its ambitious goals.[33]

Inadequacy of Equipment   Although many PTV and public radio stations have modern plants every bit as good as, or better than, the commercial stations about them, an even larger number have obsolete equipment, inadequate power, and marginal capabilities.

In their 1975 study of the state and future of public broadcasting for the U. S. Office of Education, Tressel, Buckelew, Suchy, and Brown surveyed a sample of the PTV and radio stations in the United States, and found inadequate production and distribution equipment in the stations:

While not all problems of educational television are caused by the marginal nature of equipment and by uncertainties in funding, it is our impression that many of them are. As the manager of a small California station put it: "In many respects our most important

---

[31]Gerard L. Appy, "NET and Affiliate Relationships," *in* Koenig and Hill, *The Farther Vision*, p. 101.

[32]Carnegie Commission, *Public Television*, p. 23.

[33]George W. Tressel, Donald P. Buckelew, John T. Suchy, and Patricia L. Brown, *The Future of Educational Telecommunication: A Planning Study* (Lexington, Massachusetts: D. C. Heath and Company, 1975), p. 11.

local accomplishment is just to stay on the air." With few exceptions, there is little systematic planning of capital budgets to replace worn-out or outdated equipment; most available funds go directly for operation. This situation is usually beyond the control of individual stations. Capital improvements call for large sums which must be raised through special appropriations or major fund drives. And in many organizations that fund educational television, mere obsolescence is not interpreted as a reason for replacement.[34]

Of the fifty stations surveyed by these Batelle Institute researchers, more than one fourth still did not have color studio cameras. Few had more than a single studio. The lack of adequate switching, audio, and remote equipment was mentioned frequently. In a separate study by the CPB, 20 percent of the stations surveyed did not have a color film chain (but it would appear the CPB study included repeater stations).[35]

In most cases transmitters appeared to be in good shape, but repeaters and higher power transmitters were needed in the mid-1970's to extend the range of stations; new stations were needed in many locations. In 1975, two states still had *no* PTV stations; and, as far as radio is concerned, "until recently, one could drive from Vermillion, South Dakota, to Pullman, Washington, covering the entire states of Wyoming, Idaho, and Montana, without encountering a [full service NPR] station—more than half the distance across the continent."[36]

Priorities and Funding Tressel and associates recommend that three related policies

be given high priority to improve public broadcasting transmission capabilities:

1. *Radio-Television Piggyback:* PTV stations should be required to operate a public radio station. (Although they do not say, it is assumed they would not require PTV stations to initiate a radio service where other institutions provide it in the community.)

2. *Minimum Standards of Television Transmission Facilities:* PTV stations should have full color capabilities, adequate power and antenna location, and a minimum of four color videotape recorders.

3. *Minimum Standards for Radio Transmission Facilities:* FM stations should have stereo broadcast and recording capabilities. The NPR interconnection service should be high-fidelity stereo.

For both radio and television, Tressel and his colleagues recommend that additional coverage be given high priority, "and even 'slave' transmission is preferable to no service at all."[37]

The CPB study established bare minimums for "full production stations," limited production stations, and network production centers. To bring stations up to the bare minimum standards recommended by the corporation would cost $100 million.[38]

The problem in implementing the recommendations of the U.S. Office of Education and CPB studies is, of course, *money*. The Department of Health, Education, and Welfare continues to administer the educational broadcasting facilities program begun in 1962, which provides matching grants for transmission and production facilities (Section 4.1). The dollars are limited, and priorities must be established. According to the secretary of HEW, priority in the mid-1970's

34*Ibid.,* p. 13.
35"A Report on the Minimum Equipment Needs and Costs to Upgrade the Facilities of the Public Television Stations in the United States and Its Territories," Corporation for Public Broadcasting Technical Monograph Number 2, CPB Office of Engineering Research, March 1975, p. 4.
36*Ibid.,* pp. 16–17.

37*Ibid.,* p. 58.
38"A Report on Minimum Equipment," CPB Technical Monograph.

for television had to be the "colorization" of existing stations rather than an expansion of TV service. Priorities for public radio similarly were for increased power and FM stereo rather than new stations.[39]

8.5   Getting the Public into Public Broadcasting

Despite efforts to make public broadcasting "The People's Business" (a slogan of PBS), not all "publics" are positively involved; and some groups are now making themselves heard through such adversary techniques as formal challenges to PTV licenses. As NAEB President James Fellows noted,

The success and the increase of license renewal challenges must be seen as more than an aggravation and a nuisance. Where it is not simply ego-tripping, it is beginning to be a declaration of expectations from community groups that have taken seriously the slogan that public broadcasting is the people's business. This topic needs much more hospitable and affirmative discussions and clearly needs attention at the [next NAEB] convention and at any other points where interested parties come together.[40]

At the national level, Willard Rowland sees the problem getting worse, not better:

Given the changing organizational structure of the national public television establishment . . . the problem of representativeness in local station governance is magnified. Since the Public Broadcasting Service corporate body is now dominated by a board of governors that is made up of members

elected from among representatives of the local station boards of directors, any distortions in the local governing profile are necessarily ratified and extended at the national level.[41]

Basis of the Problem   Many groups and individuals have noted that there are two reasons why large portions of the viewing public are "insulated" from the management of public broadcasting stations: (1) the way public broadcasting is financed locally through contributions and memberships; and (2) the ways that governing boards are selected. Les Brown is particularly bothered by the fact that boards are likely to be composed of members of the elite: "By and large, the boards of directors of the local public television outlets are made up of prominent representatives of industry, finance, education, and the professions—key figures in the local power structure, many of them with iron-bound political points of view and few with any real experience in communications or with a feel for the objective purposes of broadcasting."[42]

Brown obviously refers to community stations that are governed by boards of directors, citing community stations in New York City, Chicago, and Washington, D. C., as examples of stations with elite boards. He claims the problem is not restricted to community stations:

Nor were these the worst examples of Establishment stations, or of broadcasting by fear. Many, in other cities, were supported by school systems and state boards of regents, and they were even more responsive to the will of the authorities and less inclined to venture beyond the national mores and the conventional pieties because they stood in danger of losing their sources of funds.[43]

---

[39]Caspar W. Weinberger, Secretary of Health, Education, and Welfare, in a speech to the Western Educational Society for Telecommunications (WEST), San Francisco, October 1974.

[40]James A. Fellows, "Letter From . . . Las Vegas," *Public Telecommunications Review,* December 1974, p. 62.

[41]Willard D. Rowland, "Public Involvement: The Anatomy of a Myth," *Public Telecommunications Review,* May/June 1975, p. 9.

[42]Brown, *Television,* pp. 324–325.

[43]*Ibid.,* p. 327. *See also* "The Politics of Public TV," *Columbia Journalism Review,* July-August 1972, pp. 8–15.

Even Hartford Gunn, vice-chairman of PBS, acknowledges that the way in which the board of governors is composed at the national level makes it difficult to have minority viewpoints represented on the PBS Board:

> The problem of bringing minority persons and women into positions on our Board of Governors and Board of Managers is more difficult, since the Boards are elected from individual station Boards and staffs. In effect, not even PBS has control over the composition of its Boards and to assume such control would be to interfere with the essentially representational structure of PBS. The problem is further complicated by the fact that many individual station Boards are made up of the trustees of the Universities or governmental organizations to which the station is licensed and over which the local stations can exercise no control of the selection process.[44]

Necessary Action From the above observations, it is obvious that PBS and local stations should keep trying to get minority interests and the views of the silent middle-class majority represented on governing boards. Stations should give governing and policy boards a more active role in station activities. Rowland, one-time PBS staff member, concludes a thorough analysis of the problem with these general suggestions:

> At the very least, then, it would appear that the stations, the national agencies, and the publics they all claim to serve must begin a mutual re-evaluation of the terms under which they interact. The public needs to take a more active role in the affairs of its noncommercial broadcasting system, and the several local and national agencies

must find ways to open their board structures and advisory procedures to encourage such increased, and substantive, public input. The stations and the national agencies must resist the temptation to conceive of their audiences in traditional mass-society terms.[45]

The station, its board, and the community in general can encourage the public to get involved with station activities through its volunteer programs. Margo Tyler, director of communications for the National 4-H Foundation, believes there is probably no other institution in our country today offering more diverse opportunity for volunteer involvement than public broadcasting. In an article for "Voluntary Action News," Tyler noted four guidelines for those who want to become involved as volunteers in public broadcasting:

1. Find out about your local station—who holds the license, who serves on the governing board, how it is funded and staffed, and what volunteer group may already be in operation.

2. Look and listen to public broadcasting programing, and be prepared to look for ways in which additional segments of the community can be served.

3. Make an appointment with the station manager or volunteer coordinator to mutually discuss his or her needs for volunteers—and your talents and interests.

4. Follow through on that commitment with a willingness to do some of the "nitty-gritty" jobs, as well as to gain understanding of its potential.[46]

WJCT(TV): A Case Study of Community Involvement According to Charles Remsberg, a free-lance writer and critic, WJCT, Channel 7, in Jacksonville, Florida, offers a model where

[44]Hartford N. Gunn, Jr., in a letter to Senator John O. Pastore, Chairman, Senate Commerce Committee, dated August 9, 1974. Printed in *94th Congress Report 94–55,* March 21, 1975, p. 22.

[45]Rowland, "Public Involvement," p. 19.

[46]*Voluntary Action News,* Newsletter for the Center for Voluntary Action, December 1974.

"'community involvement broadcasting' is more than a promotion department's slogan." Among the activities described by Remsberg are the following:

Thorough and meaningful coverage of local issues through regular and special events programs.

An interview-telephone show called *Feedback,* aired live from seven to eight o'clock five nights a week. At times, the whole show is devoted to a single guest or issue, but generally there are three or four interview segments, with the option (peculiar to public TV) of letting the show run past its allotted hour if some segment merits it. After initial interrogation by studio interviewers, questioning of guests is turned over to telephone callers.

A "Gripe Night" once each week on *Feedback,* which allows viewers to voice any gripe they have about anything, including station policies and programing.

Complete telecasting of all Jacksonville city council meetings, and all Duval County school board meetings and "work sessions." During lulls the cameras switch to the studio where calls are taken commenting on decisions taken.

"Town Meeting" telecasts in which issues are debated in full by antagonists.[47]

One element inherent in the Jacksonville model is an intangible spirit of commitment. This attitude can be sensed, to some extent, at most PTV stations, but it seems to be most apparent at medium-sized community stations. It is a feeling that the station has a role to play in the community that extends far beyond providing programs. It is an attitude about *people* in the community and about *ideas.* This feeling of commitment—an opportunity to do something meaningful in a creative sense—attracts many persons to work in the public broadcasting arena.

Although public television and radio will not change overnight, it is almost certain that public involvement in station activity will increase and that attitudes of station bureaucracies toward public involvement will continue to change. There probably is as much chance of the public staying out of the governance of public broadcasting as there ever was in keeping the public out of formal education. Everyone expects a public board of education to govern our schools—to set policy and make decisions that are carried out by professionals. This is increasingly true in public broadcasting, and almost certainly will be even more true in the future. Public broadcasting is a relatively young institution in our culture, and it will take time before the public defines the exact role or roles it should play.[48]

8.6   Getting Education into Educational Broadcasting

The final issue in this chapter is, in part, a programing matter; and it therefore will help to serve as a bridge to the next chapter on issues in programing. The question of the roles of CPB, PBS, and NPR in providing "educational" programs and services raises fundamental questions that make *education* one of the contemporary major issues in public broadcasting.

"Educational programing" in this sense refers to those sequentially structured series of TV presentations concerned with exploring a given body of information, which may or may not be offered for academic credit. It usually is thought of as falling somewhere in the crack between good cultural/public affairs programing as seen on commercial TV and strictly instructional TV programing as seen in the classroom. The lack of emphasis on this type of programing by

---

[47]Charles Remsberg, "Jacksonville's 'Feedback' Experiment," *Columbia Journalism Review,* July/August 1972, p. 39.

[48]See also R. J. Blakely, "Rethinking the Dream: A 'Copernican' View of Public Television," *Public Telecommunications Review,* September/October 1975, pp. 30–39.

PTV outlets corresponds roughly to the demise of sequential programing over noncommercial stations (Section 7.2).

Lack of "Educational" Emphasis Since the signing of the Public Broadcasting Act of 1967, public broadcasters have been concerned more with cultural and public affairs programing than with educational programing. There are many reasons for this: The Carnegie Commission chose to concern itself only with *alternative service* programing; federal funding and corporate underwriting have provided dollars for *noninstructional* programing; and public broadcasting was anxious to demonstrate what it could do in the areas of *cultural and public affairs* programing given an interconnection and some money. The lack of attention to *educational* programing finally led one educator in 1974 to suggest that educators ought to campaign all over again for educational television channels.[49]

In passing the Public Broadcasting Act of 1967, Congress did not intend that instructional programing should *not* be a concern of public broadcasters. The act stipulates that "The Congress hereby finds and declares . . . that it is in the public interest to encourage the growth and development of noncommercial educational radio and television broadcasting, *including the use of such media for instructional purposes* . . ." (italics added).[50]

Not long after CPB and PBS were created, *Sesame Street* and *The Electric Company* were developed by the Children's Television Workshop (Section 11.7), and everyone pointed to these programs as examples of what educational programing should be like. Despite CPB's support of these programs, in the first seven years of public broadcasting there was not "a national

public broadcasting perspective of service to education, [or] a cohesive agenda for its own part in such a service."[51]

The ACNO Task Forces In 1974, CPB commissioned its Advisory Council of National Organizations (Section 6.6) to conduct a study and make recommendations to CPB regarding the role of the corporation in the relationship between public broadcasting and education. ACNO organized four task forces to deal separately with early childhood education, elementary and secondary education (including teacher education), post-secondary formal education, and adult education.

The task forces worked for a year studying the issues, and in early 1975 submitted their reports to the CPB board. The many different recommendations of the four task forces were summarized in eleven basic, general recommendations; these stated that the Corporation for Public Broadcasting should:

1. Intensify its efforts to bridge the traditional chasm between broadcasting and education, building a working partnership to serve their common purpose.

2. Recognize and support the principle of cultural pluralism, which is rooted in our common concerns as humans as well as in the differences that enhance the strength and diversity of the American people.

3. Undertake activities to assist professional development of the educators and broadcasters engaged in educational broadcasting, and encourage the application of broadcasting for the in-service education of teachers.

4. Undertake promptly certain instructional programing activities, taking into account the legal and traditional roles of other educational agencies and institutions.

---

[49]Richard Bell, during a panel discussion on the ACNO Task Forces Reports, Annual Conference of the Association for Educational Communications and Technology, Dallas, Texas, April 1975.

[50]Public Broadcasting Act of 1967, Section 396(a)(1).

[51]*Public Broadcasting and Education: A Report to the Corporation for Public Broadcasting from the Advisory Council of National Organizations* (Washington, D.C., March 1975), p. 9.

5. Assure adequate attention to the strategies, materials, and other services that are critical to effective use of educational programing.

6. Actively develop the educational programing applications of related technologies, in order to meet the educational needs of people at all age levels.

7. Assure, through its own operations and through support of others' work, an effective program of research, evaluation, and demonstration regarding educational applications of public broadcasting and related technologies.

8. Facilitate the development of new, more flexible patterns of rights clearance.

9. Encourage the development of the skills of aural/visual literacy and critical listening/viewing.

10. Recognize and support effective activities for promotion and community outreach in the educational applications of broadcasting.

11. Move at once to act upon these recommendations, initially by conducting a financial analysis, determining a calendar agenda for specific actions, and assigning responsibility for developing funding.[52]

These eleven general recommendations were then broken down into specific activities that the corporation could undertake. Among other items, ACNO suggested that the CPB should initiate the following actions: set up a number of task groups to deal with matters such as coordinating instructional TV activities of public and private agencies, bringing together publishers and others involved in educational programing, and developing a clearinghouse or library on broadcast ITV; set up a number of workshops and in-service sessions for teachers who would be using television in education; develop an "Instructional Program Cooperative" that would be an

instructional counterpart to the PBS Station Program Cooperative; encourage cooperation among postsecondary institutions for mutual development of ETV materials; pursue nonbroadcast media technology such as instructional TV fixed service, community antenna TV, closed-circuit TV, satellites, video discs and cassettes; and tackle the problems of rights and clearances.

If these suggestions seem general, it must be remembered that the recommendations of the four task forces dealt with diverse questions, ranging from the practical problems of utilization of educational broadcast material and the need for extended rights to programs, to such fundamental policy issues as the relationship of broadcasting and education and the propriety of a national organization involving itself in something that traditionally has been a local and state concern.

Reactions to the ACNO Task Forces Report
The ACNO task forces recommendations, along with a synthesis and summary by John Witherspoon (consultant to the study), were published as a pocketbook-sized report by ACNO and CPB in March 1975. About 10,000 copies of the booklet were printed, and CPB attempted to get reactions to the report and its recommendations from educators and broadcasters alike.

George Hall, director of the Virginia Public Telecommunications Council, and NAEB President James Fellows are two who question the basic premise of the report—that the CPB should become involved in formal education. Fellows quotes a letter by Hall to the president of CPB, and then continues with observations of his own:

"While representing the accumulated frustrations of hundreds of educators and technologists at the historic difficulties encountered in trying to use broadcasting for educational purposes, [they] would impel the CPB into policy sectors that are improper, unwise, unnecessary, unrealistic,

---

[52]*Ibid.*, pp. 6–7.

or ambiguous. The persons who voiced to ACNO their concerns about the limited success of educational broadcasting were almost never made aware of the fact that the CPB is manifestly not the appropriate legal instrument for implementing suitable remedies in this complex area."

This is too substantive a challenge to be dismissed easily, if at all. For it raises the question which needs first to be raised about ACNO and its recommendations: should the CPB be involved in the area of direct instruction? True, there is congressional impetus for such involvement. But the question has never been subjected to the scrutiny of formal exchanges.[53]

Other reactions have been mixed. Most who have reacted to the report agree that the needs discussed in it are real, but many suggest that the Agency for Instructional Television or some other agency that is formally tied to the chief state school officers or other institutions representing education should attempt to solve the problems.

An unavoidable problem of the ACNO report is that it is not a "program for action" as was the Carnegie Commission report (Section 4.6). A blueprint not only can be followed but also generates specific alternative designs. For example, those who disagreed with the Carnegie Commission's recommendations for an excise tax on TV sets were obliged to make specific alternative suggestions. The ACNO recommendations are seldom specific and often use terms such as "foster," "encourage," or "coordinate." They help to establish a perspective on the relationship of public broadcasting and education, but it will probably be some time before very many specific educational activities by CPB or PBS will be attributable to the ACNO recommendations.

Like many of the other questions discussed in this chapter and Chapter 9, the issue of the place of educational programing in PTV cannot be readily and simply answered. But, like these other major issues, the questions must be asked and attempts made to find the most rational and feasible courses of action.

---

[53]James A. Fellows, "Public Broadcasting and Education: Much Ado About Everything," *Public Telecommunications Review,* July/August 1975, p. 6.

**Issues
in Public Broadcasting:
Audience Analysis
and Programing**

Educational television in the United States has had an identity problem ever since its creation in 1952. Ten years after the authorization of ETV, veteran broadcaster Yale Roe examined the operations of the sixty stations then on the air and noted, "Not only do the educational stations fail to compete, but they also fail to define their goals. Who is it that they are trying to reach?"[1] Roe then looked at the audiences and program schedules of these stations and concluded:

> It would seem that the problem is that educational television has not defined its goals. If research suggests that the person with the highest intellectual pursuits and the highest education is likely to be a viewer of ETV, is not the television time wasted in programming about shorthand and skin diving? Or is this for someone else, the viewer whom the survey has not uncovered? Clearly it is, for it is the general philosophy of educational broadcasters that they should have something for everyone. One might answer, as Fred Allen did to the statement that the show must go on, "Why?"[2]

Another decade later, in 1972, twenty national leaders in public broadcasting were brought together by the Aspen Program on Communications and Society to consider a set of recommendations on the financing of public television compiled by Wilbur Schramm and Lyle Nelson of Stanford University. One of the proposals for action, based upon the Schramm-Lyle report, was that public broadcasting should "specify the long-term objectives and actual services to be provided by public television, as the basis for the long-range financial plan."[3]

[1]Yale Roe, *The Television Dilemma: Search for a Solution* (New York: Hastings House, 1962), p. 97.

[2]*Ibid.*, p. 98.

[3]Wilbur Schramm and Lyle Nelson, *The Financing of Public Television* (Palo Alto, California: Aspen Institute for Humanistic Studies, 1972), p. 57.

9.1   In Search of an Identity

Even more recently, R. J. Blakely, writing in *Public Telecommunications Review,* concluded:

> The difficulty is not simply that the total and individual audiences are small; it is also that the stations and the system have no *rationale* for their services and consequently none for the natures, sizes or relationships of their audiences. From all this, we must infer that public television has not built up a clientele, or a set of clienteles, that will give it support—political, social, ultimately financial.[4]

These pleas for some agreement on what should be the goals of public television do not mean there have not been attempts to define a mission for PTV. In fact, there has been so much debate on these questions that H. Rex Lee, while still FCC commissioner, told the public broadcasting community, "I sometimes wish that we could have a new matching grant formula: for every 100 hours of debate about who or what you are, the public could be guaranteed one hour of quality programing."[5] Despite this friendly admonishment, the debate will go on. As with the other issues we have examined, there is no simple answer to the question of *who* is to be served with *what* kind of programing.

Definitions and Legislation   The problem begins in the laws that created, govern, and fund noncommercial broadcasting. The Communications Act of 1934, the 1952 *Sixth Report and Order,* the Public Broadcasting Act of 1967, and other legislation and decisions, have all attempted to give public broadcasting some degree of freedom and flexibility along with necessary restrictions and controls. As Anne Branscomb said, "The problem begins as philosophy, but finds its expression in the law."[6] Branscomb believes the ambiguities in the laws relating to public broadcasting are the result of, and in turn add to, a general inability on the part of everyone to resolve certain basic philosophical questions:

> Consequently, there is likely to result much soul searching among public broadcasters and the other entities which serve and regulate them concerning the nature of public broadcasting and its development in the future. During this crisis of identity, answers to some tough questions must be found: What is public broadcasting? What is the justification for reserved channels? What does a noncommercial service really mean? How are noncommercial educational licensees to ascertain and serve *public needs* and *interests* as well as programing preferences? Who is responsible for what? Do "public stations" have higher or lower standards of responsibility with respect to political and other public affairs programing? How are these various responsibilities and needs to be funded? Who is to decide what?[7]

Branscomb reviews the FCC references to "noncommercial educational" broadcasting and points out that the FCC has never really interpreted what this term means. She then supports the concept of a service concerned with providing both *education* and an *alternative* to commercial broadcasting, by quoting Commissioner Frieda Hennock who spoke of "an unprecedented opportunity for education both formal and

---

[4]R. J. Blakely, "Rethinking the Dream: A 'Copernican' View of Public Television," *Public Telecommunications Review,* September/October 1975, p. 32.

[5]H. Rex Lee, "Public Telecommunications: The Task of Managing Miracles," a speech to the NAEB convention, November 1973. Reprinted in *Public Telecommunications Review,* December 1973, and in Ted C. Smythe and George Mastroianni, *Issues in Broadcasting* (Palo Alto, California: Mayfield Publishing Co., 1975), p. 342.

[6]Anne W. Branscomb, "A Crisis of Identity: Public Broadcasting and the Law," *Public Telecommunications Review,* February 1975, p. 10.

[7]*Ibid.*

informal," and who further said that ETV stations should

> supply a beneficial complement to commercial telecasting. Providing for greater diversity in television programing, they will be particularly attractive to the many specialized and minority interests in the community, cultural as well as educational, which tend to be by-passed by commercial broadcasters speaking in terms of mass audiences.[8]

The Concept of Specialized Audiences

This idea of "specialized and minority interests" is a fundamental concept for public television and radio programing. Steve Millard believes this concept essentially *defines* public broadcasting.

> The spirit behind "serving specialized audiences" is that public broadcasting—beginning at the local level—will devote its attention to establishing, first of all, what true specialized audiences *are;* that it will try to ascertain in detail what the audiences need and want; and that its programming will be tailored to meet those expressed needs. Where commercial broadcasting aspires to give the public what it wants, public broadcasting will give the *publics* what *they* want. It is assumed, however, that this philosophy—unlike commercial broadcasting's—will, by definition, produce good programming, because the programs will be attuned to the needs of active, knowledgeable, devoted groups, not simply tossed out to a passive audience.[9]

Millard's statement implies that public television should serve either the *educational* needs of the audience or should be an *alternative* to commercial TV in serving specialized

entertainment/diversion needs of the public(s). This could lead to a somewhat misleading dichotomy in which all PTV programing would have to be justified on the grounds of meeting the criteria for one or the other of these separate categories.

Alternative Programing Criteria   Any program could be classified as "alternative programing" (and therefore justified as a PTV offering)—regardless of whether or not it was "educational" in nature—as long as it was not being offered on commercial television. For example, it would be considered irrelevant to question whether *The Forsyte Saga* or *Upstairs, Downstairs* were "educational." It is enough that they were well-executed dramatic productions that commercial TV could not "afford" to carry in its competition for mass audiences. They were therefore appropriate "alternative" fare for public broadcasting. A similar argument can be made for offering Bergman movies, *Monty Python,* or a tennis tournament (until tennis was "discovered" by commercial TV).

Educational Criteria   On the other hand, following this educational/alternative-programing dichotomy, it would not be appropriate to question whether or not an "educational" series should be carried on a PTV outlet when it can attract as large an audience as commercial programing. For example, some observers have asked why *Sesame Street* is not now carried on commercial television since it has proven its ability to attract large numbers of viewers. The same might be argued for *National Geographic Specials* such as "The Incredible Machine." However, the insertion of commercials into *Sesame Street* or *The Electric Company* (or any other "educational" series) would substantially interfere with their educational values (not to mention ethical questions that could be raised); and the programs are therefore justified purely on their educational purposes with little consideration as to whether they offer an alternative to commercial television programing (even if they attract a popular or mass audience).

---

[8]*Ibid.,* footnoting 41 FCC 148 at 591 (1952).
[9]Steve Millard, "Specialized Audiences: A Scaled-Down Dream?" *Public Telecommunications Review,* October 1974, p. 48.

Examining the Educational/Alternative-Service Dichotomy The position that public television has two distinct, although related, functions—(1) providing *education,* and (2) providing an *alternative service* for specialized audiences—can lead to an artificial "hardening of the categories." For one thing, a program can be both educational and offer an alternative to commercial programing. The Bronowski series, *The Ascent of Man,* is typical alternative fare, yet it is a sequential educational series available for college credit in many communities for those willing to enroll in a special course and do additional reading. *The Adams Chronicles,* an expensive (more than $6 million) and well-executed historical series produced by WNET for the Bicentennial, was viewed by millions solely for the entertainment value of the programs. Yet from the inception of the project, the series was to have educational uses; learning materials were developed to enable high school teachers to use the programs in classes and to assist colleges in offering home-study courses based upon the series. As pointed out in Chapter 1, this type of programing is increasingly being referred to as "educational television" (or sometimes "educational public television"). The ACNO report, *Public Broadcasting and Education* (Section 8.6), encouraged this sort of programing as meeting postsecondary education needs; it undoubtedly will continue to play a major role in PTV programing.

A second fallacy in the dichotomy is that it implies that programing should be thought of as *either* an alternative service to commercial broadcasting *or* as educational fare (or perhaps both). However, there are other ways of looking at PTV programing. Blakely suggests that we should think of noncommercial television as a "primary experience" opportunity that is an alternative to other such means of experiencing life—rather than just an alternative to other forms of entertainment and diversion.[10] Although Blakely provides few examples of primary experiences other than the Watergate hearings and similar public affairs events, he apparently would include coverage of the arts as well as observances of civic activities. These kinds of special programs do appear on PTV stations, but Blakely's main point is that when they are telecast they are thought of as *programs* rather than as the principal mission of public broadcasting. They are offered as alternative entertainment and diversion, or as education, rather than as needed access to activities.

> At the very least people should be given access and help to see how and in what ways their interests are concerned. Not to try to give such access and help would be inexcusable. To try is justifiable, regardless of the sizes of the audiences at first. If the attempt fails, much more than public television will have failed. If public television tries merely to serve the entertainment and diversion interests and tastes of special groups—if it tries and fails, nothing very important will have failed (particularly in the light of needs that are not being met).[11]

The debate will continue—and public television's identity problem is something that PTV will have to learn to live with. We should keep in mind the existence of this problem of purpose and identity—and the simple fact that there *are* different rationales for programing noncommercial radio and television stations—as we consider another of the current issues in public broadcasting, the ascertainment of community needs.

9.2 Ascertainment of Community Needs

On February 23, 1971, the FCC released a *Report and Order* following a year-long study

---

[10]Blakely, "Rethinking the Dream," p. 36.

[11]*Ibid.,* p. 37.

on the obligations of broadcasters to ascertain and meet specific needs in their communities. The *Report and Order* provided to broadcast applicants (including those renewing their licenses) a "Primer on Ascertainment of Community Needs," which spelled out the ways in which the commission expected community needs to be determined. The more specific requirements— which were originally applied only to commercial broadcasters—were then incorporated in application and renewal forms.

The Ascertainment "Primer"   The roots of the ascertainment question go all the way back to the Radio Act of 1927 which first used the phrase "public interest, convenience and necessity."[12] The current interest in ascertainment—in determining public interest and necessity—goes back to 1960, when the FCC ruled that broadcast licensees are obligated "to make a positive, diligent and continuing effort, in good faith, to discover and fulfill the needs, tastes and desires of the public in [their] communities."[13] This action by the FCC was taken in response to pressures by minority groups that felt their unique programing needs often were not being satisfied. Continuing pressure on the FCC and on individual stations during the 1960's made it obvious that clear guidelines were needed to spell out for broadcasters what was intended by the 1960 policy statement. Therefore, in 1971, the FCC issued a *Primer* on ascertainment that spelled out exact procedures to be taken by commercial broadcasters in meeting the FCC's objectives for ascertainment.[14]

Noncommercial broadcasters were initially exempted from having to follow the *Primer* procedures—since it was assumed that they were meeting public needs by their very existence. However, considerable debate followed as to whether or not public broadcasters should continue to enjoy this exemption; in 1973 the FCC invited formal comments on the question of ascertainment procedures for noncommercial stations; and in the fall of 1975 the FCC announced that the requirements of the *Primer* would extend henceforth to noncommercial broadcasters, although noncommercial licensees would be allowed considerable flexibility in complying with the requirements.[15]

Specific Ascertainment Requirements Four basic requirements in the *Primer* must be systematically followed:

1. Interviews must be conducted by station management with community leaders.

2. At least once during each license term, and/or in the six months preceding the filing for license renewal, each licensee (except some small-market stations) must conduct a random sample survey of community needs and programing preferences.

3. A statement of needs ascertained by the two surveys is then developed; this statement of needs should interpret the data, and it is a sort of "State of the Community" consideration of the audience for the station.

4. A statement of the programing that will serve the identified needs is finally drawn up and implemented.

The FCC has said it is inappropriate to require noncommercial applicants to ascertain the *instructional* needs of their communities because most instructional programing is pre-

[12]Frank J. Kahn (ed.), *Documents of American Broadcasting,* 2d. ed. (New York: Appleton-Century-Crofts, 1973), p. 36. Kahn points out that the phrase was actually borrowed from public utility legislation.

[13]Erwin G. Krasnow and John C. Quale, "Ascertainment: The Quest for the Holy Grail," *Public Telecommunications Review,* June 1974, p. 6, citing NBMC Comments in Docket 19816, pp. 1–2.

[14]*Primer on Ascertainment of Community Problems by Broadcast Applicants,* 27 FCC 2d 650,651 (1971).

[15]Krasnow and Quale, "Ascertainment," p. 9.

sented in cooperation with educational institutions. (It also exempted 10-watt FM stations since most were licensed to colleges and high schools and were adequately supervised by those governing the institutions granted the license.)[16]

Refining Ascertainment Requirements
Even with the 1975 actions, the issue of ascertainment for noncommercial broadcasters is far from resolved. Before the 1975 FCC decision Branscomb asked six "very basic questions," which have not yet been satisfactorily answered but which must be before ascertainment procedures for public broadcasters can be more than just an academic exercise in data gathering:

(1) Are noncommercial educational broadcasters to be treated differently from commercial broadcasters? (2) Are noncommercial educational broadcasters to ascertain the program offerings of commercial licensees and demonstrate that their own programing offers an alternative to existing programs available in the market? (3) Should ascertainment be related only to the "public programing" portion of the broadcast day (that directed to general audiences) or also include instructional programing as well? (4) Are noncommercial educational licensees to be held to higher or lower or different standards than commercial broadcasters? (5) How shall noncommercial educational broadcasters meet the requirement of Section 396(a)(4) of the Communications Act, as amended, "to encourage noncommercial educational radio and television broadcast programing which will be responsive to the interests of people both in the particular localities and throughout the U.S."? . . . (6) Can the FCC or any other publicly funded entity render a judgment which involves evaluating programing choices?[17]

Partially in answer to her second question, Branscomb maintains that noncommercial ascertainment procedures require a look at what other stations are offering:

The overall purpose of such ascertainment should be to assess the programing of all stations in the market and to show how the public broadcasters are meeting the unsatisfied programing needs of the particular market. This is vastly different from what is expected of commercial broadcasters and would provide a useful yardstick by which the public might assess the performance of all stations in a market.[18]

Other Requirements to Determine Community Needs
Actually, for years stations have had to determine audience needs and to demonstrate how they would meet them. For one thing, any station that applies for federal assistance from HEW must show the necessity for educational broadcasting in that particular community. Also, like commercial stations, each noncommercial station must renew its license every three years; it therefore must continually reassess community needs no matter what precise ascertainment procedures are prescribed by the FCC. For example, a challenge of all Alabama State Network stations in 1974 resulted in nonrenewal of all nine TV licenses in the state for a failure to meet the needs of the black majority in the state—even though the PTV system had corrected the problem by the time the issue was heard by the FCC (Section 5.2).

Whether or not there are legal requirements for ascertaining community needs, there is universal agreement that community needs must be identified in order to make intelligent decisions about programs. As one public broadcaster put it, "I believe . . . that ascertainment is as vital to programming as medical diagnosis is to treatment. Treatment, or programming, without

[16]*PBS Newsletter,* August 28, 1975, p. 1.
[17]Branscomb, "A Crisis of Identity," p. 16.

[18]*Ibid.,* p. 23.

1976 PROGRAM QUESTIONNAIRE

It's time to begin the process of selecting and acquiring programs for the 1976-77 season. KPBS wants and needs your help in doing this.

Production costs for many fine programs such as MASTERPIECE THEATRE are paid for by grants. But a major portion of our programs are "purchased" through the PBS Station Program Cooperative (SPC). The SPC is a complex democratic process by which groups of public TV stations throughout the country collectively underwrite the programs they feel will best serve their various audiences. Whether a program is actually purchased or not depends on how much it will cost and what the station can afford.

We're asking you to do three things. First, give us your *general feelings about subject areas and program formats*. Second, express your *degree of interest in specific programs* which are now or have recently been on our schedule. You certainly don't need to feel you must check every program — just those you've seen. Finally, we'd like your *suggestions, comments and criticisms* about anything you feel we should know. What you say is extremely important and influences our decisions about programs and scheduling. We promise to read and absorb all of the comments and to respond to your questions and suggestions as much as possible —*Brad Warner*
TV Program Director

9-1 Representative SPC Audience Questionnaire (Part I)

Part I

The following is a list of broad subject areas and formats in which programs will be available. Your expression of interest will be helpful in decisions regarding both national acquisition and the development of programs to be produced by KPBS. Please CIRCLE the number that best reflects your degree of interest. (1) is low or no interest, (5) is highest interest.

	Low			High			Low			High
LITERATURE AND POETRY	1 2 3 4 5				GOVERNMENT AND POLITICS	1 2 3 4 5			
THEATER	1 2 3 4 5				CANDIDATES AND ISSUES	1 2 3 4 5			
HISTORY	1 2 3 4 5				ECONOMICS	1 2 3 4 5			
VISUAL ARTS	1 2 3 4 5				SCIENCE AND HEALTH	1 2 3 4 5			
SELECTED FILMS	1 2 3 4 5				COMEDY/SATIRE	1 2 3 4 5			
MUSIC, CLASSICAL	1 2 3 4 5				AGING	1 2 3 4 5			
MUSIC, JAZZ	1 2 3 4 5				CONTEMPORARY WOMEN	1 2 3 4 5			
MUSIC, COUNTRY/FOLK/REGIONAL	1 2 3 4 5				CONSUMER CONCERNS	1 2 3 4 5			
OPERA	1 2 3 4 5				PARENT EDUCATION	1 2 3 4 5			
DANCE	1 2 3 4 5				COOKING SHOWS	1 2 3 4 5			
LIFE STYLES	1 2 3 4 5				HOW-TO-DO-ITS	1 2 3 4 5			
PETS AND ANIMALS	1 2 3 4 5				CHICANO PROGRAMS	1 2 3 4 5			
SPORTS EVENTS	1 2 3 4 5				BLACK AMERICAN PROGRAMS	1 2 3 4 5			
PHYSICAL FITNESS	1 2 3 4 5				INTERNATIONAL CONCERNS	1 2 3 4 5			
PHILOSOPHY/ETHICS	1 2 3 4 5				NATIVE AMERICANS	1 2 3 4 5			
CREDIT COURSES	1 2 3 4 5				INTERVIEWS	1 2 3 4 5			
TALK SHOWS	1 2 3 4 5				INFORMAL INSTRUCTION	1 2 3 4 5			
SPECIAL EVENT COVERAGE	1 2 3 4 5								
THE LAW	1 2 3 4 5								
PUBLIC AFFAIRS DOCUMENTARIES	1 2 3 4 5								

diagnosis or ascertainment, becomes pure mockery, quackery, or pretense."[19]

Ascertainment and the SPC   The PBS Station Program Cooperative (Section 7.3) has provided a catalyst for developing "community

outreach" activities that seem to meet the spirit, if not the letter, of FCC ascertainment requirements. Three specific kinds of procedures evolved, which different PTV stations have used to get maximum community input in making their SPC programing decisions.

Special Previews: Some stations broadcast special previews consisting of excerpts of

[19]Harry J. Skornia, "Ascertainment at Work—Without Legislation: Lessons From Overseas," *Public Telecommunications Review,* February 1975, p. 15.

Part II

The following is a list of many programs which are now or have recently been on our broadcast schedule. Please circle the number that best expresses your interest in having the series *continued or returned to our schedule*. (1) is low or not interest, (5) is highest interest.

9-2 Representative SPC Audience Questionnaire (Part II)

	Low	-			High
MASTERPIECE THEATRE	1	2	3	4	5
HOLLYWOOD TELEVISION THEATER	1	2	3	4	5
THEATER IN AMERICA	1	2	3	4	5
BERGMAN FILM FESTIVAL	1	2	3	4	5
CLASSIC THEATRE	1	2	3	4	5
JENNIE	1	2	3	4	5
THE ADAMS CHRONICLES	1	2	3	4	5
OURSTORY	1	2	3	4	5
THE STRAUSS FAMILY	1	2	3	4	5
ANYONE FOR TENNYSON?	1	2	3	4	5
GREAT PERFORMANCES	1	2	3	4	5
EVENING AT POPS	1	2	3	4	5
IN PERFORMANCE AT WOLF TRAP	1	2	3	4	5
EVENING AT SYMPHONY	1	2	3	4	5
LEONARD BERNSTEIN AT HARVARD	1	2	3	4	5
OPERA THEATER	1	2	3	4	5
DANCE IN AMERICA	1	2	3	4	5
SAN DIEGO SHOWCASE	1	2	3	4	5
PICCADILLY CIRCUS	1	2	3	4	5
INTERNATIONAL ANIMATION FESTIVAL	1	2	3	4	5
YOUNG FILMMAKER'S FESTIVAL	1	2	3	4	5
SOUNDSTAGE	1	2	3	4	5
CLUB DATE	1	2	3	4	5
AUSTIN CITY LIMITS	1	2	3	4	5
PHILADELPHIA FOLK FESTIVAL	1	2	3	4	5
NOVA	1	2	3	4	5
THE TRIBAL EYE	1	2	3	4	5
NATIONAL GEOGRAPHIC SPECIALS	1	2	3	4	5
THE ASCENT OF MAN	1	2	3	4	5
SAN DIEGO SCIENCE SCENE	1	2	3	4	5
LIFE OF LEONARDO DA VINCI	1	2	3	4	5
THE JAPANESE FILM	1	2	3	4	5
JOURNEY TO JAPAN	1	2	3	4	5
IMAGES OF AGING	1	2	3	4	5
AVIATION WEATHER	1	2	3	4	5
CONSUMER SURVIVAL KIT	1	2	3	4	5
ROMAGNOLIS' TABLE	1	2	3	4	5
WHAT'S COOKING?	1	2	3	4	5
PLAY BRIDGE WITH THE EXPERTS	1	2	3	4	5
ANTIQUES	1	2	3	4	5
LILIAS, YOGA AND YOU	1	2	3	4	5
THE MAGIC OF OIL PAINTING	1	2	3	4	5
PAINTING THROUGH EUROPE	1	2	3	4	5

	Low	-			High
INVENTIVE JEWELRYMAKING	1	2	3	4	5
ERICA	1	2	3	4	5
GUPPIES TO GROUPERS	1	2	3	4	5
THE FLOWER SHOW	1	2	3	4	5
WALSH'S ANIMALS	1	2	3	4	5
FIRING LINE (Buckley)	1	2	3	4	5
SPEAKING FREELY (Newman)	1	2	3	4	5
GLORIA PENNER IN CONVERSATION	1	2	3	4	5
DAY AT NIGHT	1	2	3	4	5
KUP'S SHOW	1	2	3	4	5
LOWELL THOMAS REMEMBERS	1	2	3	4	5
BOOK BEAT	1	2	3	4	5
WOMAN ALIVE	1	2	3	4	5
SAY BROTHER	1	2	3	4	5
BLACK JOURNAL	1	2	3	4	5
REALIDADES	1	2	3	4	5
WORLD PRESS	1	2	3	4	5
KRISHNAMURTI DIALOGUES	1	2	3	4	5
BEHIND THE LINES	1	2	3	4	5
SOLAR ENERGY	1	2	3	4	5
SCIENCE INTERVIEWS (Jeff Kirsch)	1	2	3	4	5
SAN DIEGO SCIENCE SCENE	1	2	3	4	5
WASHINGTON WEEK IN REVIEW	1	2	3	4	5
WALL STREET WEEK	1	2	3	4	5
CALIFORNIA JOURNAL	1	2	3	4	5
BLACK PERSPECTIVE ON THE NEWS	1	2	3	4	5
EVENING EDITION WITH MARTIN AGRONSKY	1	2	3	4	5
ROBERT MACNEIL REPORT	1	2	3	4	5
BILL MOYERS' JOURNAL	1	2	3	4	5
MARK RUSSELL COMEDY SPECIALS	1	2	3	4	5
BALLOT '75 (Candidates & Issues)	1	2	3	4	5
DISCUSSIONS WITH OUR LEGISLATORS	1	2	3	4	5
THE WAY IT WAS (Sports)	1	2	3	4	5
TENNIS COVERAGE (General)	1	2	3	4	5
CONTEMPORARY CALIFORNIA ISSUES	1	2	3	4	5
COURSE OF OUR TIMES (Abram Sachar)	1	2	3	4	5
MISTER ROGERS' NEIGHBORHOOD	1	2	3	4	5
SESAME STREET	1	2	3	4	5
THE ELECTRIC COMPANY	1	2	3	4	5
BIG BLUE MARBLE	1	2	3	4	5
VILLA ALEGRE	1	2	3	4	5
CARRASCOLENDAS	1	2	3	4	5
ZOOM	1	2	3	4	5

pilot programs proposed through the SPC, and viewers were urged to provide feedback—through newspaper questionnaires, letters, phone calls, or return-mail inserts (in the monthly program guide)—on desired programs.

Studio Previews: Some stations invited viewers to special closed-circuit previews at the station, where viewers were asked to respond to SPC offerings through questionnaires and evaluation sheets.

Program Guide Questionnaires: Many stations included return-mail questionnaires in the regular monthly guide, through which viewers could react to the more than 100 SPC program proposals. Figures 9-1 and 9-2 are examples of a questionnaire used by KPBS-TV, San Diego.

The biggest problem with these procedures is that they generally use membership lists to obtain reactions to SPC proposals; and, as noted earlier (Section 8.5), those who donate to public television tend to be the same sort of people who donate to other public institutions and to charities. If they were the only voices heard, a vicious circle would be perpetuated—elite tastes picking elitist programing, which would attract more viewers with elite tastes who would again vote for elitist programing. This is not to say that stations should not invite feedback on SPC programing, nor is it meant to imply that stations do not look for other ways to involve the public in programing decisions. As Ron Bornstein, manager of WHA-TV at the University of Wisconsin, puts it, "I don't think any broadcaster can say he knows exactly what the people want. All we can do is get the best indication we can."[20]

Other Methods for Determining Audience Needs   Several years ago, Donald Browne suggested that the "citizen's councils" that are now built-in to several foreign broadcasting systems be copied in the United States by both commercial and noncommercial broadcasting.[21] More recently, educational broadcast pioneer Harry Skornia recommended that an award should be given for the best ascertainment proposal or effort:

> I propose that public broadcasters set forth a policy statement, flexible enough to be applied to widely different local situations, which includes not only ascertainment but also "dialogue" (with the public), research (into effects), fairness (which is certainly a "need") and "access" (a bone of contention which is also leading to millions of words and dollars in litigation and regulation). Why not a good faith statement of the careful way in which a station, or network, will

> identify the principal problems of the area for which it is responsible—and then program to them?[22]

E. B. Eiselein, the "media anthropologist" at KUAT, Tucson, suggests that public affairs *programs* should be thought of as "feedback" to *messages* from the audience. He thinks broadcasters have difficulty with the matter of ascertainment because our mass communications models describe broadcasters as message *originators* and the audience as *receivers* who *"feedback"* reactions *to the broadcasters*. It should be, says Eiselein, the other way around.[23]

The whole question of formal ascertainment, of course, raises many questions about government intervention and increased federal control. Two Washington attorneys, Erwin Krasnow and John Quale, are pessimistic about formal ascertainment requirements. They note that all ascertainment requirements are quests for *certainty*.

> But where does this quest end? That question—and the answer, such as it is—applies to all regulation, not merely to ascertainment. The quest for certainty yields new rules, and further refinement of old ones—but not certainty. So there is yet another cycle of complaint, comment and temporary resolution, ending in more rules and regulations. . . .
>
> With that in mind, public broadcasters will have to begin turning away from debates about whether they should be held to "higher," "lower," or simply "different" standards, and debate the problem as one that all broadcasters—indeed, all communicators—must face. . . . And if ascertainment requirements are to be imposed by

[20]*PBS Newsletter,* August 28, 1975, pp. 1, 4.
[21]Donald R. Browne, "Citizen Involvement in Broadcasting: Some European Experiences," *Public Telecommunications Review,* October 1973, pp. 16–28.

[22]Skornia, "Ascertainment at Work," p. 4.
[23]E. B. Eiselein, "The Program as Feedback: One Station's Approach to Ascertainment," *Public Telecommunications Review,* March/April 1975, p. 11.

Congress and the FCC on public broadcasting today, why not cable tomorrow? And if the requirements are imposed now, who is to say that they will not be more specific and burdensome tomorrow? These problems are not public broadcasting's or commercial broadcasting's. They are everyone's.[24]

9.3   The Audiences of Public Broadcasting

Although public broadcasting is still wrestling with the problem of determining audience *needs,* it has researched audience *characteristics* to a considerable degree. ETV audience research spans close to a quarter century.

The Impact of Educational Television in the Late 1950's   From 1954 to 1960, ETV, and in particular the audiences of ETV, were studied by many individual researchers. Some researchers were doctoral students at universities with new noncommercial TV licenses; some studies were commissioned by the Educational Television and Radio Center; and some research was funded by Title VII of the National Defense Education Act after its passage in 1958 (Section 4.1). In 1960, Wilbur Schramm reviewed and summarized thirteen of these studies in a single volume entitled *The Impact of Educational Television.*[25]

The principal conclusion of Schramm was to be echoed over and over again in the years following the book's distribution: Viewers of ETV programs "go purposefully to the educational channel for a program they know in advance they want."

They read the station program guide or the TV programs in the newspapers. They remember the schedule, and turn to the

station at a certain time with a certain purpose. The program completed, they usually turn off the set or tune it elsewhere. This pattern of highly selective viewing makes for attentive audiences and station loyalty, but also for lower program ratings.[26]

The People Look at Educational Television in 1960   Shortly after Schramm's summary of ETV audience studies, he teamed with Jack Lyle and Ithiel de Sola Pool in the first large-scale study of the audience for educational television. Their report, published as a book entitled *The People Look at Educational Television,* tabulated data from 30,000 interviews conducted in 1959 and 1960 among the audiences of nine representative educational television stations in all sections of the United States.[27] To put this landmark study in proper perspective, note that this was the beginning of the White era at NETRC and, despite accelerated improvements in programing at this time, ETV was still concerned primarily with lectures and interviews, academic considerations of public affairs, and detailed explanations of phenomena, objects, and processes.

Schramm, Lyle, and Pool reiterate the point made by the earlier Schramm work:

Educational television asks for rather uncommon behavior. It asks for behavior one might expect of persons who, after a day's work, hurry off to a lecture or a symphony concert, or go to the library for a book on the history of Africa, or study mathematics because they feel they ought to know about modern notions like set theory. People who do those things are small minorities of their communities.[28]

But how large a minority actually viewed ETV? The study found that in the larger metro-

[24]Krasnow and Quale, "Ascertainment," p. 12.

[25]Wilbur Schramm (ed.), *The Impact of Educational Television* (Urbana: University of Illinois Press, 1960).

[26]*Ibid.,* p. 22.

[27]Wilbur Schramm, Jack Lyle, and Ithiel de Sola Pool, *The People Look at Educational Television* (Stanford, California: Stanford University Press, 1963).

[28]*Ibid.,* p. 47.

politan areas (Boston, Pittsburgh, San Francisco) 21 to 24 percent of the potential audience view the community stations at least once a week. For school, university, and state-owned VHF stations, the percentage is about 9 to 13 percent. For UHF stations (this was before legislation requiring the manufacture of "all-channel" receivers), the figure drops down as low as 3 percent.[29]

The interviews conducted by Schramm and his associates included data on demographic and selected personality variables in an attempt to find out just what kinds of people watched educational television.

> The feature of the ETV audience that first comes to one's attention is that there are all kinds of people in the audience. There are young people and old ones, well-educated people and little-educated ones, parents and childless couples, nuclear physicists and manual laborers. There are eggheads and businessmen. It is a minority of choice rather than a minority by determination.[30]

There were a few sociopsychological variables, however, which were significant in the statistical tests applied to the data.

> The main factors that distinguish ETV viewers from nonviewers we have found to be higher social status, higher aspiration level, higher cultural level, higher interest in public affairs, and higher energy. We have also noted that viewing ETV goes by families, and that ETV viewers tend to be purposeful rather than "let's see what's on" users of television.[31]

The People Look at Public Television, 1974 (and Later)   In the 1970's a vast amount of data on public television began to be generated by the Corporation for Public Broadcasting in cooperation with the National Center for Education Sta-

tistics of HEW. A CPB report issued in 1975 was authored by Jack Lyle, and (since Lyle was one of the researchers in the 1960 study) was appropriately entitled *The People Look at Public Television, 1974*.[32] According to CPB studies, the audiences for public television in the mid-seventies had these characteristics:

1. More than 80 percent of the people in the United States could receive one or more public television stations.

2. About one third of all TV households tuned to a PTV program in any given week, and one half of all TV households viewed a PTV program at least once in any given month.

3. From 1970 to 1975 the number of people viewing evening programs on PTV increased by 50 percent.

4. Among national PTV offerings, programs in the "arts" were viewed by twice as many households as "public affairs" programs, with an average Nielsen for arts programs of 1.5—representing over a million households.

5. *Sesame Street* reached nearly a quarter of the nation's TV households in a month, despite the fact that only 26 percent of American households have children under the age of six, the target audience of *Sesame Street*.

6. Most persons who watch public television programs in the evening watch only one such program per week.

7. Viewing of PTV programs continues to be related to both income of household and education of viewers, but the total audience for PTV cuts across socioeconomic and ethnic groups.[33]

---

[29]*Ibid.*, pp. 47–50.

[30]*Ibid.*, p. 59.

[31]*Ibid.*, pp. 81–82.

[32]Jack Lyle, *The People Look at Public Television, 1974* (Washington, D.C.: Corporation for Public Broadcasting, 1975).

[33]*Ibid.* Additional conclusions from S. Young Lee and Ronald J. Pedone, *Status Report on Public Broadcasting, 1973* (Washington, D.C.: Corporation for Public Broadcasting, December 1974) and from periodic "Focus on Research" contained within some issues of the *CPB Report.*

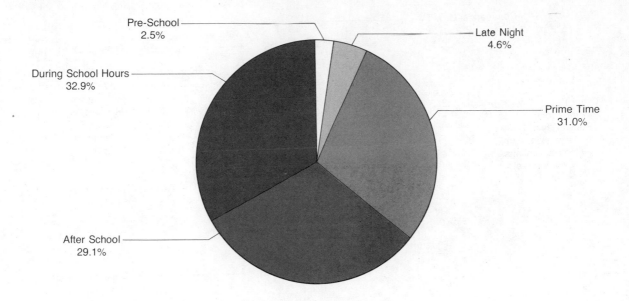

9-3 Average PTV Station Percentage of Broadcast Hours by Time of Day

Source: Natan Katzman, Public Television Program Content: 1974 *(Washington, D.C.: Corporation for Public Broadcasting, 1975), p. 21.*

Looking at the data over the years, it appears that the audiences for children's programs and programs in the arts are gaining increasingly larger shares of the audience each year. This does not mean that fewer people are viewing public affairs and special-audience programs but rather that larger numbers are tuning away from commercial television to view such programs as *Sesame Street,* the Boston Pops and musical specials. This may be because people turn on the TV set in the evening more to be diverted than to be stimulated, and the Boston Pops offers more diversion than *Washington Week in Review* —so the audiences for the latter remain relatively stable, as audiences for the former increase.

9.4   PTV Programing Balance

At about the time of the earliest studies on ETV audiences, a report was published by

Brandeis University for the ETRC on *One Week of Educational Television, May 21–27, 1961.* After 1961, a series of similar reports were issued by different organizations, the last one published by the CPB, reviewing *One Week of Public Television, April, 1972.* In 1973, the CPB decided to begin a new series of reports, which would sample an entire year of public television content. The first report in the new series was published in July 1975, reviewing *Public Television Program Content: 1974.*[34]

Total Programs and Hours   The average PTV station in 1974 broadcast 6,547 programs

---

[34]Natan Katzman, *Public Television Program Content: 1974* (Washington, D.C.: Corporation for Public Broadcasting, July 1975). Percentages of programs and program hours throughout the chapter are from this work, as are all pie-charts.

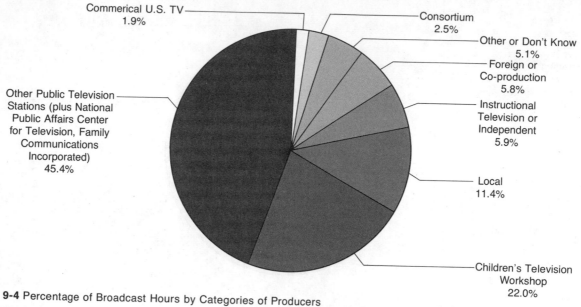

9-4 Percentage of Broadcast Hours by Categories of Producers

Source: Natan Katzman, Public Television Program Content: 1974 *(Washington, D.C.: Corporation for Public Broadcasting, 1975), p. 23.*

during the 3,872 hours it was on the air. These broadcast hours were divided fairly evenly among the three basic time periods that a station is on the air: school hours (and daytime hours on the weekend)—33 percent of total broadcast hours; after-school hours (and late afternoon on the weekends)—29 percent of total broadcast hours; and prime time—31 percent of total hours. Before school hours and late night broadcasts comprised the remaining 7 percent of all broadcast hours. Figure 9-3 graphically shows how total broadcast hours were divided by time periods.

Sources of Programs   Figures 9-4 through 9-7 relate to the sources of PTV programing, showing where programs come from—by different categories. Figure 9-4 indicates the *producers* of programs carried by the typical PTV station. Nearly half (45 percent) of the average station's program hours were produced by other

PTV stations.[35] The next largest source of programs is the Children's Television Workshop (*Sesame Street* and *The Electric Company*) which produced 22 percent of all program hours broadcast by the average PTV station.

If we stood in master control and noted where programs came from, what would we find? If we asked each minute what was punched up on the switcher, we would find that nearly half the time the PBS "network" line was feeding the transmitter (figure 9-5). Another 18 percent of the time a videotape delay or repeat of a PBS-distributed program would be feeding the transmitter. Films run on the "film chain," which once

---

[35]Included as programs "from other stations" are programs produced by National Public Affairs Center for Television and Family Communications Incorporated (producers of *Mister Rogers' Neighborhood*), since NPACT is now a part of WETA in Washington, D.C., and FCI operates as a unit within WQED, Pittsburgh, which originally produced the series.

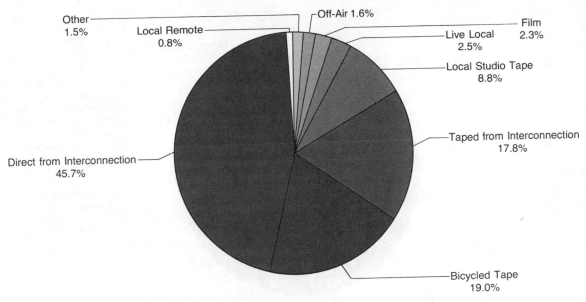

9-5 Percentage of Broadcast Hours by Transmission Technique

Source: Natan Katzman, Public Television Program Content: 1974 *(Washington, D.C.: Corporation for Public Broadcasting, 1975), p. 28.*

accounted for most of the broadcast hours, now are a small source (about 2 percent) of the broadcast hours. And about 2½ percent of the time stations are broadcasting live local programs.

If we were to ask each minute what *agency* distributed the program on the air, 62 percent of the time the answer would be "PBS" (figure 9-6). Ten percent of the time the programing comes from one of the tape libraries (7.7 percent from an ITV library and 2.1 percent from PTL or NET); 11 percent of the time the program is a local production.

Figure 9-7 shows what happens if we ask the question about the distributing agency a little differently. Instead of asking each minute, we can ask where the program came from each time a new program begins. Because PBS programs tend to be longer than local programs or those from instructional TV libraries, a little less

than half of the actual program titles carried by a station come from PBS, with all other categories assuming a larger percentage of the total.

Types of Programs Figure 9-8 indicates the types of programs that comprise a typical broadcast day, by looking at the percentage of broadcast hours devoted to each program category. One of the largest categories of programs broadcast is "instructional TV." Seventeen percent of the broadcast hours in 1974 were designed for use in schools or for formal instruction at home. These figures do *not* include *Sesame Street* or *The Electric Company* broadcast during school hours or in-service offerings for teachers — unless credit was given.

Looking just at prime-time hours (figure 9-9), we can see a fairly even distribution of program hours devoted to public affairs, music/

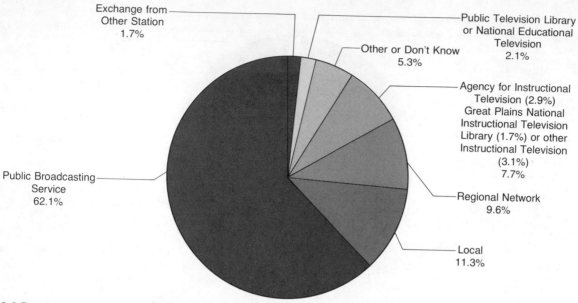

Exchange from
Other Station
1.7%

Public Television Library
or National Educational
Television
2.1%

Other or Don't Know
5.3%

Agency for Instructional
Television (2.9%)
Great Plains National
Instructional Television
Library (1.7%) or other
Instructional Television
(3.1%)
7.7%

Regional Network
9.6%

Local
11.3%

Public Broadcasting
Service
62.1%

9-6 Percentage of Broadcast Hours by Distribution Agency

Source: Natan Katzman, Public Television Program Content: 1974 *(Washington, D.C.: Corporation for Public Broadcasting, 1975), p. 24.*

dance/drama, and general information/history—with all other types of programing comprising another quarter of the prime-time hours.

All these data help to furnish some insight into the issues raised above and in previous sections: Who should produce PTV programing? What should be the balance between local and network productions? How much "educational" or ITV programing should a PTV station carry? What about the balance of public affairs, cultural, and children's programing?

9.5 Trends in Public Television Programing

In the next couple of decades, PTV will probably change nearly as much as it has in the past quarter century. Although it is difficult in the mid-1970's to foresee the influences of new technologies such as satellites, video discs, and cable

TV, we can discern some trends as we look at the different types of programing—beginning with public affairs presentations.

Controversies Over Public Affairs Programing Most people in and out of public broadcasting find incredible any suggestion that public radio or television should *not* broadcast public affairs programs. Yet, such suggestions are made; a 1975 *TV Guide* article said, "Under these circumstances, and in this tight recession year of 1975, we ought to drop the federal funding—at least for the public affairs element, preferably for PTV as a whole."[36]

Public affairs programing was a central issue in the Whitehead-Nixon attack on public

---

[36]Kevin Phillips, "Should Federal Money Support Public Television?" *TV Guide,* July 12, 1975, p. A–4.

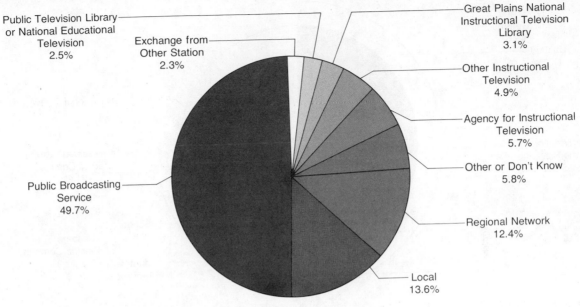

9-7 Percentage of PTV Programs by Distribution Agency

Source: Natan Katzman, Public Television Program Content: 1974 *(Washington, D.C.: Corporation for Public Broadcasting, 1975), p. 24.*

broadcasting in the early 1970's (Section 6.2). In his infamous speech to the 1971 NAEB convention, Whitehead was quite specific as to where he stood on the question of public affairs broadcasts:

> Instead of aiming for "overprogramming" so local stations can select among the programs produced and presented in an atmosphere of diversity, the system chooses central control for "efficient" long-range planning and so-called "coordination" of news and public affairs—coordinated by people with essentially similar outlooks. How different will your network news programs be from the programs that Fred Friendly and Sander Vanocur wanted to do at CBS and NBC? Even the commercial networks don't rely on one sponsor for their news and public affairs, but the Ford Foun-

dation is able to buy over $8 million worth of this kind of programming on your stations.[37]

A short time later, Whitehead said to the Macdonald committee:

> There is a real question as to whether public television, particularly . . . the national federally-funded part of public television, should be carrying public affairs and news commentary and that kind of thing, for several reasons. One is the fact that commercial networks, by and large, do, I think, quite a good job in that area. Public tele-

[37]Clay T. Whitehead, "Local Autonomy and the Fourth Network: Striking a Balance," *Educational Broadcasting Review*, December 1971, p. 4. Address to the 47th convention of the National Association of Educational Broadcasters, October 1971, Miami, Florida.

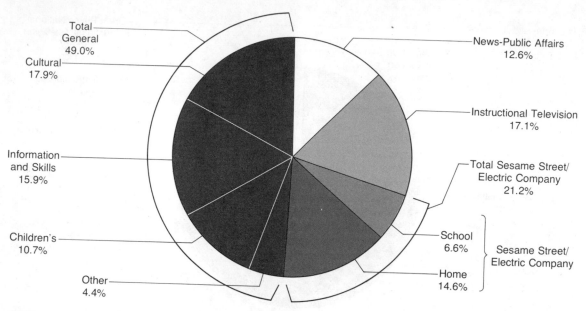

9-8 Percentage of Broadcast Hours by Programing Categories

Source: Natan Katzman, Public Television Program Content: 1974 *(Washington, D.C.: Corporation for Public Broadcasting, 1975), p. 38.*

vision is designed to be an alternative to provide programing that isn't available on commercial television.[38]

Until the question was raised by Whitehead in 1971, no one seriously questioned the propriety of offering public affairs programs on PTV or ETV. From its inception, the ETRC considered public affairs an important and essential area of programing. In a 1954 ETRC publication, three of its ten program categories were related to public affairs—"Contemporary America," "International Affairs," and "The Individual in Society."[39] A year later, the Center talked in terms of four broad areas of educational concen-

tration—one of which was "Social Studies and Public Affairs."[40]

Later the ETRC turned to a complex system of six program areas and twenty-eight themes. Three of the six main areas dealt with public affairs—political institutions, social institutions, and "the formation of public policy in domestic and foreign affairs." Under John White, NET program planning and acquisition was streamlined into three primary areas—public affairs, cultural programing, and children's programing (Section 7.2).

Clearly, the importance of public affairs programing as an integral and vital part of the programing operations of the fourth network had never been questioned. How could a viable noncommercial "public" network—licensed in the "public interest"—even consider any programing

[38]"Public Television in Transition," School of Journalism, University of Missouri at Columbia, Freedom of Information Center Report No. 301, April 1973, p. 5.
[39]ETRC, *Program Reports,* September-October 1954, p. 1.

[40]"ETRC Report to the Fund for Adult Education," October 1, 1955, p. 7.

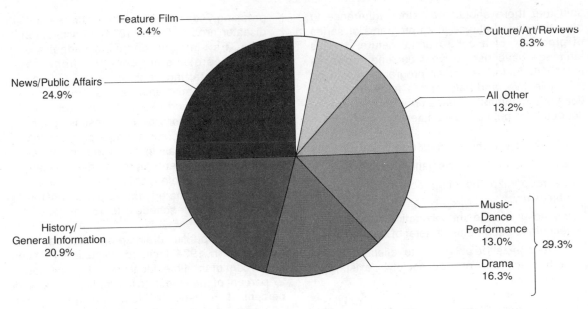

9-9 Percentage of Prime-Time Hours by Program Type

Source: Natan Katzman, Public Television Program Content: 1974 *(Washington, D.C.: Corporation for Public Broadcasting, 1975), p. 127.*

plans that excluded public affairs? As of 1967, according to White, "NET devotes at least half of its resources and half of its schedule to programs in public affairs."[41]

In the late 1960's, however, controversy had begun to develop within public broadcasting ranks over public affairs—not over whether there should be public affairs programing but over who should regulate it. At first, the debates were between PBS and NET, the principal producer of public affairs programing at the time, who wanted PBS to air NET programs without questioning their content. However, PBS was forced to field criticism from both stations and politicians about some of the programs and wanted more of a say about program content. In late 1970—still well before Whitehead jumped into the fray—two NET documentaries were sharply criticized as unfair and lacking objectivity. "Does Advertising Tell It Straight?" also involved corporate underwriting and is discussed in Section 8.2. "Banks and the Poor" attacked both banks and congressmen who had banking connections; a study of the program by a political scientist for PBS called it "an abuse of the air waves."[42]

In late 1971, PBS formed a committee to develop guidelines for news and public affairs. The committee, headed by Elie Able, one-time NBC correspondent and then-dean of Columbia University's Graduate School of Journalism, developed a set of guidelines based upon the language of the Public Broadcasting Act—which

[41]John F. White, "National Educational Television as the Fourth Network," *in* Allen E. Koenig and Ruane B. Hill (eds.), *The Farthest Vision: Educational Television Today* (Madison: University of Wisconsin Press, 1967), p. 90.

[42]"Public Television in Transition," p. 2.

said that there should be "strict adherence to objectivity and balance in all programs or series of programs of a controversial nature."[43] This language gave rise to what became known as the "FOB factor"—that all programs should be fair, objective, and balanced. The recommendations of the committee became a kind of "code of conduct" for public broadcasters:

We recognize the obligation to be fair;

we pledge to strive for balanced programing;

we recognize the obligation to strive for objectivity;

we acknowledge the obligation not to let technique become the master of substance;

we recognize the obligation to reflect voices both inside and outside society's existing consensus.[44]

Current Status of Public Affairs on PTV

Public affairs programing consists primarily of studio-type talk shows produced by local stations and regional networks, plus specials and "reports" produced by the National Public Affairs Center for Television (NPACT) and other national producers. The national programing distributed by PBS has increasingly become a principal issue in public broadcasting. As a result of the Station Program Cooperative (Section 7.3), there had been a reduction of public affairs programing on PBS. Paul Duke, who in early 1974 left a high-paying prestigious job as NBC News congressional correspondent to become the key reporter and personality for NPACT, summed up the scene in the 1970's this way:

But if things are looking brighter for the other programming divisions, they are de-

pressingly darker for public affairs—the disaster area of PBS's 1974–75 schedule. NPACT's staff and operating budget are down approximately one-third from the bureau's starting year of 1971. . . . Such weekly PBS programs as *The Advocates, Bill Moyers' Journal* and *Washington Connections* are, for various reasons, gone. There is, in fact, not a single program left that regularly deals in depth with major national issues other than the studio-type talk shows. As a whole, public affairs programming had plummeted from 30 per cent of the weekly network schedule to 16 per cent.[45]

The public affairs programs that were on the air in 1974 focused on national issues 75 percent of the time, local/state/regional issues 27 percent of the time, and international affairs 45 percent of the time (the totals are greater than 100 percent since there is some overlap in the content of many of the programs). About two thirds of all public affairs/news programs are in-studio presentations (discussions/newscasts/interviews), and only 13 percent of the 1974 program hours—national and local—were actual coverage of events including governmental bodies at work, despite the interest generated in this sort of programing by the Watergate hearings in 1973 and despite the extensive coverage of state legislatures in some states (such as Florida and Nebraska).[46]

News programs at all levels will probably increase in number and popularity in the coming years for two reasons. The concern for meeting ascertained community needs (Section 9.2) will often focus on informing the community about local problems and various efforts and activities being undertaken to answer those problems. At the same time, portable low-cost slant-track

---

[43]Public Broadcasting Act of 1967, Public Law 90–129, Section 396(g)(1)(A).

[44]John Macy, Jr., *To Irrigate a Wasteland: The Struggle to Share a Public Television System in the United States* (Berkeley: University of California Press, 1974), pp. 71–72.

[45]Paul Duke, "Public Affairs: The Commitment We Need," *Public Telecommunications Review*, October 1974, p. 27.

[46]Katzman, *Public Television Program Content: 1974*, pp. 94–103.

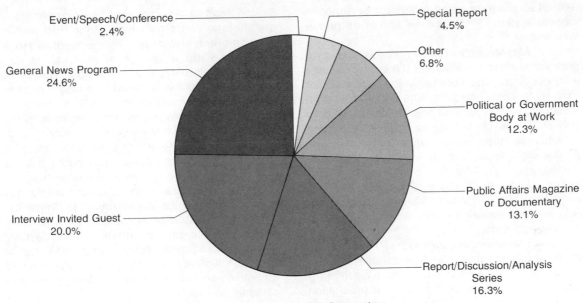

Event/Speech/Conference
2.4%

General News Program
24.6%

Interview Invited Guest
20.0%

Special Report
4.5%

Other
6.8%

Political or Government
Body at Work
12.3%

Public Affairs Magazine
or Documentary
13.1%

Report/Discussion/Analysis
Series
16.3%

9-10 Percentage of Local Public Affairs Programing by Format Categories

Source: Natan Katzman, Public Television Program Content: 1974 *(Washington, D.C.: Corporation for Public Broadcasting, 1975), p. 119.*

recorders and smaller and more sophisticated electronic news gathering (ENG) equipment make it increasingly easy to move into the community with cameras and microphones. With ENG facilities, it is feasible for a station to get into extensive on-location and remote coverage without the inconvenience and expense of a fully equipped mobile van.

Public broadcasting certainly will continue to produce national public affairs programs and just as certainly will continue to face criticism for tackling tough issues. The question is not how to avoid criticism but how to live with it. The danger is that public broadcasters could decide it is not worth trying. As Bill Moyers, one-time press secretary for President Johnson and sometime public broadcasting commentator, put it, "It would be painful irony if, in trying to get out of the poorhouse, public broadcasters convinced

themselves—privately, of course—not that it is dangerous to take risks, but that it is wise to avoid them."[47]

Cultural Programing: Drama   In the mid-1970's, drama on PTV consisted of three basic sources—plus specials. Public TV continued to carry the British dramatic productions, primarily as adapted by Boston's WGBH-TV, under the general title of *Masterpiece Theatre*. The Los Angeles PTV station, KCET, continued to produce the *Hollywood Television Theater* and was starting the ambitious *Visions* series. WNET in New York produced several series and specials such as *The Adams Chronicles*. Series and programs such as these, and occasional offerings produced by other PTV stations, comprised about 10 per-

[47]Quoted by Duke, "Public Affairs," p. 28.

cent of all program hours broadcast by noncommercial stations in 1974 (one fifth of all prime-time hours).[48]

Masterpiece Theatre consists of programs created by the BBC and ITA in Great Britain, with openings and commentaries produced for PBS by WGBH. The programs are produced on videotape with a care and precision that is seldom achieved in the United States. In addition to being entertaining, they set a standard in acting and production that can only be healthy for American commercial TV. (One public broadcasting BBC series, Upstairs, Downstairs, inspired a short-lived commercial TV network imitation, Beacon Hill, which—however—was unable to sustain a mass audience.)

The American dramas on PTV also have used videotape rather than film, in relatively conventional fashion. In the future, there quite probably will be more experimentation in production styles. The staffs of KCET in Hollywood and WNET in New York proved in the early 1970's their ability to produce quality conventional drama. In the 1980's, the challenge will be to explore the qualities of intimacy, involvement, and immediacy—as well as experimental production techniques and innovative dramatic formats—so that drama on PTV will take on new dimensions clearly different from what is available on commercial TV.

Cultural Programing: Music and Dance

With 13 percent of all broadcast hours in 1974, music and dance programs make up a smaller proportion of programs than most other types— yet they are the most popular programs on PTV, capturing twice as many viewers as public affairs and most information programs.[49]

The Boston Symphony programs and the Boston Pops series—both produced by WGBH— are the standards by which other PBS music programs are measured. These programs are straightforward coverage of musical performances, reminiscent of the Firestone Hour that in the late 1950's was "must" viewing (on commercial TV) for music lovers.

In this category also are the many short series and specials that encompass every musical interest from avant-garde to bluegrass, jazz, and rock. Dance programs have been mostly ballets, although some modern dance has been tried.

Improvements in television audio are sure to continue the popularity of performance programs on public broadcasting. There already have been musical performances simulcast by PTV and public radio stations; and as NPR converts to full stereo networking, the opportunities for stereo-sound simulcast performances will certainly be explored.

Children's Programs

It is sometimes difficult—on first consideration—to understand why it took public broadcasters so long to discover Sesame Street. In the 1950's and 1960's, educational television station managers and critics talked about quality children's programing, but— with few exceptions such as the NET What's New? series started in 1960—they were generally incapable of meeting the needs of children, primarily for financial reasons. Most programs were much like the pap served on commercial TV—only 4-H activities and science demonstrations, instead of cartoons, filled time between games and interviews with children in the studio.

The principles that later were incorporated in Sesame Street and The Electric Company were not unknown during the early days of PTV. A 1962 award-winning book by the Television Information Office offered the following principles and suggestions to commercial (and educational) broadcasters: Children have difficulty at times relating words to things; children enjoy repetition; fantasy and reality become confused for children; the child tends to view events as being distinct and

---

[48]Katzman, Public Television Program Content: 1974, p. 125.

[49]Lyle, People Look at Public Television, 1974, p. 25. In one four-week period of late 1974, for instance, 24 "arts" programs reached a cumulative audience of 23.5 percent of the nation's TV homes, compared with 12.3 percent reached by 25 "public affairs" programs.

unconnected (parts are not always seen as contributing to the whole); children like rhymes and rhythm and made-up words that have fascinating sounds; children like music, particularly if they can clap and sing along with it; children like nonverbal messages (a clown tripping over his own feet) or pantomime; children like simplifications, which is why cartoons are so popular.[50]

It took more than five years for the financing to be arranged so that these ingredients could be incorporated in a series for preschoolers (Section 11.7); but, when such a series finally was created, it was a winner. As noted researcher Warren Seibert wrote:

> With little doubt, the most important development in educational broadcasting in recent times has not been color, or satellites, or CATV prospects, or even the Public Broadcasting Act and the Corporation for Public Broadcasting combined, but "Sesame Street." It demonstrates powers that educational broadcasters have ascribed to the media and to their services from the beginning, except that now these things seem credible. It has, in the words of one writer, revived a demoralized field. In the midst of abundant superlatives and favorable comment, it seems especially important to pause and consider both the unfavorable reactions to "Sesame Street" and the evidence, if any, which might maintain belief in the series' importance and success.[51]

Since 1972, *The Electric Company* has proven to be a series that runs a close second to *Sesame Street*, but there are virtually no other examples of this caliber. The greatest need is programing in the evening for preteenagers.

Zoom, the preteenager series broadcast in the early 1970's, proved too expensive for the PBS Station Program Cooperative (SPC-1) in 1974, and the series has been struggling ever since. With support from corporate underwriters the series has continued, but only by repeating many segments from early programs. The failure of *Zoom* in the SPC suggests that some other daily preteen series should be developed for PBS. Perhaps some combination of *Zoom, The Electric Company,* and *The Mickey Mouse Club* is what is needed—as evidenced by the renewed popularity of the old Mickey Mouse Club programs syndicated to independent commercial stations and cable TV systems.

For the next age group—teenagers—PTV programing is virtually nonexistent. This, perhaps, is public broadcasting's greatest programing challenge: to create and promote the kind of constructive programing that will be of significance to a substantial portion of America's preadult population.

Educational Programing: History/General Information/How-to-do-it Materials Next to public affairs programing, the informational program was the most popular offering in the early 1970's—especially historical series such as *Civilisation, America, The Ascent of Man,* and *The Adams Chronicles*. More than one fifth of all programs broadcast by the average PTV station fell into this category.[52]

Information and how-to-do-it programs were once the basic staple of ETV programing, comprising a large percentage of the program hours in the late 1950's. The cooking and sewing programs of the 1970's are little changed from the formats used in the earliest days of noncommercial broadcasting, but the information program frequently has gone from the "talking face" to a documentary/travelogue format; and when a lecturer is used—as in *The Ascent of Man*—

[50]Ralph Garry, Fred B. Rainsberry, and Charles Winick (eds.), *For the Young Viewer* (New York: McGraw-Hill Book Company, 1962), pp. 156–172.
[51]Warren F. Seibert, "Broadcasting and Education," *Educational Broadcasting Review,* June 1972, p. 143.
[52]Katzman, *Public Television Program Content,* p. 125.

we see the lecturer only occasionally, and then he or she is more likely to be sitting on a rock in the middle of a battlefield rather than behind a podium in the studio.

There is little doubt that the general information program, particularly histories and biographies, will remain popular on public television. The interest in our heritage created by the Bicentennial, the increased technical capabilities to take us outside the studio, and the numerous opportunities to offer college and high school credit for viewing these types of programs, insure their continuing popularity.

Increasingly, how-to-do-it programs possibly will *not* be carried on broadcast PTV, but rather will be offered on video discs through bookstores and public libraries. The video disc allows the viewer to select exactly the program needed at the time it is needed and has the added advantages of being able to repeat portions that are not immediately understood (Section 15.1). The programs that are offered by PTV stations in the future will seldom be supplied on the regular interconnection service, since they are not "time-bound." As John Montgomery, then vice president of programing for PBS, remarked in a memo to PTV station managers in 1975, "Perhaps the

'How to Do It' series are more appropriate for library distribution . . ."[53]

These, then, are some of the more crucial issues and programing questions that face public broadcasting as it prepares for the 1980's: How much public affairs programing? And how shall it be financed? What kinds of dramatic presentations? What kinds of cultural productions are needed other than performances? What kinds of constructive children's programing can be designed and made appealing to large numbers of youngsters? What kinds of "educational" materials need to be emphasized? What will be the mix between broadcast PTV and nonbroadcast distribution technologies for various informational programing? This chapter has attempted to summarize some of the issues and challenges facing public broadcasting—in the all-important area of programing—as we leave the field of public television and radio to turn to the world of instructional telecommunications in Part Three.

---

[53]Memorandum from John Montgomery, PBS vice president, programing, to "All Station Managers and All Program Directors," dated August 1975, Re: SPC-3, pp. 2–3.

three

**ITC Perspectives
and Organizational
Approaches**

10

Part Three of this book takes a theoretical look at the various roles of Instructional Telecommunications—in schooling and industrial applications. We are concerned with narrowing the total field of educational telecommunications down to specific uses of television, audio materials, and related media that are applied for particular *instructional purposes* —teaching/learning situations, connected with identifiable courses of instruction—whether for credit or not, in the classroom, at home, in the training center, or anywhere else.

This chapter introduces several theoretical frameworks for organizing ITC considerations, various perspectives on the total scope of instructional applications of "telecommunications used for educational purposes."

10.1   Some Basic ITC Considerations

Several basic concepts were introduced in the first chapter—concepts that should shape the thinking of any practitioner involved in instructional telecommunications. If you were an ITV administrator—placed in charge of a fairly complex ITC program—how would these theoretical considerations affect your organizational thinking, the way you would structure your program?

A Means of Communication   It must be stressed that any telecommunications medium, specifically television, is basically just a *medium* — a means of communication, a channel for sending information from one spot to another (Section 1.4). We can shape it and use it for many different communication purposes. We are not limited to a preconceived notion of what a particular medium can and cannot, should and should not, be used for. If it can be useful to accomplish a certain instructional/communication purpose, use it. If it is inappropriate for some other job, do not use it (even though a school system down the street may be using it for that exact job). Although it may be appropriate for a particular need, there may be another, less expensive, tool to do the same communications job. Use the most efficient means—not the one that is newest or most glamorous.

This perspective gives rise to our extended analogy of print and television (Section 1.6). Just as there are countless individual tasks that "instructional print" can help us with, so are there numerous applications for "instructional television." This comparison is explored further in Section 10.3. Television—as a carrier of signals—is not in itself "educational," any more than a printing press is inherently "educational." Both media do, of course, have tremendous potential as channels for educational messages, as means to accomplish certain instructional tasks.

Integration of Media As a corollary to the above, it must be recognized that television, or audio tapes, or filmstrips, or any other medium, cannot be viewed in isolation. Television must be viewed as part of the total media/method picture in any learning situation (Section 1.7). Telecommunications media must be integrated with other nonelectronic learning aids (including print materials), personnel, grouping techniques, and resources into a total "systems" approach to a given learning problem.

Too often, school or business training directors view instructional TV as a "project." They confuse the presence of the medium with the purpose of a specific learning task. "Television is something that must be used for a given project." We then can refer to "our ITV project." True, television can be an important part of a particular learning project; but the medium itself is not "A project."

With this kind of overview we can plan in terms of a "functions" organization—a management system that incorporates all media into an integrated approach that does not discriminate among various media in its organizational table. Rather, the system would be organized into functions such as the following: instructional planning, acquisition of materials, local production, storage

and retrieval and cataloging operations, distribution alternatives, utilization, and evaluation and feedback (Section 13.1).

Media as Part of the Curriculum Another consideration—specifically in formal schooling situations—is that media, television particularly, must also be viewed as a proper part of the curriculum. Television is a subject area. From this perspective, it *is* more than "just a means of communication." What do schools need to do to prepare students to be intelligent consumers of television fare, advertising messages, political propaganda, violence, and fantasy?

Much more could be done with formal courses in mass media, television production, visual arts, propaganda and persuasion, criticism and analysis, and the like. Also, undoubtedly, substantial educational benefits could be derived from extracurricular activities and projects: production clubs, studio crews, and other broadcasting activities. Section 10.7 is concerned with these considerations.

Television as a Facilitator of Change The new media must also be seen as change agents—elements that, because of their very existence, tend to facilitate other changes. For instance, cannot ITC actually enable schooling administrators to aspire to new curricular objectives in other subject areas? What can we now do—using the new media—that we were not even able to conceive of before the media were developed? What new training programs might an industry inaugurate using video cassettes or discs? What fresh directions might be possible in the arts or humanities making use of new telecommunications technological techniques?

Further, how might television and related media be utilized in order to accelerate desirable changes in traditional systems? Some changes in our conventional schooling systems are inevitable: changing roles of teachers, different grading patterns, more flexibility as to the times and places of formal education, and the like. How will media help to accelerate some of these changes? These are some of the areas that we will look at in Chapter 12.

10.2   Traditional Organization of ITV

In an attempt to organize an approach toward instructional television—or School TV— there are several varied paths. This section is a look at one means of categorizing the uses of television and related electronic media in a formal schooling context as well as for informal educational viewing. Television can be considered within at least four different educational settings: *direct instruction, informal education, in-service teacher education,* and *administrative purposes.* This section merely outlines some considerations in each of these areas.

Direct Instruction   The most obvious use of school ITC is that of beaming instructional materials directly to students—usually in the classroom setting. There have traditionally been three or four categories of this direct instructional use.[1]

Total Teaching   This relatively little-used category refers to situations in which television is utilized for the total instructional job. No classroom teacher assumes any instructional responsibility for the actual presenting of material. A teacher or monitor or aide of some sort is usually present, especially at the lower-grade levels, for obvious administrative and supervisory purposes. This use seldom is appropriate for elementary or even secondary uses. It has some applications at the college level, and its use at this postsecondary level appears to be expanding. It is most often employed within business, industrial, or military training programs—in which the learners are mature and self-motivated, with awareness that their advancement (and possibly their very lives) could depend upon learning the material.

Major Resource   This category refers to the use of school TV in a scheduled, routine manner, with instructional TV sharing the instructional load with the classroom instructor. There is heavy reliance upon the structure of a televised ITV series, for instance, but the teacher still is in control of the total learning situation. In this category, TV could be compared with the textbook, which has traditionally been called the major resource in the classroom. This use of STV is also sometimes called "team teaching television" (referring to the team of the on-camera teacher and the classroom teacher).

Supplemental Viewing   Occasional or irregular viewing of a few ITV telelessons during a course falls into this category. The television portion of the course certainly is not integral to the total curriculum. Supplemental viewing may include *semiregular* (maybe weekly) viewing of a given series of programs—or it may include only occasional viewing, on an *enrichment* basis, of a few programs throughout the course. One could make a comparison here with occasional use of traditional audio-visual films. The term "enrichment viewing" is sometimes used synonymously with "supplemental viewing"—although "enrichment" implies even more of an occasional, irregular use, less directly tied in with the curriculum.

As we examine other means of thinking about STV, we will find that these traditional categories of direct instruction may be useful for reference purposes.

---

[1]These classifications, usually attributed to I. Keith Tyler (of JCET fame, Section 3.4), are not the only ones in use. Authors have outlined many variations in this traditional categorization. For example, Philip Lewis *(Educational Television Guidebook* [New York: McGraw-Hill Book Company, 1961]) uses the categories of direct (total) instruction, supplementary teaching by the TV teacher, enrichment, and other teacher or support services (p. 12). Martha Gable ("Television in the Secondary School" *in* Robert M. Diamond (ed.), *A Guide to Instructional Television* [New York: McGraw-Hill Book Company, 1964]) listed the categories of total teaching, direct teaching (major resource), supplementary teaching aid, and enrichment (pp. 126–128). Several other patterns exist.

Informal Education   In this area we might include the entire realm of *public television*—the general, nonsystematized, evening programing of virtually every noncommercial station—as well as *commercial* programing of educational merit.

This encompasses cultural programs, public affairs presentations, and out-of-school children's programing. The scope of this category includes everything from the best of PBS to the simplest local panel discussion—from a series of lectures on local history to a national exposé of consumer fraud. Some may be presented within a structured series format; others are "one shots." Some may be designed to be viewed in discussion groups; most will not.

Within a schooling context, many of these evening programs can be assigned for homework viewing. An astute classroom teacher can frequently capitalize upon an occasional outstanding "special" from the previous evening—even though it was not assigned as required viewing.

Evening television also presents many opportunities for more systematized adult instruction (Section 8.6). Courses ranging from reading *(Operation Alphabet)* to music *(Folk Guitar)* to cooking *(The French Chef)* may be available. A complete high school curriculum *(TV High School)* makes it possible to obtain the equivalent of a high school diploma by watching television at home. With the *Chicago TV College* (Section 3.9), we have come almost full cycle; the line between public television and School TV becomes somewhat hazy. In Chapter 12, we will look at other considerations that muddy the distinction between formal and informal education even further.

In-Service Teacher Education   The use of television for in-service training and education of teachers could be discussed in two basic categories. One large area, of course, is the production of *series-oriented formal "lessons."* These may be produced on a local, regional, or national level. They may consist of a college-credit course of dozens of telecasts. They may be broadcast

in the early morning hours, during school, or after school. They may be delivered by means of a closed-circuit system at the convenience of a group of teachers viewing a certain series. Broadcast in-service programs possibly may be recorded off the air and then played back in more suitable configurations such as after-school seminars, concentrated workshops, and informal evening viewing sessions.[2]

Another whole category of in-service use of television includes various *"microteaching" applications* and related approaches—using television as a mirror for self-confrontation. The concept of recording a classroom teacher's presentation of a lesson and then playing back the recording for a critique has many possibilities. A short microlesson may be recorded, or a lengthy segment of the teacher's classroom activity may be recorded. Many configurations are possible. This approach is shown in illustration 10-1. Many other in-service applications could be detailed. In essence, the distinction between the two categories of in-service uses of the media discussed above is similar to the distinction between "produced ITC" and "audio-visual" uses discussed below (Section 10.5).

Outside of the schooling context, the medium can also be used as an in-service tool for many other business and professional groups: medical, law-enforcement, sales, financial, industrial, and the like.

Administrative Purposes   Finally, the use of television as a practical educational tool must include various administrative functions of the medium. The television system can be used for many different kinds of administrative communications—intraschool, intradistrict, and interdistrict.

Television is an observation tool. Many special areas can be monitored by television cameras: playgrounds, study halls, cafeterias,

---

[2]For a detailed case study see William Hansen, "Television, Kinescopes, and In-Service Education," *in* Diamond, *Guide to Instructional Television*, pp. 141–148.

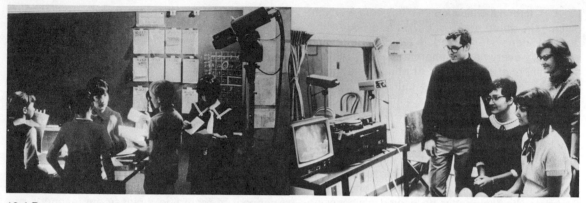

10-1 Representative Microteaching Application: Recording the Classroom Teacher (left) and Teacher Self-Evaluation (right) (Courtesy: University of California, Santa Barbara, and San Jose State University)

corridors, and so on. This conceivably could free teachers from these menial assignments. The medium could facilitate observation of classroom situations by student teachers.[3]

The public information uses of television cannot be overlooked: programs ranging from filmed documentaries over open-circuit stations (commercial or noncommercial) to simple demonstrations recorded for a PTA meeting; from blatantly obvious public relations attempts to orientation programs introducing parents to new school projects.

Schools may establish their own "library" uses of the television medium—limited interschool exchange of STV programing to meet specific needs. Recordings and archives of memorable school activities or outstanding guests could be preserved. Perhaps the distribution of popular audio-visual films and other materials using a school's closed-circuit TV system (instead of traditional distribution arrangements) could be included in this category.

Television can be used to prepare instructions and simple demonstrations for substi-

tute teachers: (a) when authorized absences can be anticipated, specific substitute lessons can be prepared in advance; (b) substitute programs can be recorded and held on a standby basis to be used when unexpected absences occur.

Still other possible administrative uses fall into that broad area of research and development: administering standardized quizzes and examinations; obtaining immediate feedback (through a multiple-channel response system) to test items; validation of new curricular programs; and the initiation of pilot programs on a controlled basis.[4]

As with the other traditional categories outlined above (direct instruction, informal education, and in-service training), many parallel uses of television for administrative purposes in business and industry are also suggested: communication between company headquarters and regional offices, demonstrations of new products and services, advanced supervisory practices, and the like.

The list could be continued; it could be organized along any number of approaches. The

[3]*See* John M. Hofstrand, "Television and Classroom Observation," *in* Diamond, *Guide to Instructional Television*, pp. 149–157.

[4]*See* John C. Woodward, "Standardized Testing by Television," *in* Diamond, *Guide to Instructional Television*, pp. 135–140.

four traditional categories discussed above serve, simply, to exemplify one practical framework of considering ETC as an instructional/schooling tool. School TV is not a single project, approach, production, or delivery system. It is a flexible, unique, ubiquitous, immediate means of communication that can be used for virtually unlimited practical applications.

10.3   Print-Inspired Approaches to ETC

There are many different ways of gaining a perspective on instructional telecommunications. One approach would be to examine instructional ETC from a print perspective. This is a somewhat tongue-in-cheek extension of our print analogy introduced in Section 1.6, but using print terminology to describe various functions or applications of electronic media.

Televised Textbook   One obvious application would be a television counterpart to the standard printed textbook. In essence, this would involve the construction of a major series of telelessons—each one probably twenty to forty-five minutes in length—which would form the core of instructional materials for a particular class. This would apply to the "major resource" and "total teaching" categories described above.

Programed TV Workbook   Just as a printed programed workbook can be designed to facilitate mastery of specified objectives, so can television be designed according to the same principles of programed instruction. This would involve extensive testing and revision of small units of information, but this careful validation process certainly can be (and has been) adapted to visual media as well as to print.

Televised Pamphlet   Many materials lend themselves to brief print treatment in an inexpensive pamphlet format: state guidelines, curricular ideas, new programs, short subjects, summaries,

limited treatises, and so forth. Similarly, many ITC materials can be produced in a like manner—with inexpensive equipment, designed for limited circulation, perhaps only for short-term use. In fact, if we can compare the structured ITV series ("televised textbook") with a textbook printed on a high-speed quality printing press, perhaps we can also compare the "televised pamphlet" with materials run off on a school spirit duplicator or mimeograph machine.

Memo-by-Tube   As this term implies, there are many administrative applications for television and related media (see above). Audio transmission, slow-scan television, facsimile reproduction, advanced applications of the telephone, all are instances of various telecommunications media being used for specific management and information-dissemination purposes: announcements by a company president, audiocassette briefings, instructions to regional offices, and the like. In fact, more and more large corporations are relying heavily upon video-display devices and other applications of the "memo-by-tube" to carry out their everyday administrative functions.

Telecopier   In the early 1970's, inexpensive photocopying technology began to revolutionize many aspects of the way we communicate and dispense printed information. Similarly, the relatively inexpensive slant-track video recorder introduced in the late 1960's has had a comparable impact upon the copying and dissemination of video information (with many parallel copyright and clearance problems).

Videocyclopedia   The encyclopedia and related reference works have certainly become an integral part of traditional educational print resources. In a like manner—and with the recent advances in cassette, disc, and electronic retrieval technologies—video information also is rapidly being adapted for reference and referral applications. It is becoming increasingly easy for a teacher to tell a student to look up such-and-such a work in the video library or to check out a specific 10-minute program and report on it to the class. It certainly will soon be commonplace to have video reference files that are parallel resources to the printed encyclopedia of today.

Video Chalkboard   The chalkboard cannot be overlooked as a relatively inexpensive, readily accessible print resource available, in one form or another, to teachers all over the globe. It enables instructors to spontaneously manipulate written verbal symbols and present them to a sizable number of students at one time. Similarly, inexpensive camera equipment can be used for a parallel purpose in a single classroom for such things as image magnification, close-up views, and detailed diagrams—allowing instructors to manipulate visual symbols and realia spontaneously, presenting them to a large number of students simultaneously. The overhead camera and the permanently mounted microscope camera are but two examples of how a "video chalkboard" can be applied in the lecture hall.

TV Looking Glass   This category incorporates uses of TV apparatus for self-confrontation and self-evaluation purposes such as in-service "microteaching" applications (discussed above) and analysis of various student performances. The camera and recorder in the back of the classroom or on the athletic field have often been referred to as a mirror for self-confrontation. Possibly, other applications of television equipment for observation/monitoring uses could also be considered (as a one-way mirror) to be part of this category.

Magazine-of-the-air   Certainly one of the largest sectors of the total print industry is the magazine. Consider just the "information" magazine (or newspaper)—printed on a regular basis, distributed to our homes or places of business—bringing current news of general interest or specialized professional information to our attention.

Are there not parallel "information" applications of the telecommunications media—daily news broadcasts, documentaries, special in-depth reports, instantaneous live coverage of important events—which have (or should have) important functions in many facets of our educational programs?

Home Study TV Course   Correspondence courses and home-study programs have long played a specialized role in our educational picture. Similar telecommunications potentials are being realized in a variety of parallel operations, such as the Chicago TV College (Section 3.9) and Nebraska's S-U-N program (Section 5.5). In virtually every major metropolitan community, it is now possible to take a limited amount of credit course work by enrolling in early-morning college courses offered over television (either commercial or noncommercial stations)—inspired by *Continental Classroom* and *Sunrise Semester* and, more recently, by Britain's Open University.

TV Literature and Composition   In every literate culture, man has included the study of his own literature as part of his curriculum. Written composition, starting with basic literacy, has always been at the core of the three R's. Should not the study of television literature and composition/production skills be included in today's curriculum as a vital counterpart to written literature and composition? This is explored below (Section 10.7).

These eleven "print-inspired" approaches to looking at ETC should underscore the necessity of trying to examine the realm of "telecommunications used for educational purposes" from as wide a perspective as possible.

10.4   Fifteen Components of ITC

Just as we drew up a list of the "twelve components of a public broadcasting station" (Section 5.1), so might it be helpful to pull together a similar list for instructional telecommunications.

At an international conference in 1966, Jack McBride presented a paper listing twenty elements of ITV.[5] Borrowing freely from his ideas and consolidating a few other concepts, the following compilation of fifteen components is presented for examination.

To some extent, each of these fifteen components is present in virtually every educational situation that utilizes instructional telecommunications. The discussion below is presented in terms of a formal schooling situation; but, of course, the same components are necessary in any business training program. These elements may be present in varying degrees; certain projects may stress one item more than another; but, in one way or another, they are vital considerations in any operation. In fact, by substituting some terminology, these concepts are the components necessary for *any* instructional situation—whether telecommunications media are involved or not.

1. Concept of the Media: First of all, one must start with a basic understanding of the philosophy of telecommunications media—what they are, how they can be used, when they should not be used, and so forth. Many of the items discussed in Chapter 1, and covered quickly above in Section 10.1, must be considered at this point. Television is basically just a means of communication. Electronic media can be compared to print. All media should be considered in an integrated program. School TV is not a project. Media should also be viewed as proper subjects for the curriculum. ITC will also serve as important change agents. Et cetera.

2. Educational Needs: The starting point for any educational program must be rooted, theoretically, in some demonstrated need. One does not initiate a solution without having a clear picture of the problem to which the solution is addressing itself. One must determine goals and objectives before looking for the specific tools that will be used for achieving those objectives.

Several elements must be considered at this point. What is the total educational situation—including the broad goals of education, the specific community and societal needs, the finances of the local schooling system? What are the formal (vocational, general, and college preparatory) and informal (social and personal) needs of the students? What are the specific characteristics of the students in this learning situation (age, economic background, academic abilities, and the like)? What is the status of the present schooling situation—plant and facilities, special programs, community involvement, strengths and weaknesses of the teachers? (Section 13.5.)

3. Administrative Support: Regardless of the importance of an instructional program or the validity of a specific tool used in the program, the project will not succeed unless it has the support of the school administration. Regardless of how convincingly research indicates that the media work, the facts are meaningless unless the decision makers think it is a good idea to use the media.

There are essentially three different levels of necessary administrative support. First, there is the top-level *policy support*. Usually this is at the board level (board of education, trustees, regents). Second, there must be support at the *top administrative level*—the chief executive (superintendent, president, principal) and his immediate subordinates. Third, the project must have support from the local or *immediate management* level (department chairman, dean, curriculum supervisor). (Section 13.2.)

4. Curriculum Planning and Educational Support: Plans must be outlined for the specific uses of various media. Exactly what jobs can the media do? How can they best be utilized? First of all, a careful review of the goals and needs is

---

[5]Jack McBride, "The Twenty Elements of Instructional Television: A Summary," paper presented at the Seminar on Educational Television Sponsored by the Centre for Educational Television Overseas, Cairo, Egypt, February 6–12, 1966.

necessary. There must be careful consideration of the broad functions of education—those things we can expect a schooling system to be able to do (imparting information, shaping attitudes, instructing in skills, stimulating creativity and analytical thinking, and the like).

We must think in terms of the specific advantages and limitations of the media. The nature of the learning situation must be considered: large-group instruction, small group patterns, individual instruction. Specific curriculum planning procedures must be initiated with the team members who will be involved: curriculum administrators, consultants, teacher committees, the TV teacher or content authorities, and the media production staff. It is of the utmost importance that any STV project have the full support of all faculty members who are in any way connected with the project. Many of these considerations are discussed further in Chapter 13.

5. Acquisition of Materials: Once the needs have been identified and instructional programs have been planned, the next step is to determine what materials may already exist that could meet those needs and plans. If there is one general rule-of-thumb that can be applied to the obtaining of materials, it is this: It is cheaper to rent or purchase copies of existing materials than it is to produce your own! So, the obvious first step should be a thorough search of national and regional libraries, local stations or projects, and other sources of existing STV materials to see what might already be available that could fit in with your plans (Section 14.2).

6. Producing Personnel: Of course, the chances are that much of what is needed will not be found in a ready-to-use format. Some amount of local original ITC production will be required. This sixth component gets into the first of four considerations concerning production of instructional telelessons. All of these factors are considered further in Chapter 14.

The first necessary production item is the personnel to put the programs or materials together. There obviously must be an overall content authority, a producer, someone with the final decision-making responsibility. It is often assumed that there will be a television teacher, an on-camera instructor to do the presenting. However, this assumption should be carefully and conscientiously examined. Must there always be a live body talking directly to the camera? Also, there must be arrangements for the studio producing personnel: the TV director, the associates, the floor crew, the engineers, the supporting artists and craftsmen. Which ones are to be a permanent part of the STV staff? Which positions can be filled on a temporary or contract basis? Are there any positions that can be filled with students?

7. Television Lesson: An obvious component is the telelesson itself. Many questions arise as to the nature and format of this element. What is the nature of the purpose of the telelesson? How will the programs be designed? How long will each telelesson be? How many units are needed for the series? What will be the format of each? Illustrated lecture? Filmed field trip? Interview/panel discussion? Perhaps the question of validation should be raised at this point. Will the lesson be tried out, tested, validated as to its reliability and ability to achieve the stated purposes?

8. Ancillary Materials: In addition to the telelesson itself, there must almost always be ancillary, or accompanying, materials produced to go along with the lesson: classroom teachers' guide, student handouts and worksheets, charts and diagrams, other printed items, audio materials, testing materials, and whatever else might be appropriate to complete the learning situation. How will these materials be produced? Does the school have an adequate print shop? What materials will have to be contracted out? What kinds of lead time are necessary?

9. Production/Origination Equipment: This final production component is the obvious category of the hardware. Depending upon the nature of the STV operation, what equipment is necessary? A single inexpensive vidicon monochrome camera or an elaborate multicamera switching installation? A desk in a corner of a classroom, or an extensive studio facility? How much can be rented? Depending upon the specific needs of the learning situation and the extent of the television contribution to the entire learning system, the total price for this component of ITC could run anywhere from a couple of thousand dollars to a multi-million-dollar investment.

10. *Distribution System:* Once the television presentation is recorded (or produced live for immediate dissemination), it must be delivered to the learners who will be using it. Basically, there are three broad overall categories of distribution. The first is *over-the-air* open-circuit systems—VHF-UHF stations, microwave, instructional TV fixed service (ITFS), satellites, and so forth. This first category is often divided, pragmatically, into two parts: station broadcasting and limited "closed" open-circuit systems (ITFS, microwave, and so forth). The second broad category would be closed-circuit or *through-the-wire* distribution—intraschool closed-circuit TV, statewide closed-circuit projects, cable TV, remote-access systems. Finally, we have the category of direct *physical delivery*—hand-carried videotapes, cassettes, or discs, mail delivery, audiovisual carts and trucks, library retrieval, and the like.

Criteria must be considered for determining which delivery system is most appropriate for a given program. The distribution of ancillary materials—guides and program schedules— must also be worked out. These elements are considered in Chapter 15.

11. *Reception Facilities:* Once the televised materials are delivered to the proper locations, they must be received and displayed for the learners. There are several aspects of this all-important component. First, the TV receivers or monitors must be considered. Will a standard

receiver be sufficient, or should high-quality video monitors be used? Will there be separate audio speakers? What special safety considerations should there be? How about special convenience features? Will the receivers be permanently mounted (ceiling, wall, floor) or placed on portable stands? What are the specifications for the movable stands?

Another consideration involves maintenance. Does the school system have engineering personnel trained to handle maintenance? Should you have a maintenance contract, or service the receivers on a per-call basis? These considerations—plus the next three components—are discussed in Chapter 16.

12. *Viewing Arrangements:* The less tangible counterpart to the reception hardware would be the actual viewing arrangements themselves. There must be coordination of four physical elements: the television receivers, a suitable location or room for viewing, the classroom teacher, and the learners themselves. Getting all of these elements together at the right time is an obvious necessity; yet in many school situations it involves quite a bit of logistical juggling.

There must also be consideration of the personnel involved—coordination of various teaching positions, perhaps a team teaching effort, resource persons, supervisory personnel who might be involved. The actual physical arrangement of the viewing room must be considered. Will a regular classroom be used? Or perhaps a special AV viewing room? Where should the receiver be placed? What are proper viewing angles?

13. *Classroom Teacher:* At this point, the classroom teacher or receiving instructor certainly must be considered. Often underplayed by administrators—and sometimes by classroom teachers themselves—the crucial functions of the classroom teacher must be stressed. The teacher must be prepared for the media utilization functions—preservice instruction in utilizing ITC should be included in teacher education

curriculums, and in-service training during the ITC project should be conducted by the school or STV system.

Several crucial roles of the teacher must be performed before, during, and after the actual media presentation. The teacher must be both an active participant in the telelesson as well as an observer and evaluator. (Several aspects of the changing roles of the classroom teacher will be included in Sections 12.6 and 16.4.)

14. *Utilization:* This component is concerned with the concrete learning period itself—the actual interaction of the learners and the televised materials. Included in this period is all of the preparation that precedes as well as the follow-up that follows the STV portion of the lesson. Conventionally, this utilization period is broken down into three periods—herein referred to as the "Three R's of Utilization"—the preparation or "readiness" period, the actual viewing of the telelesson or "reception" period, and the follow-up or "reinforcement" period.

15. *Feedback, Evaluation, and Research:* The final chapter of the book is concerned with this most important component—finding out whether or not, and to what extent, the desired learning has taken place as a result of the ITC program. This can be attempted on several different levels. The simplest level is direct and *informal feedback*—random information from the learners and classroom teachers. This feedback may be in the form of program evaluation forms and sheets, anecdotal reporting, and/or classroom visitations. A second level of evaluation might be more formal *project evaluation.* This could include a more systematized collection of data regarding learning achievement, test scores, and the like. It might include special consultants or objective observers and analysts. It could certainly include cost-effectiveness considerations to see if the results were worth the dollars spent. A third aspect of evaluation could be *program validation*—the application of the principles of programed instruction, involving exten-

sive testing and revision, in order to be able to "guarantee" the results of a specific learning program. Finally, this component should also include theoretical *pure research* undertakings—special projects set up to test specific approaches or uses of the media. This category would include specially funded programs, more carefully designed controlled learning situations (laboratory conditions), more involved statistical treatment of data, and the like.

These fifteen components, then, comprise the essence of any educational undertaking involving instructional telecommunications. They are not listed in any special priority ranking; they are presented for consideration as guidelines or criteria for any ITC project. By evaluating any instructional project in terms of these fifteen components, one can obtain a good idea as to the comparative structure of the project. As should be apparent, these fifteen components generally form the outline for Part Four of this book.

10.5   One Basic Distinction

If there has been any single theme of this chapter up to this point, it has been simply this: instructional telecommunications must be viewed from as broad a perspective as possible. There are no convenient divisions and pre-designated pigeonholes into which one can easily stuff various categories of STV projects. In fact, school ETC is *not* a project. It is simply a means of communication.

However, one basic distinction should be made at this point. Two somewhat separate categories can usually be applied to most uses of ITC. (Even here, we qualify the distinction somewhat; there undoubtedly are uses of television that do not fall easily into one of these divisions or the other.) In general, most applications of electronic media for schooling purposes can be seen as falling into one of these two categories.

The first category would be *produced*

ITC, referring to *those STV applications that involve the use of software materials produced, packaged, and distributed with identifiable titles and program lengths.* The other category would be *audio-visual/hardware ITC,* including *those applications of telecommunications hardware that involve no identifiable programing materials— where the tools of the medium are used as an extension of the teacher's immediate presentation.*

The first category, produced ITC, roughly incorporates our traditional classification of total teaching/major resource/supplemental applications. Indeed, it may be convenient to think in terms of two separate parts of this category: *curriculum-structuring ITC* and *enrichment/ reference ITC.* Both are "produced ITC" in that they consist of definite software units, but the former is firmly tied in with the curriculum of a subject area, while the latter is more of an appendage—nice, but not indispensable.

Produced ITC: Curriculum-Structuring
The first subdivision is concerned with uses of television that have a substantial impact on the curriculum and structure of a given course. It typically is identified with a series of programs (anywhere from five or ten up to sixty or more individual telelessons), which progress in an orderly sequential fashion through a given content unit or entire course. The individual lessons or programs are designed to be viewed on a regular basis (from one per day to one per week); therefore, they more or less dictate the structure and content of a course—hence the term "curriculum-structuring."

Returning to our analogy to print, the obvious corollary would be the textbook. Once a teacher has decided upon the use of a given text for a course, the direction and limits of the course have been fairly well determined. The teacher obviously can adapt the text, use portions of it, rearrange the chapters, skip parts of it, and supplement it with other material—but the text is instrumental in partially shaping the structure of that course.

The same is true when a teacher elects to use a given ITV series for a class; the TV lessons dictate the direction and substance of a good portion of that course. Again, the ITV series may be adapted, supplemented, possibly rearranged, and partially edited, but still it determines—to some extent—the structure of the course. The telelessons, presented in sequence at regular intervals, form the outline for the course—dictating the basic content and pacing. It is curriculum-structuring.

Produced ITC: Enrichment/Reference
The other subdivision of produced ITC involves non-curriculum-structuring applications, in which the medium is used to help illustrate a certain point, to enrich a concept, to highlight one portion of the lesson, to serve as reference for a point, or whatever. These occasional/enrichment/ reference uses usually are under the control of the classroom teacher (as far as scheduling and decision to use the materials are concerned) and often are locally produced.

Examples of enrichment/reference non-curriculum-structuring produced ITC would include preparing and recording a chemical demonstration that might be dangerous or expensive to attempt in repeated class sessions, recording a field trip, checking out a video cassette from the library to illustrate a point, using a newscast from the previous evening in a government class, leasing a single program from a national ITV library to reinforce one particular concept, recording a special guest lecturer, and similar applications. In fact, any use of produced ITC software that does not dictate what the overall content or structure of a course will be falls into this category.

The best comparable print analogy might be, say, a particular pamphlet or supplemental reading assignment that helps to reinforce or give details about a particular item in a course. The extra reading assignment does not dictate the outline or structure of the course; it merely helps to fill it out. It is an adjunct, an appendage. It is enrichment/reference material.

10-2 Two Representative Overhead Camera Installations: Dual Overhead Magnifier Cameras for Biology (left) and a Mini-Studio Installation to Facilitate Prerecording of AV Demonstrations (right) (Courtesy: University of California, Santa Barbara, and San Jose State University)

Audio-visual/Hardware ITC   In this major category of *nonproduced ITC,* there is no content as such. Television becomes pure hardware without any produced software. Since there is no content, it should be considered an instrument employed to extend some other medium of communication. These uses of television are often called audio-visual uses, since they are similar to using an opaque projector or an overhead projector or some other audio-visual aid in class presentations.

Specific examples of this last category of ITC would include the following: using a mounted (overhead) camera to enlarge drawings in a book or show the dissection of a frog for large-group viewing (illustration 10-2); recording a student's performance in a speech class and playing it back for self-analysis; connecting a TV camera to a microscope for easier viewing by a biology section (illustration 10-3); or connecting three rooms together with a closed-circuit hook-up to hear a guest lecturer. The common element in these uses of television is that *there is no separate* *software—there is no produced TV presentation as such*.

Of course, the recorded speech or the dissection of the frog may do such a good job of illustrating some concept that the recording is

10-3 Students Viewing a Laboratory Demonstration (Courtesy: San Jose State University)

kept; it becomes "produced" software, which can be titled, cataloged, stored, and used in the future. The videotaped student speech or frog dissection would then be used as enrichment/reference produced ITC (or might possibly be eventually incorporated within a series of programs that would be curriculum-structuring produced ITC).

Comparing this category of audio-visual/hardware ITC with print, we have such familiar items as the chalkboard, the overhead projector, and even scratch paper—none of which have any content, or software, until they are utilized for a particular communications job.

As mentioned above, these categories are not entirely exclusive; some STV uses can overlap and have dual applications. For instance, a history teacher may take half a dozen major historical programs (each of which was developed as a separate enrichment/reference telelesson) and structure a course around them; another teacher may use a short series of four or five programs (normally a curriculum-structuring resource) as an enrichment to his traditional curriculum without actually letting the additional programs dictate any change in course structure, content, or pacing; a biology teacher may have an overhead camera installed (normally an AV tool) and then substantially modify the structure of the course in order to take advantage of the resulting new opportunities. However, generally, it will be helpful to think in terms of these two basic categories—produced ITC and AV/hardware ITC—as various possible applications of school ETC are explored.

10.6   Comparison of Various ITC Models and Perspectives

So far in this book, the distinction between produced ITC and audio-visual/hardware applications has been implied several times. In explaining figure 1-1, the "ETV Circle of Instruction/Appreciation," the audio-visual uses were first introduced in the inner circle. The next three

rings are curriculum-structuring applications of produced ITC, and the last three circles are enrichment/reference examples of produced ITC.

In discussing the eleven print-inspired approaches to ETC (Section 10.3): *Curriculum-structuring produced ITC* would include the televised textbook, the programed TV workbook, and the home study TV course; *enrichment/reference produced ITC* would include the televised pamphlet, the videocyclopedia, and the magazine-of-the-air; and *audio-visual/hardware ITC* would include the memo-by-tube, the telecopier, the video chalkboard, and the TV looking glass. (The eleventh approach, the study of media as a subject area, falls into its own category.)

It is interesting to note that, historically, the development of ITV and the evolution of traditional audio-visual centers have moved from opposite directions, as shown by the table (10-4). The audio-visual film movement clearly evolved around our so-called audio-visual and enrichment uses of films and other media (Section 2.2). Only within the past decade or two have the traditional AV operations begun to move in the direction of developing major curriculum-structuring materials and complete course packages. This movement is probably best typified by the trend of audio-visual centers to become transformed into *instructional design* centers, concerned with a total *instructional systems* approach (Section 13.4).

Conversely, instructional television started out as a curriculum-structuring tool from the beginning. The first clear-cut examples of School TV, dating back to 1947 (the Philadelphia schools using commercial TV stations to transmit instructional series), were using the medium to outline and design new curricular patterns of instruction (Section 2.5). Virtually all of the classic case studies of STV usage throughout the 1950's—Hagerstown, Chicago TV College, *Continental Classroom,* and the like (Section 3.9)—were examples of a curriculum-structuring approach.

10-4 Examples of Various Applications of Media in Three Classifications

Categories of applications	Books/magazines/ educational print	Films and related traditional AV	Instructional television
Produced Media Curriculum-structuring	Textbooks Programed instruction Syllabuses	Instructional design and "systems" approaches	Course-related, sequential, ITV series
Produced Media Enrichment/reference	Libraries Magazines Reference works Pamphlets	AV films Charts Slides Audio tapes Filmstrips	Single programs Specials Documentaries One-shots
Audio-visual Hardware uses	Chalkboards Scratch pads Notebook paper	Overhead projector P.A. microphone Microscope	Overhead camera Single-room CCTV Microteaching

This curriculum-structuring emphasis in produced ITC was coincidental with much national curriculum revision that was initiated in the 1950's—the "new math" projects such as the School Mathematics Study Groups, new curriculums in the sciences such as the Biological Sciences Curriculum Study and the Physical Science Study Committee, and later programs in languages and social sciences. These projects were developed, in part, in response to the recognized needs of the 1940's and 1950's—a serious teacher shortage, a phenomenal knowledge explosion, and a substantial growth in student enrollment. Curriculum-structuring applications of ITV were started for the same reasons.

In the late 1960's and early 1970's, the teacher shortage became a surplus, and teachers no longer required major assistance with new curriculum projects. At the same time, inexpensive vidicon cameras and slant-track recorders made it possible for schools and teachers to plan simpler single-classroom uses of the medium. Many audio-visual/hardware applications of TV tools were initiated during this period.

10.7   The Subject Is Television: Media in the Curriculum

In this chapter we have attempted to place telecommunications in a broad theoretical framework—to view electronic media from as many perspectives as possible. In another context, we can examine television as a proper area for study within the curriculum—to look at the media as an academic subject.

Many times, when schools get involved with various projects using telecommunications media as tools of education/communication, they find that the students (and teachers) also get involved with the *study* of the tools they are using. The same facilities (cameras and receivers) that are used to produce and deliver telelessons concerned with history and science can also be used to study the media themselves. The school ITV administrator—who is primarily concerned with scheduling TV facilities and delivering various ITC materials to various classrooms—may also be asked to teach a unit on "television" for an English class. Even the 1975 report of the Advisory Council of National Organizations, advising the Corporation for Public Broadcasting as to its role in educational broadcasting, included as one of its recommendations that the CPB should "encourage the development of the skills of aural/visual literacy and critical listening/viewing" (Section 8.6).[6]

Why Study Telecommunications?   A school's decision to offer media studies is based upon some impressive data and hard facts about our involvement with television and radio. Numerous writers have mentioned that children typically have watched 3,000 to 5,000 hours of television before they begin school; that school-age children spend well over one third of their waking hours in front of a TV set; that, by the time he or she graduates from high school, the average student has watched 15,000 hours of television but has spent only 10,800 hours in school; that the average high school graduate has watched more than a quarter of a million TV commercials; that adult viewers spend more of their waking time watching television than any other activity except working. The TV set is obviously a popular piece of furniture: Surveys indicate that more than 96 percent of American homes have at least one set, and that set is typically turned on five, six, or more hours per day. As Theodore W. Hipple has put it:

> Few homes, even those in the nation's pockets of intense poverty, are without television. The two-television set family is as common as the two-car garage or the two-bath home. Numerous evening meals are eaten in the company of the network newscaster. . . . Series performers become instant heroes to the viewing public. The nation's business all but halts as the popu-

lace sits glued in fascination before the gripping spectacle of an astronaut's televised moonwalk. Television has shaped Presidential elections, changed family sleeping habits, altered preschool education, indeed, transformed American society.[7]

We would not think of designing a school curriculum that did not include the study of literature or communication skills or social studies. Does it make any more sense to leave out the study of the grammar and analysis of television and its impact upon society? John Culkin succinctly summed up the argument for training young people to be critical viewers: "To fail to prepare kids to be intelligent consumers of television films and the whole schmeer today is to fail to prepare them to live intelligently in real life in the real world."[8]

Approaches to the Study of Mass Media
The mass media are complex activities/processes, and there are many different ways that telecommunications can be studied in schools. We can approach media as theoretical phenomena, as institutions, or as social forces. We can examine their messages, their literature, their effects, or their methods of creation. Media may be studied as a complete formal course of instruction or as a unit of instruction included within some other subject area. Several of these different approaches are outlined below:

Communications Theory   Theorists have often approached communications as a complete process. A source of a message encodes some thought into symbols (such as words) and then transmits that coded message through some

---

[6]*Public Broadcasting and Education: A Report to the Corporation for Public Broadcasting from the Advisory Council of National Organizations,* (Washington, D. C., March 1975), p. 7.

[7]Theodore W. Hipple, *Teaching English in Secondary Schools* (New York: Macmillan Company, 1973), p. 238.

[8]John M. Culkin, "Unstructuring Education," in *New Relationships in ITV* (Washington, D.C.: Educational Media Council, 1967), p. 27.

medium or channel to receivers. The receivers decode the message and then respond to it in some way, either by feeding back reactions or by taking some action (Section 1.4). Where does television fit into this theoretical model? When should it most appropriately be used? What kinds of messages are suited to different kinds of channels? What are the feedback processes involved in mass media? The student could profit from a detailed study of all aspects of telecommunications media by means of this theoretical perspective.

Visual Literacy   The concept of visual literacy includes the study of media characteristics and the elements of visual messages, which enables an individual to perceive the subtleties in the messages and to understand the effects the visual messages may have in their entirety or in their parts. Visual literacy approaches the study of audio-visual messages in the same way that beginning studies in literature look at written prose and poetry. In more technical language, the National Conference on Visual Literacy has proposed this definition of the field of study:

Visual literacy refers to a group of vision-competencies a human being can develop by seeing and at the same time having and integrating other sensory experiences. . . . Through the creative use of these competencies, he is able to communicate with others. Through the appreciative use of these competencies, he is able to comprehend and enjoy the masterworks of visual communications.[9]

The Grammar of Television   Just as visual literacy attempts to teach how to "read" visual messages, we also can help students analyze the grammar of a medium such as television. Detailed analysis of the structure of television—its grammar, syntax, and rules for complete statements—should be included in a broader study of visual literacy, media literature, or in the study of the practical aspects of television.

Elements of the language or grammar of television include items such as editing (the distinctions between film cutting and TV studio editing), pacing, the use of camera movement, camera transitions, fades and special effects, audio perspective, depth-of-field, subtleties of lighting, the montage, use of multiple images (both superimpositions and inserts), and other production cues that help to transmit the totality of the message to the viewer/reader.

Mass Media and Society   Another approach to a media curriculum would consider the role of telecommunications in twentieth-century America. Students, if only as consumers of television, should understand the effects of mass media upon society and upon the individual. Among the general topics suggested for study are the following: broadcast history and regulations; the way commercial TV programs are selected and produced; the types of programing on commercial and noncommercial TV; the role of advertising in commercial television. Most such study proposals also emphasize the importance of establishing criteria for judging the worth and trustworthiness of a specific program or type of program.

Many course proposals emphasize the study of advertising messages, political programs, editorials, and other persuasive program-

[9]Roger B. Fransecky and Roy Ferguson, "New Ways of Seeing: The Milford Visual Communications Project," *Audiovisual Instruction,* April, May, and June/July, 1973, p. 12. Everyone does not agree on the definition of visual literacy or its importance. Michael L. Lasser ("A Humanist Looks at Visual Literacy," *Audiovisual Instruction,* September 1975, pp. 31 ff.) sees both a danger and a promise, a too-easy way past difficult problems of teaching living, breathing people and a potentially useful way of reaching TV-reared children.

Dean R. Spitzer and Timothy O. McNerny ("Operationally Defining 'Visual Literacy': A Research Challenge," *Audiovisual Instruction,* September 1975, pp. 30–31), striving for an operational quantifiable definition of visual literacy, see film comprehension (the aspect they chose to concentrate on) as composed of film content, film conventions, sound, and measurement instruments for judging the degree of verbal skill necessary for understanding and answering questions.

ing. They stress the importance of recognizing emotional appeals, correct applications of evidence and logical reasoning, and various other approaches and combinations. The effects of televised and filmed violence is also a prime area of concern in this academic approach.

Television Literature Although much "good literature" is presented on television, there is also much of questionable or even negative value paraded before us on the tube. Television also has developed its own literature, much of it evolved from radio and the films: situation comedies, detective/police action; quizzes and game shows, and so forth. The commercial TV Saturday morning cartoon line-up *(Hong Kong Phooey, Speed Buggy, Pebbles and Bamm Bamm,* ad nauseam) certainly represents a category of original children's literature for television. It unquestionably plays a much larger role than Mother Goose or *Grimm's Fairy Tales* in the lives of most youngsters. Yet, what are we doing to prepare children to cope with this kind of literature? What critical viewing patterns are established? Where are TV literary standards being taught?

Additionally, we must also be concerned with the creation of literature that *is* television. No less a literary personage than John Steinbeck made the following observation:

> I run into people who seem to feel that literature is all words and that words should preferably be a little stuffy. Who knows what literature is? The literature of the Cro Magnon is painted on the walls of the caves of Altamira. Who knows but that the literature of the future will be projected on clouds? Our present argument that literature is the written and printed word in poetry, drama, and the novel has no very real basis in fact.[10]

Television as an Art Form Americans seem especially slow to consider visual media

as art. In European countries, film has long been recognized as an art form, and, in countries such as France and England, television is increasingly perceived the same way. Americans have been slower to put film, television, and radio into the same category as the more fashionable intellectual popular arts of opera, symphony, and drama—but the study of the mass media as art forms is slowly gaining respectability.

An even more extreme form of television as an art form is the manipulation of TV sounds and pictures to transmit abstract impressions—producing images for their artistic effect apart from any message. Some of these experimental effects include colorization, solarization, reversed polarity, overbeaming, feedback, multiple and combined images, mirror effects, and a host of other electronic techniques. Some advances in the study of television as an art form have come out of the work of the National Center for Experiments in Television at KQED, San Francisco. Even at the elementary and high school levels—using basic television facilities—similar artistic experimentation can be encouraged; and the students can discuss and analyze television art in terms parallel to those used to critique elements of opera, painting, sculpture, music, dance, and drama (balance, mood, tone, intensity, color, harmony, texture, and so forth).

Program Construction and Production Techniques Knowledge of, and practice in, the ways TV programs are made can add depth to a student's understanding of media. We stress the four skills of language arts—reading, writing, speaking, and listening. But what about camera work? Or shooting and recording? Editing? Microphone techniques? The camera and recorder may be as basic tomorrow as the pencil and paper are today.

By manipulating a camera, a setting, lighting, and characters, the student can get an idea of the illusionary reality that comes across so well on the TV screen. By studying the structure of a TV program—from the production stand-

---

[10]Quoted by Neil Postman, *Television and the Teaching of English* (New York: Appleton-Century-Crofts, 1961), p. 40.

nt—and by being aware of the subtle effects
lighting, sound, editing, and camera angles
and movements, the student can analyze television with some sophistication.

Extracurricular Media Activities   In addition to formal courses or academic work in the
study of mass media, extracurricular television
and radio activities also provide a healthy opportunity for many students to learn how to work positively with the media. These activities can take a
variety of forms: broadcast clubs, work as audio-
visual assistants, after-school informal seminars,
field trips to broadcast stations, operation of a
campus "station," presentations by visiting professionals, and other projects. Some of these
activities may be basically recreational or designed for personal growth or vocational training
or intended as service functions for the school or
for other worthwhile purposes. From such types
of involvement, students also gain insights into
the functioning of the medium in society.

Alternate Media   In addition to schooling
applications and formal study of media, alternate-
media groups are appearing across the country—
exploring nontraditional uses of television, particularly using portable video equipment. These
groups, known also as "guerrilla television" and
"underground video," form an important element
in the total fabric of noncommercial telecommunications. There are at least a hundred of these
operations from coast to coast—such as Rain-
dance and Alternate Media Center (New York
City), April Video Cooperative (Washington), and
Media Access (Menlo Park, California). The emphases of such groups usually fall into two basic
categories: artistic expression and/or community
dialogue.

Alternate television used for community
dialogue affords its practitioners with useful
lessons in pragmatic communication. Such media
activists often become involved in alternate television without much experience in professional
or academic television production, so they essen-

tially teach themselves about the interplay of
television and messages and audiences. When
they meet in organizations devoted to production
or to finding outlets for their messages, their discussions of techniques and problems may lead
them to discoveries about the effects of television.
They form new concepts about the media and
their interaction with society. "Videotape can be
to television what writing is to language. And
television, in turn, has subsumed written language as the globe's dominant communications
medium."[11]

Whether we are concerned with students
in a formal schooling situation or activists in an
alternate media group, whether we are dealing
with extracurricular broadcast activities or academic courses of media study—the important
concept is that young people, as well as older
individuals, begin to look *into* television, as well
as *at* it. In the age of telecommunications, we
should do no less.

One of the most frightening and urgent
possibilities lies in the area of preparing students
for the day when reading and writing may actually
play a much less prominent role than they do
today. Malcolm Crowley raised this question, in
a somewhat unsettling manner, in his essay, "The
Next Fifty Years in American Literature":

A final possibility must be considered that
printed literature in the future will be written
for and read only by scholars. For the public
at large it might give way to picture books,
or to spoken and tape-recorded stories, or
else to dramas and serials composed for
television or the new medium that will come
after it.[12]

---

[11]*Radical Software 1*, Summer 1970, preface.
[12]Quoted by Postman, *Television and the Teaching of English*,
p. 13.

**Scope and
Structure of
School TV**

In examining various approaches to the uses of television specifically for formal schooling purposes, one can profitably consider the several different geographical levels or scope of instructional telecommunications applications. One can examine projects ranging from a very limited, one-person operation to vast programs having world-wide applications. How widespread can a given project be? What different kinds of administrative patterns might be established for programs of different scope?

Of course, many School TV programs incorporate elements of more than one scope or level: a single-school project uses instructional TV materials produced by a national consortium; a district closed-circuit TV system redistributes programs originally transmitted by a statewide network; a regional cooperative venture relies heavily upon administrators at the district level for actual implementation. However, it should be beneficial to become familiar with some of the considerations and administrative patterns common to schooling operations encompassing various geographical levels.

This chapter will examine eight different levels of STV involvement: *single-classroom applications, school-level projects, district administration, metropolitan ITV associations, statewide operations, regional activities, national programs,* and *international developments*. In considering these eight levels, we will look at several case studies (a couple are hypothetical)—some of which may be representative of that particular scope or level and others of which may be unique to one particular project.

11.1   Single-Classroom Applications

At the most basic level are uses of television that are limited to a single classroom. Virtually all users of the medium at this level fall into the categories of audio-visual/hardware or enrichment/reference produced ITV (as opposed to curriculum-structuring applications of ETC).

Some of the most widespread uses at this level are in the medical and dentistry education programs. "For many years, medical education has led in the application of television for training in clinical practices and nursing techniques. In fact, some of the most sophisticated color television facilities may be found in teaching hospitals."[1] The heaviest use has been in this audio-visual category: close-up images shown to a large class, self-evaluation by medical students at various stages, viewing diagnostic techniques while assuring patient privacy, recording surgery procedures, and the like.

Other specialized—and highly visual curriculums—have also made extensive use of television in this category. However, for our purposes, let us examine a hypothetical instance of a possible application at a regular high school.

Tenth-Grade Biology Chuck Jameson is a fictional biology teacher at Central High School, a large general educational high school in a suburban area of a major metropolitan city. Most of the 2,000 students take general biology—either from Chuck or one of his two biology colleagues—in order to fulfill their science requirement. Chuck teaches four sections of general biology and one elective class in physiology.

In 1959, as part of a research project under the National Defense Education Act, Central High School purchased its first television equipment, two inexpensive viewfinder vidicon cameras and four TV receivers. After completion of the research project (comparative methods of teaching Spanish verbs), the television equipment was stored in the English department (because they had more storage space) and was largely forgotten.

In 1966, with the encouragement of a new principal, Chuck obtained some additional funds for lens adapters and was able to perma-

nently attach one of the cameras to a microscope. After some experimentation and expenditure of a little more money for additional cabling and better lighting and optical equipment, he was using the camera-microscope combination quite effectively in all his biology classes.

Chuck was able to show to an entire class many special slides and delicate displays he had painstakingly prepared. Previously he had shared much of this material only with a few interested students on a one-at-a-time basis. As Chuck made more use of the camera-microscope hookup, he discovered that he could cover a lot of material much faster than he had previously been able to; he could see exactly what picture every student was looking at. It saved tremendously on the amount of time that he had to spend checking individual student microscopes.

In 1969, Chuck requested—and eventually was able to obtain—another inexpensive monochrome vidicon camera. With the assistance of the metal-working shop class, he put together a stand so that he could use the camera as a permanent overhead camera. This set-up proved invaluable to show close-ups of various table-top displays, small specimens, charts and pictures, and various dissections.[2] He found that he was making so much use of the two cameras (the microscope and the overhead) during the same class period that he was losing too much instructional time connecting and disconnecting the camera cables to his two classroom receiver/monitors. So, with the help of the electronics class, he was able to build a crude switching device that enabled him to switch instantaneously from one camera to the other (with just a momentary "glitch" or picture breakup).

At this point, Chuck Jameson had less than $1,500 of school funds invested in his two-camera facility. He was easily able to justify this

---

[1]Charles S. Tepfer, "Still Needed—Quality in Medical Programming" (editorial), *Educational and Industrial Television*, November 1972, p. 6.

[2]For a detailed discussion of the possibilities of the overhead camera, *see* Rudy Bretz, "Overhead TV—the Ideal Visual Aid," a nine-page monograph duplicated by Santa Clara County Board of Education (San Jose, California, 1963).

to his principal and to the school board of education. However, his next request was for a portable ½-inch slant-track recorder, which would enable him to prepare certain materials and demonstrations ahead of time. This was another $1,500 item, and it took stronger justification. However, the board finally approved the 1972 purchase of a video recorder—to be shared by the entire science department. As it worked out, Chuck was able to schedule the recorder two days a week in his classroom, so he could arrange all of his prerecorded materials and demonstrations to be shown during those days. Chuck now has a library of about twenty prerecorded demonstrations, special slides, and simple experiments that he has prepared himself.

Chuck's main ITC project right now is the establishment of a self-pacing instructional approach consisting of various individualized learning modules that students would work through at their own rate. He had been inspired by the pioneering work undertaken by Samuel Postlethwait at Purdue University,[3] and he wants to set up a similar program at the high school level. He has already produced some of the self-teaching modules, which consist of audio cassettes and slides as well as television materials (he has received considerable support from the district's audio-visual center in producing these materials). The idea is to package a number of these individualized learning units, which would be independently checked out by students, who would view and listen and do the required lab work for each unit and then take the examination for that particular unit.

Chuck has been developing this "competency-based" approach for his physiology class. He feels that these students are more motivated and could more easily adapt to the self-teaching concept. Of course, this usage of the various related media would take Chuck definitely

into the curriculum-structuring applications of ITC; it is an exciting project that has occupied most of his spare time during the past two years. This year, the board indicated approval of the purchase of a video-cassette recorder in order to facilitate this project (matching funds are available from the state for innovative projects of this nature). Chuck has every hope of being able to start his newly designed, individualized, self-teaching, audio-tutorial physiology class the year after next.

USC Interactive ITFS System   One of the most extensive amplifications of the single-classroom approach—actually it should be labeled an "extended classroom" application—is the interactive instructional TV fixed service (ITFS) installation (Section 4.5) at the University of Southern California. The heart of the system is the main campus ITV Center which has four studio classrooms, each of which is equipped with three remote-controlled cameras—one overhead camera shooting the demonstration-and-scratch-pad area in front of the teacher, one camera at the rear of the room shooting the teacher, and one camera in front of the room to pick up shots of the students (illustration 11-1). The three cameras are operated by a single technician in the rear of the room.

Each studio classroom is normally patched into one of the four ITFS transmitters, which beam the ongoing nonrehearsed classroom lecture-discussions throughout the greater Los Angeles region. There are fourteen receiving locations in the area, most of which are at various corporation sites. The basis of the ITFS system is to provide academic opportunities for in-service courses in engineering, business, and so forth, at numerous business locations. Full-time employees can thus enroll in regular campus courses (which are taught in the four studio classrooms) and participate through the outlying viewing centers without having to attend classes on campus.

The term "interactive" refers to the fact that each receiving location is equipped with

---

[3]*See* S. N. Postlethwait, J. Novak, and H. Murray, *The Audio-Tutorial Approach to Learning* (Minneapolis: Burgess Publishing Company, 1969).

11-1 Interactive ITV System at University of Southern California: Teaching Console (left) and Student Viewing Stations in the Studio Classroom (right) (Courtesy: Patrick Loughboro)

microphones and an audio transmitter that can relay questions and comments back to the campus studio classroom. Thus, students from several different receiving locations can participate simultaneously in the class proceedings. This is still basically an audio-visual/hardware ITC application (in that there are no recorded materials, no studio procedures, no scripts and rehearsals, or whatever)—but quite an extended "classroom" installation.

11.2   School-Level Projects

Moving from the classroom to the school level, one finds an amazing diversity of various applications of instructional telecommunications. There are many varied patterns combining school-level productions, implementation of district programs, distribution of metropolitan and regional materials, and utilization of national materials.

From Pacoima to Beverly Hills   It may be instructive to look briefly at two actual school situations—as a quick indication of the variety of school approaches to ITC—and then take a more detailed look at another fictional case study.

Pacoima Junior High School, located in the San Fernando Valley, is part of the Los An-

geles Unified School District. In 1966, funds were made available from the State Compensatory Education Act to establish demonstration programs in economically deprived areas. Pacoima set up a demonstration program in reading and math and, later, a math demonstration center. In addition to classroom, library, and counseling space, this center included a very handsome television facility: three studio cameras, five 1-inch slant-track video recorders, complete film chain, and an impressive video control center and associated equipment. Among other accomplishments, during 1971–72 the school produced a weekly student news analysis series, the *Newseekers,* which was aired over the Los Angeles public TV station, KCET.

However, there were several problems with the project: the needs of the school were never accurately ascertained, there were never any adequate maintenance arrangements worked out, and the leadership for the program was divided among state, district, and school levels. Eventually, the original project was disbanded; and in 1975 the facility was converted to an area radio-TV media center for the Los Angeles public schools (illustration 11-2). From 1975, it functioned as a production center for other schools in the San Fernando Valley, as a training/curriculum operation for the students at Pacoima, and as a satellite for the school district's own ETV station, KLCS, Channel 58.

A few miles away, another system was started in the same year as the Pacoima project; and—although it was supported by a school district at the other end of the economic spectrum—ran into some similar problems. Beverly Hills High School, although located close to the center of Los Angeles, is part of the relatively small (and well-budgeted) Beverly Hills school district. It started out with a very ambitious dial-access retrieval system in 1966, with the high school operating top-quality production and distribution facilities. The main problems centered around the lack of adequate ascertainment of needs, infatuation with an ambitious hardware installation, and lack of guaranteed commitment to adequate maintenance. As at Pacoima, there

was an overemphasis on equipment and facilities, and relatively little thought was given to the securing and production of software programing that was actually needed by the school. In the case of Beverly Hills, the expensive-to-operate (and relatively underutilized) dial-access system was dropped; and, by the mid-1970's, the high school was operating a rather elaborately equipped closed-circuit system—primarily as a student-training operation and as a programing source for a local community cable channel (illustration 11-3).

If there is one lesson to be learned from operations such as those at Pacoima and Beverly Hills, it is simply that any STV program—regardless of how much money may be initially available for the construction of facilities—is off to a shaky start unless several criteria are met: the project must be in response to a real need; there must be

11-2 Pacoima Junior High School Radio–TV Media Center (Courtesy: Robert K. Shattuck)

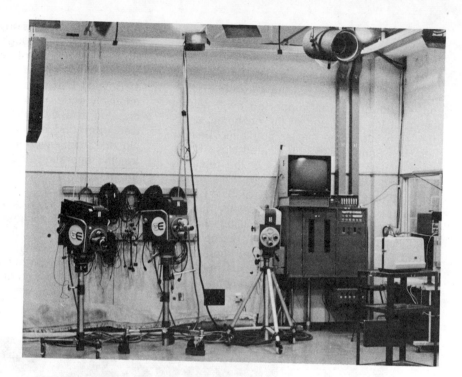

11-3 Beverly Hills High School Television Studio (Courtesy: Robert K. Shattuck)

widespread support from the faculty in general; and there must be adequate provision for technical maintenance of the program (all of which are mentioned among the fifteen components listed in Section 10.4).

Let us turn from the real world of Pacoima and Beverly Hills to a composite case study of a hypothetical school situation.

Middleborough Elementary School Located in a pleasant Midwestern, fictional, middle-sized community, Middleborough Elementary School has an enrollment of approximately 650 students in grades kindergarten through six. The school has no television production facilities— although it does use ITV fairly extensively. Three out of every four schools in the United States are equipped with at least one television receiver; Middleborough is equipped considerably above the average.

The main source of television programing is the noncommercial TV station, WZZZ, which is part of the state public TV network. Some program series are also rented directly from national ITV libraries. Although the school does not produce any of its own materials, two of the teachers have, at different times, taken groups of students down to WZZZ in order to participate in programs being produced by the station.

The school has a master antenna system, which receives WZZZ and three local commercial stations. Station WZZZ is on Channel 48, so the signal has to be down-converted (to Channel 5, which is otherwise unused in Middleborough) and distributed to connected classrooms so that it can be viewed on the VHF channel on the room receivers. Fifteen of the twenty-two classrooms are connected by coaxial cable to the school's distribution system. There are television receivers in eleven of the rooms. A twelfth receiver is kept, unassigned, on a standy basis—to be immediately wheeled into service whenever there are maintenance problems with any of the other sets. The school has a maintenance contract with a local television service and repair agency, which assures the school of routine annual preventative maintenance as well as priority on any trouble calls. Middleborough School also has a ½-inch slant-track video recorder, which enables teachers to request certain programs to be recorded off the air and played back, at a more convenient time, through the master antenna/distribution system. So, in actuality, the school has an embryonic closed-circuit system.

Joyce Loomis is a fifth-grade teacher at the school. However, she has been given two released periods during the day because she has been assigned to serve as the school's "ITV coordinator." As such, she is the one person most responsible for the success of the continuing program. She receives no extra compensation (except the released time); however, the responsibility and leadership position are rewarding in themselves. She also has attended a summer workshop on TV utilization, and she is usually able to attend the statewide media (AV-TV) conference held every spring.

Among Joyce's responsibilities are the following: meeting with the school faculty to determine instructional needs and specific ITV series that should be scheduled; ordering catalogs, classroom teachers' guides, and other materials to accompany the television lessons (thirteen teachers are using a total of ten different series); serving on the WZZZ in-school advisory council to work on selecting and scheduling broadcast ITV series over Channel 48; ordering additional ITV series from the Great Plains National Instructional Television Library, the Agency for Instructional Television, and other libraries and STV producers; scheduling the use of the school's video recorder (located behind the librarian's office, it is actually operated by a scheduled group of sixth-grade "television aides," who provide the manpower to record and play back the pro-

grams); contacting the television repair agency for needed service calls; coordinating the scheduling of television receivers when they must be moved from room to room (not every teacher who uses television has a receiver in his or her own room); and handling all program evaluation forms and providing feedback to the station and STV distributors.

Joyce Loomis has a hectic schedule. However, fortunately, she also has the support of her school principal and fellow teachers, and she has the satisfaction of knowing that Middleborough has a well-planned, well-organized STV program.

11.3   District Administration

The heart of the American educational system is the local school district—more than 17,000 individual, autonomous, independent school districts. The federal government may be handing out more money each year and the state legislatures may issue more guidelines and curriculum policies each session, but it is still the local board of education that raises the tax dollars, builds the school plants, sets the goals and objectives, hires the administrators and teachers, and actually runs the local school system. The local board sets the budgets; the local board makes the decisions.

Districts and ITC Organization   Administrative patterns vary tremendously from one district to another; and an observer can tell a great deal about the role of STV in a district by examining where the responsibility for the television operation is placed. Does the head of the ITC system have direct access to the superintendent? (Some districts have a separate administrator—ITV manager or assistant superintendent for television—who reports directly to the superintendent.) Does the ITV director or coordinator report to the chief officer in charge of *curriculum-structuring* functions (assistant superintendent for instruction or curriculum)? Does the ITV section come under

someone in charge of *audio-visual/reference* materials (director of AV center, librarian, manager
of educational materials, or the like)? In many
school districts—those with minimal interest in
ITC—some lower-echelon staff person (perhaps
a curriculum supervisor or some instructional
materials specialist) is designated to handle any
television matters that happen to come up.

   School districts own and operate many
different types of ITC installations. Hundreds
operate closed-circuit systems (of varying descriptions); Hagerstown, Maryland, (Section 3.9)
is the prototype. Many have CCTV systems that
have evolved into instructional TV fixed service
(ITFS) operations; Anaheim, California, is a good
example. Scores have ITFS stations that were
started as such from the beginning; Plainedge,
Long Island, was the first; New Trier Township
(northern Chicago suburb) has long been another
outstanding ITFS model. A couple of dozen school
districts own and operate their own noncommercial TV stations; Denver and Atlanta are two of
the largest. At least one school district—Dade
County public schools in Miami, Florida—owns
and operates both a VHF station and an ITFS
installation.

   While examining district ITC operations,
one might also include the larger private school
networks—specifically the Catholic establishments. The archdioceses of the larger cities in
the country have initiated several of the leading
ITFS stations in such centers as New York City,
Chicago, Los Angeles, Detroit, and Boston.

   Samoa: "Cooperative Instruction"   American Samoa is, of course, a territory of the United
States and is not—strictly speaking—a school
district as such. However, in terms of organization
and size (approximately 12,000 pupils), Samoa
can be studied in comparison with U.S. school
districts. Samoa's use of television, however,
truly is unique. No place else under American
jurisdiction comes close to a similar program.

   When H. Rex Lee was appointed governor of American Samoa in 1961, he inherited

the results of sixty years of "benign neglect" by
American authorities. Among other problems, the
island's educational system was virtually non-
functional—by U.S. standards; yet the products
of the "schools," the native Samoan young people,
were trying to go out and compete in the global
marketplace. Almost immediately, Lee determined that to upgrade the schooling system by
traditional means would be almost impossible:

> To have brought in the 300 to 400 teachers
> that would have been needed to staff the
> schools on a level comparable to that in
> the U. S. would have required from three
> to four million dollars. More importantly,
> it would have meant the dismissal of the
> native teachers, many of whom stood high
> in tribal circles [although their formal train-
> ing was comparable to a sixth-grade educa-
> tion]. Everywhere you looked, it became
> obvious that the physical, transportation,
> cultural, and human resources in Ameri-
> can Samoa made the traditional organiza-
> tion of a school system impractical if not
> impossible.[4]

   Governor Lee brought in a study team
from the mainland, headed by Vernon Bronson
(executive consultant to the NAEB), which developed the plans for a system of "cooperative
instruction making maximum use of television."
(It was decidedly *not* "a television project"!) Seldom has there ever been devised such a plan to
thoroughly and radically overhaul a complete
system of education in an immediate "crash"
manner.

   Twenty-six new elementary school buildings were constructed. Four new high schools
were built. A complete modern television facility
was built with four production studios and, eventually, ten broadcast-quality videotape recorders.
In October 1964, the program was initiated. Tele-

---

[4]Harry J. Skornia, "Some Lessons from Samoa," *Audiovisual
Instruction,* March 1969, p. 56.

11-4 American Samoa Television Installation: Studio Production (left) and Classroom Utilization (right) (Courtesy: KVZK-TV, Samoa, and D. N. Wood)

vision lessons were beamed over six VHF channels from transmitters on Mount Alava, 1,600 feet above Pago Pago Harbor.

Approximately thirty television teachers were brought in from the mainland—along with producer/directors and complete production and engineering crews. Lessons were produced to provide the core of instruction for every subject taught at every elementary grade level (less use was made of television at the secondary level.) Most of the native Samoan teachers were retained as classroom teachers—implementing and utilizing the television lessons. The program had a definitely planned peripheral effect of serving as a means of in-service training to upgrade the native teachers. Each individual elementary class (of approximately thirty pupils) would utilize as many as nine different short telelessons during a typical day.

At its peak, the system was broadcasting up to sixty individual lessons per day. As the competence of the native teachers increased—

and as the students progressed through the system—there was less need to rely so heavily upon the television portion of the system. By 1974, about forty-five separate lessons were being broadcast each day. By 1976, the system was utilizing only three broadcast channels, with a greater emphasis being placed on community programing—as opposed to classroom TV.

Without question, the Samoan system has served as a worldwide model for other countries and underdeveloped areas that need to upgrade educational programs on a crash basis.

11.4   Metropolitan ITV Associations

Most of the other levels of organization we are outlining in this chapter have some discernible physical or political boundaries such as a classroom, a state, or the nation. There is one organizational level, however, that has no existence except in an ITV context—one type of association that is unique to school uses of ETC. This is what we have labeled the "metropolitan ITV association."

The boundary of this type of association is congruent with the coverage area of an open-circuit noncommercial TV station in the center of the service area. To explain the function and raison d'être of this type of association, consider that the transmission pattern of a typical PTV station will include most of a major metropolitan area (with maybe parts of one or more cities and numerous smaller suburbs and towns), portions of three or four to possibly twenty different counties, anywhere from a dozen to over a hundred individual school districts, and maybe even overlapping into two or more states. The television signal knows no political or jurisdictional boundaries.

Therefore, we are faced with the situation of hundreds of individual schools—under various city, county, district, and state authorities—having one thing in common: they can pick up the broadcast signal of station WZZZ. Now, however, comes the question of how do these schools—with their numerous constituencies and political boundaries—communicate with the station personnel as to their needs and desires. Certainly we cannot expect the station to deal individually with a hundred or more separate entities. The answer: The schools themselves come together and form a loose confederation or consortium to determine their own curriculum consensus and programing priorities—a "metropolitan ITV association." See the chart (11-5) for a hypothetical situation. (The term, of course, varies from place to place; these organizations are called "regional" or "area" "authorities," "commissions," "associations," "councils," or whatever term seems convenient.)

The metropolitan ITV association organizes itself, sets up committees to determine various priorities, and establishes officers. (Remember the "WZZZ in-school advisory council" that Joyce Loomis belongs to? Section 11.2.) The metro ITV association has a designated president or executive secretary (either paid or, often, donated by one of the larger school districts), who

will be the chief contact with the PTV station's corresponding officer—the director of educational services, the ITV coordinator, or some similar title (Section 5.3). Most major metropolitan areas that have a noncommercial station have some sort of association that corresponds roughly to this general model. For a specific case study, let us turn to the Los Angeles area.

Regional Educational Television Advisory Council (RETAC)   In 1959, under the leadership of the Los Angeles County superintendent of schools, a number of school districts formed the (Los Angeles) Regional Educational Television Advisory Council. This consortium subsequently grew to include representatives from the eight Southern California counties that surround metropolitan Los Angeles. Leadership for RETAC has always been furnished by the Los Angeles county schools, division of educational media (formerly audio-visual education)—which has provided office space, certain administrative services, and the salary for a staff member who serves as executive secretary of RETAC.

The services and functions of RETAC are similar to those supplied by most other metropolitan ITV associations: determining needs of member schools that can be met by ITV; working with the metropolitan PTV station(s) (and/or commercial stations) on securing and scheduling appropriate series; serving as a production consortium to produce locally those ITV series that are not available from existing sources; serving as a coordinating forum to present a unified voice before local and state governmental agencies on budgetary and legislative matters; buying or producing and distributing classroom teachers' guides and other ancillary materials; conducting in-service workshops and utilization seminars for member schools; and similar services.

In order to carry out these functions, RETAC works through an executive committee, which consists of representatives of the county schools offices and the larger city school systems. The RETAC curriculum committee is the working

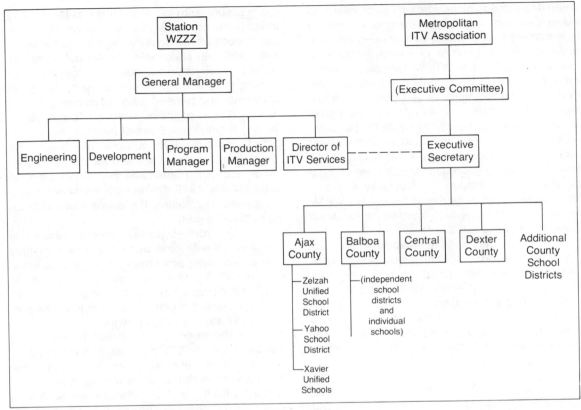

11-5 Relationship of PTV Station and Metropolitan ITV Association

committee charged with working most closely on selecting and implementing the appropriate ITV materials. RETAC was created before Los Angeles had a noncommercial TV station, and during its first five years it worked with commercial stations to arrange for broadcast time when school programs could be transmitted. Today RETAC works with community station KCET and station KLCS (the Los Angeles city schools station), as well as continuing with some commercial stations that continue to program a few ITV series.

RETAC reached its largest scope in the mid-1960's when almost a hundred different school districts—representing approximately 750 individual schools—were members of the council.

A decade later, membership had dropped to approximately sixty school districts, representing 500 individual schools. Most of the decrease in membership came about as a result of budgetary problems of the member schools. For its remaining member schools, RETAC continues to perform a vital function in coordinating many aspects of a metropolitan ITV association.

11.5   Statewide Operations

At the state level, ITC operations may be conducted in several different ways (Section 5.5). Georgia and Tennessee are good examples of a statewide open-circuit network operated by the state department of education. Nebraska and North Carolina were mentioned as models of university-operated state networks—with consideration being given to in-school programing. Texas and Indiana both have statewide microwave hook-ups to tie institutions together. However, the best example for our purposes would probably be a state network operated by a state commission or authority. Most of these statewide operations are concerned to some extent with public broadcasting, but—especially in the South—are really geared for specific instructional purposes. For example, the Kentucky Authority for Educational Television runs the statewide ETV network, but stations are licensed to the state department of education.

Two states initially established interesting statewide closed-circuit networks. One of these, Delaware, was discontinued by its state legislature in 1970 (Section 13.2). The other has evolved into a state-supported institution operating three related telecommunications networks: a six-channel closed-circuit operation, a ten-station open-circuit network, and an emerging FM radio network.

South Carolina ETC Networks   In 1958, the South Carolina General Assembly supported a demonstration project to try out closed-circuit TV in one Columbia high school. Less than twenty years later, the system was reaching more than a third of the state's 650,000 pupils, bringing TV signals into virtually all of the state's 1,100 schools.[5]

The South Carolina Educational Television Commission was created in 1960. The original mandate to the commission was to up-

---

[5]Data for this section have been derived primarily from four sources: *Annual Report of the South Carolina Educational Television Commission* (Columbia: South Carolina Educational Television Commission, 1975). Preface to the 1975–76 SCETVC budget request to the General Assembly. John P. Witherspoon, "State Public Television—A New Tool for the States," *State Government,* Autumn 1971. South Carolina 1975–76 Instructional Television Resources (Columbia: South Carolina Educational Television Commission, 1975).

grade school instruction, and this was the primary thrust of the commission's programing. However, other needs were also pressing, and the system—with additional state support—set out to initiate college credit courses, business and industrial training programs, medical and law enforcement programs, and general public programing.

Some of these needs could be met by the expanding closed-circuit network, but others called for reaching the public in their homes. So the state commission initiated its first two open-circuit stations (Greenville and Charleston) in 1963–64. By 1976, the ten-station network was completed, blanketing the entire state with an open-circuit signal.

In order to provide special services to the blind, as well as instructional audio materials and public radio programing, the state commission put its first FM radio station on the air in 1972. Two others have since begun operations, and two more are planned to complete the projected five-station radio network.

The open-circuit television stations are used for beaming STV materials to the elementary schools. The secondary schools—with more complex scheduling programs—receive their materials from the closed-circuit network. A total of six CCTV channels are used in some areas, with about 260 high schools connected to the statewide closed-circuit system. During 1975, a total of seventy-one different ITV courses were transmitted to the public schools (all grades). Almost a quarter of a million students were involved; the average television student viewed three separate ITV series during the year.

In addition to the elementary and secondary programing, the commission has also been involved at the college level. In 1975, fifty-eight college-credit courses were telecast; five of them were designed strictly for at-home viewing. The most successful college program has probably been the master's degree in business administration, which serves more than half of the MBA enrollment at the University of South Carolina. Combining closed-circuit lessons to thirteen centers throughout the state, a telephone talk-back system, and weekend seminars, the MBA-ETV program has served as a model for similar programs in nursing, engineering, and education.

A primary feature of the total South Carolina ETC system has been the commitment to a wide variety of business and industrial in-service and training programs. Both the open-circuit stations and the closed-circuit system have been utilized in this effort. In addition to high schools, more than a hundred other types of receiving facilities have been tied into the multichannel CCTV network: hospitals, university regional centers, police departments, technical education centers, and similar agencies. South Carolina has pioneered in using ITC for statewide law enforcement training (dating back to 1965). Numerous ambitious programs have been established for medical education and nurses' training (the state's ten largest hospitals are tied into the system). Many other programs have been established for special groups, such as the deaf, the illiterate, the unemployed, and prison inmates. A total of 85,000 to 90,000 adults are enrolled in specialized courses or programs every year.

As far as public broadcasting is concerned, the open-circuit network also carries a full schedule of PBS programing—contributing several series to the national service. A 1973 study indicated that 60 percent of the South Carolina citizens consider themselves regular viewers of ETV—this is twice the national average.[6]

The organization that runs this combined operation—including one of the best-equipped production facilities in the United States—is composed of about 170 full-time professional staff members as shown in the organizational chart (11-6). The annual state budget has been upped to over $8 million per year. As an article in the *New York Times* summarized it:

It will come as a surprise to a great many northerners to learn that South Carolina

enjoys probably the most outstanding—and considering its relative wealth, certainly the most generously funded—educational television network in the entire country.[7]

11.6   Regional Activities

Beyond the state, the next organizational level is the regional groupings. This category includes all types of telecommunications endeavors that involve multistate arrangements—for example, production compacts, distribution and networking arrangements, or other special services.

The STV functions of the regional networks (Section 5.6) certainly should be included in this category. Both Eastern Educational Network (EEN) and Southern Educational Communications Association (SECA) have been especially active in this area. In fact, EEN, with the Northeastern Regional Instructional Television Library, is undoubtedly the largest supplier of ITV materials that is still "regional" in character.

Another type of regional STV activity might be classified as the "supertransmitter"—some sort of method of getting a transmitter high enough in the air to cover a multistate region. The most obvious example, today, would be satellite transmission, and, indeed, the Satellite Technology Demonstration (using the ATS-6 satellite) was first used for regional purposes. Because of the international implications, however, this will be discussed below (Section 11.8). The case study we shall look at is an older model of the "transmitter in the sky"—one with propellers.

MPATI: Supertransmitter   In the 1940's, Westinghouse had experimented with the possibility of relaying TV signals from an airplane (this was before the transcontinental coaxial cable was laid). In the late 1950's, the idea was revived—and with a $7 million grant from the Ford Foundation (and contributions from a few other sources), the Midwest Program on Airborne Tele-

---

[6]Jan Collins Stucker, "South Carolina's Educational TV Leads the Class," *New York Times,* March 2, 1975.

[7]*Ibid.*

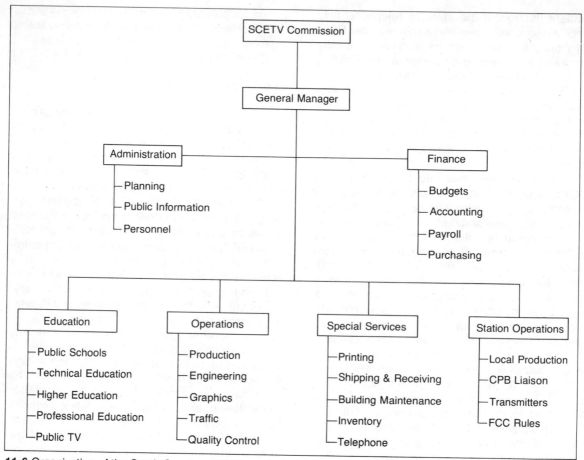

11-6 Organization of the South Carolina ETV Commission

vision Instruction (MPATI) was created. Basically, the purpose was to see if the schools in a multi-state area could work together in a regional compact to support the idea of a flying transmitter, sending ITV materials into the classrooms of a large area of the United States.

The Midwest was chosen for the project because the terrain was flat enough to make it possible to send the signals over a great distance; the population was dense enough to make the project feasible economically; the area included

extremes in cultures, economic conditions, and educational levels; and, finally, there was relatively little activity at the time in getting ETV stations and CCTV systems operational (therefore, there was more of a *need* for the supertransmitter). A preliminary study indicated that at least thirteen ground-based stations would have been required to cover the same territory that one flying transmitter could cover—from an altitude of five miles. The region finally selected for the project included most of Indiana, eastern Illinois, south-eastern Wisconsin, southern Michigan, western Ohio, and northern Kentucky.

Office space and airport facilities were leased from Purdue University, and work was started on converting two DC-6 aircraft (one would remain on the ground each day as a standby plane) into flying transmitters that would fly a tight figure-eight pattern over northeast central Indiana (illustration 11-7). Two UHF channels (72 and 76) were allocated for experimental use by the FCC, and—with a carefully designed installation—a school building could pick up the signals as far away as 200 miles. The six-state MPATI region included about 17,000 schools and colleges (more than 7 million students) in its 130,000 square-mile coverage pattern.

During 1960–61, extensive preparations were under way on both the production and plans

11-7 MPATI Flying Transmitter: Modified CD-6 with Antenna Extended (top) and Video Recorders and Transmitting Equipment in the Plane (bottom) (Courtesy: MPATI)

for implementation of ITV materials. Layers of committees, representing states, districts, and schools in the six-state area selected the curriculum—determining where TV could best be utilized. After a nationwide "TV teacher talent search," sixteen teachers were finally chosen as the original MPATI teaching group; and production was started in nine studios (university and station operations) from Wisconsin to New York (there was no reason why the programs needed to be produced in the same locale where they were to be distributed).

Also, twenty universities in the six-state area were selected as "resource institutions" for MPATI; and an area coordinator from each university was assigned to work with local schools in his area—helping them with equipment problems, holding utilization seminars, setting up summer workshops, and the like.

In September 1961, the project literally got off the ground, and the most expensive and best-produced ITV telelessons created up to that point were transmitted into classrooms across the region. At its peak, MPATI enrolled close to 2,000 individual schools and was for a time virtually self-supporting. The airborne program did run into problems, however, which eventually brought down the planes.

First, the project could not get the four to six permanent UHF channel assignments it requested because, reasoned the FCC, committing that many channels would keep ground-based facilities in the region from developing. In fact, the FCC ruled that MPATI could not continue to use even the two experimental channels beyond 1970; it would have to become an ITFS operation.

Second, MPATI succeeded in stimulating enough interest in ETV throughout the six-state region that it helped to put itself out of business. Numerous stations were started, many schools turned to their own CCTV systems (and started renting copies of the MPATI tapes), and several schools started experimenting with ITFS (which was announced in 1963). So, by the mid-

1960's, there was much less need for the flying transmitter than there had been five years earlier (partly because of the stimulus of the flying transmitter itself).

The successful solution to any specific temporal problem, in eliminating the problem, also eliminates the need to continue the solution. (The same observation may be made about the Samoan ITV operation.)

Thus, in 1968, the airplanes came down for the last time. MPATI remained as a production and library organization for another three years, however, serving during the late 1960's as one of the three major sources of ITV materials in the country, along with Great Plains National Instructional Television Library (GPNITL) and National Instructional Television Center (NITC). Finally, in 1971 the entire MPATI assets and library operation (twenty-seven telecourses consisting of 1,700 individual videotaped lessons) were incorporated into the GPNITL.

MPATI originally stated that it had two operational aspects to be evaluated: "(1) the technical system of transmitting airborne television signals, and (2) the televised instructional material."[8] To these, a third should be added: (3) the concept of a multistate consortium of independent school districts. In all three areas, MPATI could be considered a success. First, the plane delivered a usable signal with greater than 96 percent reliability; and "should the prospect of direct-to-the-school satellite transmissions become possible over a national or regional area, the MPATI experience could be considered a very apt demonstration."[9] Second, the ITV materials—though sadly dated today—were an example of the high quality that can be produced when several

[8]*This is Airborne* (information brochure) Lafayette, Indiana: Midwest Program on Airborne Television Instruction, 1961), p. 16.

[9]Ken Winslow, "The Adoption and Distribution of Videotape Materials for Educational Use," in Sidney G. Tickton (ed.), *To Improve Learning: An Evaluation of Instructional Technology,* Vol. 1 (New York: R. R. Bowker Company, 1970), p. 409.

agencies combine resources to put out a mutual product. "The result was . . . the best example of an agreed-upon body of inter-institutional curriculum materials for use at the public school level."[10] Finally, MPATI succeeded in organizing hundreds of autonomous districts to work on a common project that was far beyond the reach of any individual district. This concept was to prove valuable in the successful pursuit of future consortium projects by other agencies.

11.7   National Programs

At the national level of ITV activity, there are two main functions or categories of activities: first, production centers and consortiums; second, agencies concerned with distribution and library arrangements.

The Production Center: CTW   The Children's Television Workshop (CTW) is much more than a production center, obviously; but it can be examined from the perspective of a production enterprise—on a grand scale. At a dinner party in 1966, Joan Ganz Cooney, then public affairs producer for WNET (now president of CTW), and Lloyd N. Morrisett, then vice-president of the Carnegie Corporation (now president of the John and Mary N. Markle Foundation and chairman of the CTW board of trustees), discovered they shared a common vision: the idea that enough money and time would make possible creation of a really outstanding children's series that would be both highly entertaining and soundly educational. At that moment, the Children's Television Workshop was born.

The money (approximately $8 million for the first two years) came from a combination of foundation and federal funding; and the time came from the amount of research that was undertaken during the developmental phase. The result, *Sesame Street,* was launched in late 1969—featuring a cast composed of humans and muppets (illustration 11-8).

The secret ingredient in the success of *Sesame Street* was research. "Joan Cooney set out to achieve a working collaboration between the programmer, the educator and the researcher—and succeeded."[11] Never had a project been so thoroughly tested and revised during the developmental period as *Sesame Street.* The same careful research and development went into the creation of the CTW elementary reading series, *The Electric Company,* two years later.

Although there had been some earlier attempts at program validation and pretesting of ITV materials, never had the research-and-revise concept been so thoroughly and successfully applied as in the CTW products. "It is difficult to envision any major television-education undertaking from this point on which will not incorporate as an essential element the concept of the integration of production and research into one self-correcting creative effort."[12]

The Distribution Library: GPNITL   There are several models for nationwide distribution libraries. Major audio-visual centers (such as Indiana University) furnish one kind of service. Specialized libraries exist such as the Medical Television Network (operated by UCLA) and the Network for Continuing Medical Education (supported by the Roche Laboratories), which distributes programing to more than 700 hospitals and medical centers. Many other specialized commercial and nonprofit libraries and production centers operate throughout the country; but the best example for our purpose is probably the Great Plains National Instructional Television Library (GPNITL).

---

[11]Herman W. Land, *The Children's Television Workshop: How and Why It Works* (Jericho, New York: Nassau Board of Cooperative Educational Services, 1972), p. 5 *See also* Gerald S. Lesser, *Children and Television: Lessons from Sesame Street* (New York: Vintage Books, 1974).

[12]Land, *Children's Television Workshop,* p. v.

---

[10]*Ibid.*

11-8 Children's Television Workshop Series: *Sesame Street's* Big Bird (left) and *The Electric Company* (right) (Courtesy: Public Broadcasting Service)

Formed originally as part of the tripartite National Instructional Television Library in 1962 (Section 4.3), GPNITL was self-supporting—and operating on a national level—by 1966. Today, with more than 150 individual series, GPNITL has course offerings for every subject area. Virtually all the programs were acquired from ETV producers—stations, school districts, regional production centers, and colleges and universities. Several series were acquired in blocks—such as the twenty-seven MPATI series and eighteen series from the Chicago TV College (Section 3.9).

GPNITL materials are available on a variety of formats: transverse-scan (quadruplex) videotape, helical-scan videotape (about ten different formats), video cassettes, and films.

Leasing costs for these materials are based upon the number and length of lessons, the number of transmission points (for example, from several stations in a state network), the time span during which a lesson is used and repeated, and whether or not the user supplies his own tape. The size of the potential audience (Philadelphia or Paducah) and the means of broadcast (open-circuit, CCTV, or ITFS) do not affect the price. For example, in 1975, if the user supplied his own videotape, a 30-minute color program, to be broadcast from a single transmitter, with unlimited replays during a one-week period, would lease for $67.50.[13]

---

[13]Data are taken from *Recorded Visual Instruction: 1975 Edition* (Lincoln: Great Plains National Instructional Television Library, 1975), p. A-6.

In addition to its ITV library, GPNITL also has been increasingly active in other professional areas. It has published books about legal problems and research and instructional design (to some extent verifying the above observation that production and research increasingly will be integrated "into one self-correcting creative effort").[14] Also, GPNITL has been active in cosponsoring professional conferences and other cooperative projects with the Nebraska S-U-N project (Section 5.5).

The Agency for Instructional Television
Generally throughout this book we have tried to stress that the various functions of educational telecommunications should be considered separately and examined as distinct processes. For example, production and distribution activities (although related operations of any broadcast station) are really clearly distinguishable processes. We pointed to CTW as an example of a production enterprise; and we looked at GPNITL as a distribution agency. Both of these functions, however, are combined in the Agency for Instructional Television (AIT).

No ETV organization has undergone more changes than this one. Formed as the National Instructional Television Library (NITL) in 1962 under a U.S. Office of Education (USOE) grant, it was administered originally by NET (Section 4.3). In 1965, it became the National Center for School and College Television (NCSCT) and moved to Bloomington, Indiana—becoming a project of the Indiana University Foundation. The center shortened its name to the National Instructional Television Center (NITC) in 1968 and by 1970 was entirely a self-supporting institution.

As the center became increasingly involved in production consortium projects in the 1970's (see below), it found itself working more and more closely with state agencies (for example, departments of education and ETV commissions) and with Canadian provincial governments. As a consequence, a new institution was formed in 1973, the Agency for Instructional Television (AIT), which was governed by a board of directors appointed by the U.S. Council of Chief State School Officers and by the Canadian Council of Ministers of Education. The AIT absorbed all of the staff and operations of the former NITC. With headquarters still in Bloomington, AIT has regional offices in Washington, D.C., Atlanta, Milwaukee, and San Francisco.

As a distribution library, AIT handles about half as many series as GPNITL. The AIT telecourses, generally, have been carefully selected and meet fairly high standards. In many instances, AIT has participated in the planning and production of a series, often contributing to a project in the planning stages in order to help acquire a product of higher quality. Leasing arrangements for AIT series are somewhat similar to GPNITL rental procedures, with one notable exception: AIT rate structure takes into account the size of the potential audience. For example, in 1975, to lease a 30-minute program (on the user's tape), for one week, for a school district with 50,000 students, would cost $55.50; the same program leased to a station covering a school enrollment of 500,000 pupils would cost $96.00.[15]

Turning from distribution of existing materials to the creation of new materials, AIT has had its greatest impact with the concept of the "production consortium." Basically the idea—pioneered to a parallel extent by MPATI—is that school systems can get a much higher quality product if several systems pool their resources through some sort of cooperative arrangement. Rather than have ten schools each pay $10,000 to produce its own five-program series in third-grade science (winding up with ten mediocre

---

[14]C. Edward Cavert, *An Approach to the Design of Mediated Instruction* (Washington, D.C.: Association for Educational Communications and Technology, 1974).

[15]Data are taken from *Television Instruction: Guide Book, 1975* (Bloomington, Indiana: Agency for Instructional Television, 1975), pp. 24–25.

series), why not combine the dollars and produce one $95,000 series (including costs for administration and a duplicate set of tapes for each participating school)?

The first step was to assess accurately the need—and to determine what was currently being done (in the way of existing TV materials) to meet that need. From 1965 to 1968, NCSCT held a series of eight assessment conferences—one each in art, foreign languages, health and physical education, language arts, math, music, science, and social studies. Each two- to three-day conference involved a group of seven to sixteen participants—divided between content/curriculum specialists and ITV practitioners. At each conference, the specialists viewed and evaluated from seventy to more than a hundred telelessons (or excerpts) from representative series being produced around the country. Altogether, a total of ninety-seven top content and

media specialists viewed portions of about 550 individual programs—by far the largest critical screening ever attempted. The consensus for virtually every subject area was identical: Very little programing of real merit was being produced by individual agencies at the local levels.

The recommendation that came out of most of the subject-area conferences was that there was need for some sort of national program, a pooling of resources and utilization of the best talent (subject matter experts and media specialists). Thus was born the idea of the nationwide production consortium. Several much needed, long-term projects were outlined in the next year or two. Each of the consortium projects would involve dozens of content authorities, working committees, and months of planning and trial programs.

The first completed consortium series was released to schools in 1970; *Ripples,* which

11-9 Two Series from the Agency for Instructional Television: An Elementary Physical Education Series *Ready? Set . . . Go!,* Produced by WHA-TV, Wisconsin (left), and a Scene from *Self Incorporated,* an AIT Consortium Series Dealing with Adolescent Problems (right) (Courtesy: Agency for Instructional Television)

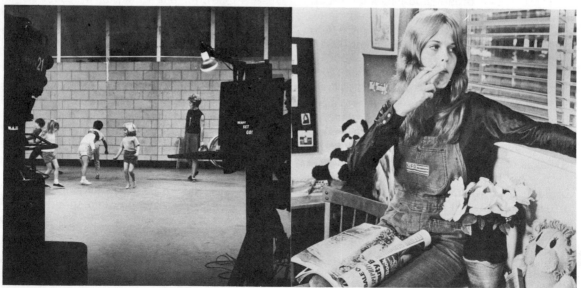

11-10 Scope of NITC/AIT Consortium Projects

Year of release	Subject area	Series title	Target age	Total project cost	Number of participating agencies
1970	Early Childhood	*Ripples*	5– 7	$256,000*	14
1971	Art	*Images & Things*	10–13	508,000*	27
1973	Life-Coping I	*Inside/Out*	8–10	779,000*	34
1974	Career Development	*Bread & Butterflies*	9–12	604,000	35
1975	Life-Coping II	*Self Incorporated*	11–13	555,000	40

A portion of the financial support for the first three consortium projects came from foundations and other sources. The last two projects were completely self-supporting.

Data are taken from Self Incorporated, *an informational brochure prepared by the Agency for Instructional Television, 1975.*

was concerned with early childhood education, involved some fourteen separate agencies contributing to the design and support of the project. Each succeeding consortium series (five had been completed by 1975) involved more participants than the preceding one; *Self Incorporated,* which deals with life-coping skills at the intermediate level (figure 11-9), involved forty state and provincial (Canadian) agencies, as shown in the table (11-10).

As the AIT 1975 catalog described its own perception of how ITV programing has progressed, "After long, slow development, school television appears to have entered a new stage. At first, it produced unimportant things badly; eventually, it did a few unimportant things well. It then produced some important things badly; now it is doing important things well."[16]

11.8   International Developments

As might be expected, there is less activity at the international level than at the other levels examined. However, due to signs of increasing interest and program exchange in multi-national ITC circles, the prospect is good for increasing activity in the near future.

International Associations and Projects Several organizations and associations are concerned with international ITC programing and projects of one kind or another. The AIT, of course, was reorganized specifically as a two-nation association. Both of the two professional associations that have played a prominent role in the development of ITC in this country—the National Association of Educational Broadcasters and the Association for Educational Communications and Technology—are international in that they have individual members from other nations (primarily Canada).

The Broadcasting Foundation of America (Section 3.8) was established in 1955 primarily as an agency devoted to the acquisition of overseas radio programing and its distribution in the United States. Headquartered in New York City, it has expanded its operations—at one point serving as the nucleus for the International Division of the National Educational Television and Radio Center (NETRC).

Other agencies have long been active in international circles. Both the European Broadcasting Union and the Asian Broadcasting Union

---

[16]*Ibid.,* p. 5.

have had extensive involvements with noncommercial educational programing. Other smaller agencies exist for specific international projects and specialized programs; the International Broadcasters Society and PTV International both are located in New York City.

American ETV associations have occasionally been involved in specific international programing projects. One of the earliest and most significant of these was perhaps Intertel, the International Television Federation. This was a cooperative programing venture, formed in 1960, by networks and production centers in four major English-speaking countries: Australia, Canada, Great Britain, and the United States. NETRC and the Westinghouse Broadcasting Corporation were the American participants. During the ten years of its existence, Intertel produced and distributed in its member countries a total of sixty-three programs.

The United Nations has played an important role in promoting the consideration of broadcasting for educational purposes. At a 1965 UNESCO conference in Paris, for example, extensive consideration was given to the possibility of satellite usage for instruction in underdeveloped countries. Many of the possibilities discussed at this conference are being explored and tested today (see below).

There also are international ITV festivals and competitions. The annual Institute for Education by Radio and Television (IERT) awards, held at Ohio State University every spring, has taken on an international flavor with many entries from other countries. The most representative international ITV competition is possibly the Japan Prize. Started in 1964, it is the first specifically designed international contest for educational radio and television programing.

Satellite Transmission   The technological device with the greatest potential for international transmission is, of course, the communications satellite. The greatest drawback to the

development of international educational programing is, of course, man himself.

> In brief, technology has provided us with the capacity for global instantaneous communication, but our inability to modify traditional modes of thinking and to adjust established social, economic, political, and legal institutions is seriously hampering effective use of this wondrous new communications extension.[17]

Nevertheless, man perseveres and progresses— demonstration by demonstration. The first demonstration of an intercontinental satellite transmission for any school-related purpose was a 1965 hook-up between schools in West Bend, Wisconsin, and Paris, France (basically for an hour of conversational practice).[18]

The first satellite launched specifically for educational purposes was the ATS-6 (applications technology satellite number six) in 1974, shown in the illustration (11-11). During its first year, ATS-6 was positioned so it could focus and relay signals to the North American continent. Three Satellite Technology Demonstration (STD) projects were undertaken—funded by the National Institute of Education, U. S. Department of HEW, the National Aeronautics and Space Administration, and the Federation of Rocky Mountain States—one in the Appalachian region, one in eight Rocky Mountain states, and one in Alaska.[19] (Symbolically, the ATS-6 satellite was used for the space-to-earth relay of the historic Apollo-Soyuz rendezvous in mid-1975.)

[17]Russell B. Barber, "The Role of Space Communications in ETV," *in* Allen E. Koenig and Ruane B. Hill (eds.), *The Farther Vision: Educational Television Today* (Madison: University of Wisconsin Press, 1967), p. 311.

[18]*Ibid.*

[19]*Satellite Technology Demonstration Federation of Rocky Mountain States, Inc. (Goodbye to the Great Divide)* (Denver: STD/Federation of Rocky Mountain States, 1974).

The satellite was then positioned over India for a demonstration of direct satellite-to-school transmissions. It was returned to the Western Hemisphere for continued schooling purposes and demonstrations at the completion of the India project.

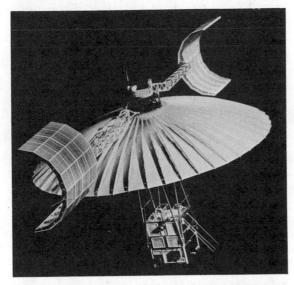

11-11 ATS-6 Satellite Used in Several Domestic and Foreign Educational Applications (Courtesy: Fairchild Industries)

In early 1975, the Public Service Satellite Consortium (PSSC) was incorporated in Washington, D.C. The goal of the consortium (organized by former Samoa Governor and former FCC Commissioner H. Rex Lee and headed by former CPB Director of Television John Witherspoon) is to guarantee that the technology pioneered in the STD demonstration is used for the establishment of a permanent operational system. Although the PSSC membership is initially American (CPB, PBS, JCET, and numerous stations and state agencies are among the incorporators), the eventual international implications are obvious.

The technology has been developed for substantial international ITC cooperation, and man is gradually learning how to organize himself for these endeavors. It is important that the future impact of international ITC not be underemphasized. Koenig summed up the importance of these international considerations as ambitiously as anyone:

> The future development of ETV lies in its acceptance as an international resource. The next step in its development should be the world-wide sharing of knowledge and culture. It is hoped that the current satellite capability over both the Atlantic and Pacific oceans will encourage electronic "people to people" exchanges. This may be the source of educational television's future history.[20]

---

[20]Allen Koenig, "The Development of ETV," *in* Koenig and Hill, *The Farther Vision*, pp. 8–9.

**Issues in ITC:
Media and the Future
of Education**

This chapter may be considered as a parallel to Chapters 8 and 9, which deal with "Issues in Public Broadcasting." We want to examine in this chapter several issues concerned with instructional telecommunications—issues that are somewhat related to those in public TV and radio. *Funding:* Who pays for ITC? Federal taxes? Local boards of education? Should parents or students pay directly? *Controls:* Who determines how ITC should be used? Learners? Teachers? Bureaucrats? Taxpayers? *Producers:* Who designs and distributes ITC materials? Schools? Universities? A federal curriculum agency? Commercial instructional materials makers? *Audiences:* At whom are formal ITC materials aimed? Children in school? Out of school? Preschoolers? Adults? *Programing:* Should we be primarily concerned with enrichment programing? ITC used as a major resource? Total teaching?

In examining these issues, however, we will not look at them separately (as we did in Chapters 8 and 9). Rather than discuss one issue at a time, we will try to place them in perspective by examining the overall question of the proper role or roles for telecommunications in the educational structure. In what areas can instructional media most effectively be employed? Where should they *not* be used? What jobs can they do best? By tackling these questions we will be examining most of the issues raised above. In short, the one question we want to examine is *"Where do telecommunications best fit into the overall educational picture?"*

A "look to the future" such as this often would be used as the final chapter in a book. However, in the structure of this text we are using it as an integral chapter of the main part dealing with "Theories and Roles of ITC" (Part Three); it may be worth while to examine these issues and do some crystal-ball gazing before proceeding to a practical consideration of how instructional telecommunications are currently organized and utilized in our schooling systems (Part Four).

In trying to answer our one broad question ("Where do telecommunications best fit into the overall educational picture?"), we first need to look at the very nature of our schooling systems. Also, we need to take a closer look at the impact of telecommunications media upon society and some of the hidden realities that may affect our future thinking about educational patterns.

12.1   Impact of Media and the Hidden Realities

To begin with, we should accept Father John Culkin's observation that most educational leaders are P-O-B's—"print-oriented bastards."[1] The present generation of educators and administrators lack the visceral "gut" understanding/orientation of television within which today's generation has been raised. It is difficult to empathize with those who emerged from the womb to perceive the world through a kinescope tube. Our understanding of the television environment/medium/experience/message can never be more than academic. As pointed out above (Section 10.7), the pervasive impact of television is statistically overwhelming and culturally overpowering; yet we do not know how to properly react to the medium. Maloney aptly described our generation's dilemma when he observed, "Television is still a new medium, and educators have not yet learned how to think about themselves in relation to it. . . . Learning a workable and comfortable stance to take with respect to television is likely to be a continuing project, both for teachers and for students."[2] What implications does this hold for television as part of the curriculum? For other subjects in the curriculum? For accelerating other changes within our schooling systems?

Television and Its Hidden Realities   This, then, is one problem of trying to come to grips with the ways in which telecommunications can be most effectively used in the schooling systems; we simply have not yet had enough experience with the media to feel intuitively comfortable with them. We are not yet sure about the "messages" of the mediums. It is still too early to conceive of the vast, tradition-shattering, cultural upheavals that will be caused by the advent of television. Marshall McLuhan several years ago pointed out that every new communications medium—especially television—has introduced new societal and interpersonal relationships that eventually have led to new social structures and patterns of individual perceptions of the environment.[3] Vietnam and Watergate may have given us some small glimpse of the political and social realities of the mass media—but it was only a glimpse. We have yet to experience any real impact upon our educational institutions due to the advent of educational telecommunications—but the potential is there.

We might refer to these potential cultural upheavals as the hidden realities of the new media. John Paul Strain points out some of the hidden realities that accompanied the print revolution—although no contemporary of Gutenberg could have foretold what was to happen.

> Therefore, it is not merely the practical results of technological additions that become important to society, but the realities of the hidden influences in the course of new developments. An excellent illustration can be found in the technological advance of print, which brought about the collateral

[1]John M. Culkin, "Unstructuring Education," *in New Relationships in ITV* (Washington, D.C.: Educational Media Council, 1967), p. 29.
[2]Martin J. Maloney, "A Philosophy of Educational Television," in Allen E. Koenig and Ruane B. Hill (eds.), *The Farther Vision: Educational Television Today* (Madison: University of Wisconsin Press, 1967), p. 24.

[3]*See especially* Marshall McLuhan, *Understanding Media: The Extensions of Man* (New York: New American Library, 1964).

learnings of solitude and individual initiative for success. Printing which produced books forced a conflict with the customs, mores, social classes, and medieval institutions. Printing of books began as a means for teaching better the old routines and traditions. In the course of time, its hidden realities aided in producing a wholly different society.[4]

Neil Postman reinforces this by pointing out that, "Print . . . changed the very form of civilization. It is not entirely coincidence, for instance, that the Protestant Reformation was contemporaneous with the invention of movable type."[5] He later adds, "The printing press provided the wide circulation necessary to create national literatures and intense pride in one's native language. Print thus promoted individualism on the one hand and nationalism on the other."[6]

Within this context, it is easier to understand McLuhan when he speaks of the electric technology (for example, television) leading back to tribalization, or when he states that electric media result in an "implosion" of civilization that is making of the world, for each of us, the Global Village. Within a broader social/philosophical context, we can safely state that very few of us are aware of the hidden realities of the television environment/medium/experience/message.

We have just begun to scratch the surface of an understanding of electronic communications. Vladimir Zworykin invented the iconoscope tube in 1923—somewhat more than half a century ago. Johann Gutenberg is usually given credit for inventing the printing press sometime around 1450. By 1500, how many scholars were able to foretell what impact printing would have upon the world? This is roughly where we stand with telecommunications today.

---

[4]John Paul Strain, "Educational Technology and Its Hidden Realities," *Educational Leadership,* May 1968, pp. 722–724.

[5]Neil Postman, *Television and the Teaching of English* (New York: Appleton-Century-Crofts, 1961), p. 9.

[6]*Ibid.*

12.2   Nature and Purposes of Schooling Systems

To begin an examination of our schooling systems, we should first look at the reasons for which the schools exist. Exactly what are they supposed to do? Well, teach, of course. Teach what? Why, reading, writing, arithmetic, biology, and driver education—along with a hundred other specialized and general courses—of course. However, as William Clark Trow points out, merely to list subject areas does not begin to explain what schools should be doing with these subjects nor does it explain the basic nature of the schooling experience.

But if the further question is asked, Why teach these subjects? the answers begin to vary: to transmit the cultural heritage, to impart knowledge, to train the mind, to enable people to satisfy their needs, to serve the state. In any case, the objective seems to be to help children (and adults) to learn, even though there are differences of opinion as to what they should learn and how they should be taught.[7]

Societal and Individual Needs   The goals of formal schooling can be divided into two broad classifications—those designed to meet the needs of society (or community needs), and those goals designed to meet individual needs of the learner. Societal/community goals include national defense, preservation of the democratic system, strengthening of the economic institutions, cultural preservation, environmental education, public health, and so forth. Some observers also include "babysitting"—the need to keep young people off the streets and out of the employment market (or unemployment lines) until they reach a prescribed age.

Individual needs include large categories such as vocational training, general liberal arts

---

[7]William Clark Trow, *Teacher and Technology: New Designs for Learning* (New York: Appleton-Century-Crofts, 1963), p. 7.

(equipping an individual to fit into society), and preparation for further schooling (learning to learn, to research, to think analytically, to use other specific tools of communication and computation). Trow lists these individual needs as the "human imperatives"—man must live (including being able to earn a living), man must live with others (socialization and citizenship), and man must live with himself (personal goals and mental health).[8] Other educators, administrators, and psychologists have compiled many different kinds of lists of educational goals and purposes (some of these are discussed in Sections 13.5 and 13.6).

The important concept is merely that the schooling system must start with a clear-cut idea of the functions it is to serve. The ends must come first; the means must not determine the ends. Yet all too often this is what has happened. Schooling purposes have been determined by what means or facilities were at hand. Irrelevant courses have been offered merely because a teacher had an interest in a particular area. Useless specialized science courses have been developed merely because a local firm donated the expensive laboratory gear. Many ill-conceived media projects have been inaugurated because the initial grant money was available to buy the hardware.

ITC as Facilitator of New Curricular Objectives Yet, at the same time, it is possible that television and related media can play important roles as facilitators of new curricular objectives— when appropriate.

Lewis A. Rhodes, former director of the NAEB's National Project for Improvement of Televised Instruction, used an analogy between television and the elevator. For untold centuries, most architectural structures were limited to two or three stories. As a practical matter, people simply did not erect any building with more stories than they could comfortably walk to the top of. Then the elevator was invented! And for many years, what did the architects do? They installed ele-

vators in two- and three-story buildings—merely so that people could get to the top floors of the traditional low buildings in a more efficacious manner. The possibility of erecting a twenty-story or a hundred-story building did not immediately occur to them; the elevator was seen simply as a more efficient means of getting where we had traditionally gone. Recall Strain's statement (above) that "Printing of books began as a means for teaching better the old routines and traditions. In the course of time, its hidden realities aided in producing a wholly different society."

We are at a similar stage with television at the present time. We are using television (comparable with the elevator) to get to the top of existing structures (traditional educational goals and objectives). The fact that we can use telecommunications to build new educational structures has not yet fully dawned upon us.

It would be a fascinating exercise to design a school curriculum around the television tube rather than around the printing press. Or, as Trow put it, "It would be interesting to speculate as to what would have been the result if film projection had been invented in the fifteenth century and the printing press in the nineteenth. The book would have had a race to overtake the picture and the spoken word."[9]

In many subject areas, for example, cannot television be used to make the curriculum concrete and vivid? Cannot the arts be brought alive? Cannot sciences and mathematics be brought up to date? Cannot areas of humanities and social studies be made meaningful and relevant? In fact, cannot the artificial compartmentalization of curricular subject areas possibly be eliminated?

Why, for example, must teaching be organized around such fixed disciplines as English, economics, mathematics or biology? Why not around stages of the human life cycle: a course on birth, childhood, ado-

[8]*Ibid.,* pp. 8–11.

[9]*Ibid.,* p. 66.

lescence, marriage, career, retirement, death. Or around contemporary social problems? Or around significant technologies of the past and future? Or around countless other imaginable alternatives?[10]

Toffler points out other priorities and imperative objectives that the schools of the future will have to deal with.

For education the lesson is clear: its prime objective must be to increase the individual's "cope-ability"—the speed and economy with which he can adapt to continual change. And the faster the rate of change, the more attention must be devoted to discerning the pattern of future events.

It is no longer sufficient for Johnny to understand the past. It is not even enough for him to understand the present, for the here-and-now environment will soon vanish. Johnny must learn to anticipate the direction and rate of change. . . . And so must Johnny's teachers.[11]

And so forth. Other critics and social observers have their own statements of needs and priorities that the schools should list as prime objectives. The important concept is that we start with a clear-cut idea of what we want the schooling system to accomplish. Then let us look at the tools and procedures for doing the things we say the schools should do. Where can ETC help to meet these goals and purposes? In our attempt to provide a meaningful educational experience for all of our citizens, we may find that the nineteenth-century institutions and traditions we are still clinging to may not be the most efficient or most humanistic way to do it.

12.3   Scrutinizing the Sacred Cows

Turning from the functions and purposes of the schooling systems, let us examine the institutions themselves. Television, and other electronic media, will help to accelerate many inevitable changes in our traditional schooling patterns. Over the years many educational practices have become institutionalized conventions merely as a result of historical precedent and/or administrative convenience. How many of these sacred cows really make sense?

Roles of Teachers   How long can we continue to consider the teacher a pedagogical jack- (or jill-) of-all-trades? Can the teacher realistically be expected to fulfill the role of information giver–counselor–manager of learning situation–tutor–discussion leader–curriculum planner–writer–evaluator–demonstrator? These are just some of the jobs related broadly to the teaching function! Trow points out the overwhelming array of nonteaching responsibilities that many teachers also must carry: custodian, clerk, foster parent, disciplinarian, examiner, audiovisualist and technician, librarian, student adviser, therapist, recreation leader, and responsible citizen.[12] How can technological devices assume more of a central place in the dispensing of information, thus enabling the teacher to concentrate on those areas in which individual contact and human leadership are needed?

Student-Teacher Ratios   The classical arrangement of housing twenty-five or thirty students in one room with one teacher has long been accepted as the conventional norm. It is administratively convenient; it is financially feasible; and it is manageable from a supervisory and disciplinary point of view. But does it make any sense pedagogically? Is a thirty-to-one ratio justifiable from a teaching/learning standpoint?

---

[10]Alvin Toffler, *Future Shock* (New York: Bantam Books, 1970), p. 410.

[11]*Ibid.*, p. 403.

[12]Trow, *Teacher and Technology*, p. 45.

How often would a one-to-one, an eight-to-one, or a 500-to-one ratio make more sense? For more efficient use of human energy, how can media be used in various-sized groups? Individualized instruction? Large-group presentation of information?

Academic Schedules   How long should students spend in school? In studying a given subject? Should academic schedules be determined by the clock and the calendar or by a student's motivation and ability to master a given set of objectives? What is sacred about a fifty-minute period for each subject area? A six-hour school day? A nine-month school year? A twelve- or sixteen-year schooling span? These time designations are convenient administratively, but do they make any sense educationally? What can they tell us about what Johnny has learned and is learning? Education is not a faucet that can be turned on and off at six-hour or nine-month intervals. We will see all of these conventional time patterns modified, and many of these changes can be based upon television and other media applications.

Locations of Learning   Where does learning take place? In school buildings, obviously—classrooms, laboratories, the auditorium, and the library. But what about the home? Church? Neighborhood centers? What can be done to recognize and structure learning experiences that take place within the context of the museum, the zoo, the park, the theater? What about the mass media—the TV tube, the radio, the newspaper, the popular magazine, and the movies? At higher education levels, how do we acknowledge and give credit for learning that has accrued from hard experience—holding down a job, travel, leisure pursuits, unstructured reading and research, "street education"? We can give a form of due credit, of course, for military experience or for some vocational apprenticeships—but how do we really give full credit for learning picked up from informal sources?

Measurement and Grading   Perhaps the most sacred cow, whose worship results in much of the inconsistency and arbitrary nature of our present educational structure, is the widespread subjective nature of grading or marking. The practice of having a teacher award a letter grade in one or more subject areas for each student one or more times each semester leads to considerable confusion and lack of standardization. For instance, how does one equate letter grades from different schools—a highly regarded prep school with high standards and a rural school in an economically depressed area? How about the difference between teachers—one with a reputation for tough grading and one who is known to be an easy marker? How does one compare a C in an honors class with an A (at the same school) in a regular class of the same subject? What exactly do subjective letter grades measure anyway—neatness of work turned in, speed with which a task is accomplished, ability to participate in a class discussion, attendance, willingness to undertake extra projects for the teacher's favor? Or, ideally, should we not simply be awarding credit for the achievement of a given set of learning objectives in a particular academic unit, with separate recognition for the amount of time spent in formal instruction?

12.4   The Role of Technology in Education

The question raised at the onset of this chapter was, *"Where do telecommunications best fit into the overall educational picture?"* In examining this question, we have looked first at some of the conventions and sacred cows of the present "overall educational picture." In examining the role of telecommunications, it may be well to turn next to the broad concept of "instructional technology."

Instructional Technology Defined   Telecommunications media and ITC generally can be viewed in the larger context of instructional technology. Generally, people loosely equate "technology" with machines and "hardware." This may be one commonly accepted definition; but in an instructional sense we are concerned with much more. "Technology" (from the Greek *technē,* meaning "art, science, or skill," and from the Latin *texere,* "to weave or construct") refers to the totality of putting together a system or process scientifically and skillfully.

The best simple definition of "instructional technology" probably comes from the Commission on Instructional Technology in its report, *To Improve Learning:*

> Instructional technology is . . . a systematic way of designing, carrying out, and evaluating the total process of learning and teaching in terms of specific objectives, based on research in human learning and communication, and employing a combination of human and nonhuman resources to bring about more effective instruction.[13]

The emphasis is not upon machines but upon the total system designed to facilitate the learning process. In this sense, telecommunications media can contribute to the "nonhuman resources to bring about more effective instruction."

Glaser points out that, within the instructional technology concept, there are four areas in which significant changes can be expected:

> (a) the setting of instructional goals will be recast in terms of observable and measurable student behavior including achievement, attitudes, motivations, and interests; (b) the diagnosis of the learner's strengths and weaknesses prior to instruction will become a more definitive process so that it can aid in guiding the student along a curriculum specially suited for him; (c) the

techniques and materials employed by the teacher will undergo significant change; and (d) the ways in which the outcomes of education are assessed, both for student evaluation and curriculum improvement, will receive increasingly more attention.[14]

Both Glaser's four points and the commission's definition of "instructional technology" are explored further in Chapter 13 as we deal with rational problem-solving patterns (Section 13.3), instructional design (Section 13.4), and determining precise learning objectives (Section 13.6).

The Place of Media in an Educational System   In context of Glaser's four areas of potential change, ITC would be considered in the discussion of his third point, "the techniques and materials" to be used in the actual teaching/learning process. Telecommunications media should not be considered as an end in themselves; neither should they be considered synonymous with "technology"; nor should they be considered in isolation—as a distinct and separate part of the total instructional/learning situation. Rather, telecommunications media should be viewed as an integral part of a total systems approach to the educational process—one of the tools to be used in a carefully constructed learning experience. In fact, the other elements of the learning system—a clear statement of objectives, a careful analysis of the learners' characteristics and prior achievements, and a firm plan for the assessment and evaluation of the learner's progress—must be considered before we can begin to examine the specific teaching/learning activities (in which media may play a part).

Therefore, we cannot determine the proper role for ITC by simply looking at the individual ITC media and trying to adapt them to an existing schooling set-up. As Trow has stated,

---

[13]*To Improve Learning: A Report to the President and the Congress of the United States,* Commission on Instructional Technology (Washington, D.C.: U.S. Government Printing Office, March 1970), p. 19.

[14]Robert Glaser (ed.), *Teaching Machines and Programed Learning II* (Washington, D.C.: Department of Audiovisual Instruction, National Education Association, 1965), p. 804.

The question that faces educators today is not how any one of these instructional media can best be used in the schools as they now are, but rather, how they can best be fitted together, along with the school personnel, all to become not aids or adjuncts but components in an educational system. This is something more than training teachers to employ the new media—use the tools and operate the machines. The new technology requires that man learn to cooperate with the machines. He must know what each component can do, and so fit them into subsystems within the larger system.[15]

Educators and instructional designers must think of ITC tools not as extra devices or gimmicks that can be attached to, plugged into, or glued on to existing educational situations.

The usual procedure is to tinker with the most troublesome problem, such as overcrowding, and add a new wing, a new course, or a new teacher. . . . Another procedure is to experiment with some new device like team teaching, television, or teaching machines. . . . Unless frantic efforts are made to rearrange schedules, provide the staff training needed, and develop instructional materials, the inchoate disruption may cause the efforts to be discontinued. Or at best, the results may be no better than under the old methods.[16]

In too many instances, this is precisely what has happened: ITV has been "added to" an existing schooling arrangement—with no other modification of the situation—and then the results have shown "no significant difference" when test scores were examined (Section 17.4). Obviously, there will be no significant change if nothing significant was done to modify the learning situation! Adding one additional piece of hardware to a faltering system will not alter that system. Only in rare cases (such as Samoa, Section 11.3) have significant changes been made to the whole system as television was introduced; and in these instances, there was indeed a significant change in the learning system—with television being but a component of that change.

12.5 Changing Patterns: Measurement and Assessment

So, let us assume that—in any ideal system of education—we will need to make some drastic changes in our present arrangements; we will need to throw out any of our current sacred cows that no longer serve our purposes. In fact, let us start from scratch.

As John Goodlad expressed it, we can rethink our decisions:

The point I am trying to make is that men have the opportunity to remake *all* previous decisions. Every single decision governing a school was made at one time or another by a man or by men. At the time the decisions were made, less data were available than are available today. The men who made these decisions were no brighter than we are, and they were less well educated. Therefore, it behooves us to examine every decision about schooling before we make it—decisions on size of buildings, and whether or not we want one at all; number of teachers, and whether we need a certified teacher for every 28 and a half youngsters; whether there will be a library that houses real books, or one which is a computerized box.[17]

Imagine for the moment that "education" does not exist. There are no schools, no teachers,

---

[15]Trow, *Teacher and Technology*, p. 116.
[16]*Ibid.*, pp. 119–120.

[17]John I. Goodlad, *The Future of Teaching and Learning* (Washington, D.C.: National Education Association, 1968), p. 19.

no instructional materials, no systems of grading and classifying students. The institutions and paraphernalia simply have not yet been created. Then *you* are given the assignment of creating the system. Let us assume that you have been appointed consultant to design a theoretical educational system for a nation similar to the United States. Your system must be designed to meet the needs of a free society; you must take into account the myriad individual needs of learners from toddlers to senior citizens.

For the purposes of discussion and stimulation, let us try to design the most provocative and radical system possible. This may not be the ideal system that any one of us would care to defend at the present time. But—in order to stimulate our tradition-laden imaginations—let us try to come up with a totally different approach. The remainder of this chapter will be devoted to such an exploration. Where do you start? What do you build first?

"The First Priority"   In the simplest kind of "design" model (or problem-solving pattern), there are three questions to be asked:

Where do we want to go?

Where are we now?

How do we get from here to there?

Let us assume that you have already tackled the first question, that you have determined your goals and broad objectives for your theoretical schooling system. You have outlined your vocational needs and liberal arts and citizenship programs; defined societal needs, leisure-time objectives, and needs for further training and education; identified the important subject areas; and so forth. (We will not be concerned—in the context of this book—with redefining the goals of education.) Which of the three questions must you answer next?

Unfortunately, too much of our current pedagogical thinking has focused on the last

question, "How do we get from here to there?" We are preoccupied with the machinery of transportation—without giving adequate thought to defining exactly where we are before we begin. We design the teaching/learning activities—the classrooms, the materials, the laboratories—without adequately examining the would-be learners to see "Where are we now?" Let us answer the second question before skipping to the third.

Should we not be able to assess accurately where the student is before we try to get him from "here" to "there"? In designing a radical educational system from the ground up, the first priority should be to establish an accurate and reliable system of measuring and assessing the learning accomplishments of learners—on a wide variety of scales or subject matter competencies. Then, once we have these measuring devices, we can begin to map out the best travel routes for each individual learner.

Many educators have pointed out the need to devise testing and evaluation procedures that measure the absolute mastery of knowledge or acquisition of skills—rather than relying upon subjective letter marks and grade designations that purport to indicate the relative standing of pupils as compared with some norm. Glaser stated, "... tests and information from other sources will need to be developed to describe competence in a more clear-cut and absolute way than is usually the case with relativistic grading procedures where test scores take on meaning only in terms of the relative standing of students."[18] And Goodlad has pointed out that

tests for individual diagnosis of performance are hard to come by, although such tests increasingly are being constructed. Even if they were easily accessible, however, it would appear . . . that overcoming grade

---

[18]Robert Glaser, "The School of the Future: Adaptive Environments for Learning," *in* Louis Rubin (ed.), *The Future of Education: Perspectives on Tomorrow's Schooling* (Boston: Allyn and Bacon, Inc., 1975), p. 132.

and group norms in seeking to use them properly constitutes a formidable obstacle.[19]

Individual Competency Scales Let us suppose, for example, that we would first establish a really comprehensive measuring system for determining precisely what competencies the learner possessed in a number of discrete knowledge and skills areas—abilities which are basic to learning in many other areas. This evaluation system would deal with distinct skills in a basic subject separately. For example, instead of measuring "verbal skills," perhaps we would have *individual scales* for "grammatical theory," "spelling," "word choice," "sentence structure," "paragraph construction," and so forth. Each learning area might be rated on what could be termed, for example, an "individual competency scale" (ICS).

These ICS examinations would be available on a flexible basis (see below), and anyone could apply for any exam he or she felt ready to take at any time. These exams would not be related to any particular curriculum or schooling program. The ICS tests would simply be a widely accepted measurement device to determine how much a learner has achieved in any specific subject area. Each ICS area might be scored on a range of, say, 500 points—with zero representing absolutely no ability or knowledge and 500 representing a theoretical total mastery of a subject.

As individuals periodically take various ICS exams, we could begin to build a fairly accurate picture of each individual learner's abilities and interests. Within the language skills area, for example, a certain student might have ICS scores of 147 for grammar, 128 for spelling, 209 for word choice, 232 for sentence structure, 186 for paragraph construction, and so on. The individual learner could then work on those areas he wanted to improve (for instance, grammar and

spelling)—by enrolling in a course, engaging in independent study, hiring a tutor, following some programed instruction, participating in an ITV series, or whatever other means he might choose individually or with the advice of a teacher/counselor in a formal schooling context. When ready, the learner would retake the particular ICS test he has been preparing for.

Most ICS exams would probably be paper-and-pencil objective tests. Others would be actual physical performances and demonstrations. Some would be subjective evaluations based upon written essays or compositions or even subjective assessment of creative works or interviews with ICS examiners. What is especially important in the more subjective areas, the teachers would not assume the role of official examiners (as opposed to today's schools where the teachers are evaluating and grading their own products).

There might be a very large number of ICS areas in your new educational system—a set of scales for any curriculum area that would be required by any other commonly pursued subject, through the entire public schooling system including higher education. ICS exams would cover acquisition of knowledge, reasoning and creative abilities, and various skills areas. There would be some basic ICS exams that everyone would be expected to take (such as language and math skills); examinations in specialized areas would be taken only by those who were interested in those particular areas.

Principles of Continual Assessment There would be several principles behind the use of the ICS scores. First of all, they could be accepted at all levels, in all parts of the country, by all schools, for all types of educational purposes. They would represent a system of nationally recognized achievement measuring.

Second, the ICS system might be used to replace—or at least substantially supplement—the current system of marking and placement by grade levels. Educators have long theorized

[19]John I. Goodlad, *Looking Behind the Classroom Door: A Useful Guide to Observing Schools in Action* (Worthington, Ohio: Charles A. Jones Publishing Company, 1974), p. 86.

about the advantages of nongraded schooling; and many quasi-nongraded programs have been instituted. However, without a really comprehensive system of absolute achievement measurement, such nongraded systems have been destined to remain only partially effective.[20]

Third, to be functional the ICS approach would have to be flexible and accessible in its implementation. Testing operations would have to be readily available to learners of all descriptions. Arrangements would have to be flexible enough to allow certain basic tests to be administered at virtually any time the learner desired. Learners would be allowed to repeat most examinations as often as desired (within certain practical limitations and with a reasonable delay between tests). Many basic exams might be administered free through the school system; other—more specialized—exams would be available for a service fee.

To some extent, the ICS system could be built upon the foundations established by the various achievement and aptitude tests designed by existing comprehensive testing agencies such as Educational Testing Services in New Jersey, the Psychological Corporation in New York, and the College-Level Examination Program (CLEP) of the College Entrance Examination Board. The main differences would be that the ICS system would contain sufficient discrete areas to enable it to be used as a substitute for present grading systems (especially at the elementary level and for basic subjects at the secondary level); and, also, that the ICS tests would be so frequent, widespread, and universally used that the trauma often attached to the current periodic use of existing universal testing programs—CLEP, Scholastic Aptitude Test (SAT), Graduate Record

Examination (GRE), and the like—would be greatly eliminated.

As a corollary to the ICS testing, there could also be widespread use of programed instructional materials and self-measuring devices so that the learner (with assistance of counselors, tutors, and parents—as necessary) can maintain a reasonably accurate picture of where he is.

This, then, would be the first priority of our innovative radical schooling system—development of a functional ICS approach. No more complete reliance upon report cards. No more immutable grade levels. No more periodic final exams. Measurement and academic evaluation would be based not upon how long a student sat in a classroom, whether or not he had good attendance, or what his chronological age happened to be; but, rather, continual assessment would be based upon what competencies a learner had in a given subject area.

There obviously are drawbacks to be avoided with the implementation of an extreme proposal such as the ICS system—just as there are dangers in the radical implementation of any complex plan. Curriculum planners and other educators would have to keep in mind always that ICS evaluations are for the use of *individuals* to measure competencies that are assumed (and therefore required) for certain courses or jobs. ICS mastery levels should not be used to compare schools—or individuals—since averages on ICS tests would be difficult to interpret and would not take into account the varying educational or vocational objectives of different schools and the varying abilities and goals of students. The individual competency scales should be designed solely to measure the absolute (not relative) achievement of learners in specific discrete subject areas.

Nor should ICS achievement be equated with *schooling* per se. There are many intangible benefits to be derived from attending an elementary school for six or seven years, secondary school for six years or so, and a college long

[20]*Ibid.* Goodlad, in his survey of sixty-seven elementary schools, reported that, "It appeared to us that most of the so-called nongraded schools were endeavoring to tack nongrading onto the existing structure instead of exploring the full implications of a new structure."

enough to earn advanced degrees. Many of these intangible and socialization benefits cannot be measured on any kind of specific testing instrument. (See below.)

Two Operations of "Education"   It should now be apparent that what we commonly lump together as "education" really consists of two quite separate and distinct operations—*accreditation* and *instruction*. To these two identifiable processes, others would add a third operation of education—"intangible benefits" that include all other learnings, experiences, and attitudes acquired through the process of socialization.

Accreditation   Evaluation. Assessment. Measurement. Thus we determine where we are, what our starting base is, and how much we have accomplished. Too often, this process is relegated to a secondary status. We devise a test *after* we have designed the unit of instruction. We think up a final exam *after* we have "taught" the course.

Instruction   This, of course, is the first area that most people think of in connection with an educational system—the process of imparting knowledge, inspiring wisdom, and the teaching of skills. Here we are concerned with the tangible aspects of classrooms, teachers, instructional materials, media, libraries, and the like. But instruction is only half of the picture; it answers only the question, "How do we get from here to there?"

In our theoretical system of education, it may be that these two processes or functions should be entirely separated; they should not be housed within the same agency. One institution could be concerned with the teaching/learning activities; and another agency with the accreditation/assessment functions. Of course, we already have many examples of this split system: the state driver's license test, the medical

and legal professions with their boards and bar examinations, other professional associations and accrediting organizations. Some institutions of higher education (for example, the University of London, the New York State Regents External Degree Program) have evolved systems of accreditation that can result in the issuance of degrees regardless of where or how the prerequisite knowledge was gained. The New York State Regents examinations accomplish a similar purpose at the secondary level.

Intangible Benefits   Clearly, a system of education provides more than specific learning objectives and formal instructional programs. Socialization contributes to many intangible benefits. Attitudes are developed; personalities are formed; future directions are mapped out. An appreciation for the value of education itself is encouraged. Students learn to relate to different kinds of peer groups; they develop abilities that enable them to function in group processes. Pupils grasp an understanding of the institutionalization of society; they learn how to recognize and work with authority figures—et cetera. There is much more to the formal schooling process than simply academic learning programs and evaluation systems.

It is probable that, in any future system of education you might devise, schooling structures will continue to play an important role in answering these impalpable needs. Perhaps two different kinds of certificates or documents will be awarded when a student has reached a specific educational plateau: one (a diploma) to recognize the hours or years spent in a certain schooling/socialization environment (elementary, secondary, or college); and one (a degree), based partially upon ICS exams, to indicate specific academic achievements.

This may be the avenue we want to pursue in our system of the future: separate the functions of instruction and accreditation; recognize the importance of the intangible benefits of socialization; use an ICS system both as a diagnostic tool and as a determination of credit. Then, having responded to the question of how to determine where we are, let us turn our attention to some of the specific instructional concerns inherent in the question of "How do we get from here to there?"

12.6 Changing Patterns: New Responsibilities

In designing your radically different educational system, you probably would come up with some sort of flexible approach that would not depend upon mandatory attendance. In fact, it might not even be necessary to require participation in certain basic "courses."

Responsibilities of the Learner What if the responsibility for charting one's educational programs and determining one's learning objectives could be primarily left up to the individual learner? At early ages, especially, there would probably have to be guidance and advising from parents, counseling centers, and instructional institutions. But, even at these ages, would it not be desirable (and quite exciting) to watch young learners grow and explore and master those skills and content areas where their curiosity and motivation takes them?

In order to participate—as a responsible citizen—in the running of a democracy, every person should be educated to the extent that he or she can cope with the complicated issues involved in making intelligent decisions and setting policy for the nation. Every child should master the skills of communication and computation—to the extent of his or her ability—in order to be able to cope on a personal basis. If learners were left to determine their own educational objectives, might not each one intuitively, or of necessity, eventually recognize the importance of *wanting* to learn to read, to write, to add, to understand the economic system?

In calling for reform, John Holt points out the importance of the learner's motivation in advocating that we design

schools and classrooms in which each child in his own way can satisfy his curiosity, develop his abilities and talents, pursue his interests, and from the adults and older children around him get a glimpse of the great variety and richness of life. In short, the school should be a great smörgåsbord of intellectual, artistic, creative, and athletic activities, from which each child could take whatever he wanted, and as much as he wanted, or as little.[21]

Imagine a five-year-old interested in butterflies. As he chases and collects them, he notices differences. He compares them with moths. Someone explains a caterpillar to him. As his curiosity grows, he finds out that he can get more information from picture books. Eventually the need to be able to read the words in the book becomes paramount (along with the curiosity about words in newspaper comics and TV listings). He can then be guided along certain learning programs to develop his reading ability as he feels the need and as he is nourished by the satisfaction of attainment of self-determined goals. As his interest in the insect world increases, so does his desire to be able to master fundamental mathematical computations in order to verify certain observations, make simple predictions, and eventually begin to carry out basic research projects. Simultaneously, his other interests (in wagon wheels, fire engines, favorite stories, TV programs, money, popular music, and girls) would lead him into other self-motivated learning programs—increasing his various skills as well as adding to his store of information and knowledge.

As he explores and grows, his progress would be continually monitored on various ICS examinations. Periodic counseling sessions would suggest areas in which he might want to sharpen his skills or develop new interests. However, basically, the learner would be responsible

[21]John Holt, *How Children Fail* (New York: Pitman Publishing Corporation, 1964), p. 180.

for determining his own learning program—setting his own objectives and working toward them at his own pace.

Of course, the proposal is not so simplistic that we can expect every child to become a perfect self-motivated learner as soon as we open up the world's resources to each and every one. The student must be educated and trained in the techniques and attitudes which lead to successful academic self-management. Glaser stresses the job of the schools in teaching students how to handle responsibility for their own learning.

> Because an essential part of adapting the environment to the capabilities of the learner involves adjustments made by the learner himself, students must acquire increasing competence in self-directed learning. To accomplish this, the school must provide students with models and standards of performance so that they can evaluate their own attainments and select teaching activities (with or without the help of teachers and peers) as a function of their increasing competence.[22]

Many other problems are raised with this kind of idealistic proposal that relies heavily upon the responsibility of the individual learner. This system might work well for highly motivated, intelligent students with a rich cultural background. But would it be effective with inner city children? Rural pupils? Students with subtle learning problems? What about children from homes where education is not respected? What about the economically deprived? What kinds of checks and monitoring systems would be necessary to guarantee that each learner was benefiting from his individual learning programs? Could such a system realistically be worked out? Could it be made financially feasible?

Responsibilities of the Educators   As learners accept more responsibility for the direc-

tion of their own learning programs, and as technology assumes much more of a burden for the presentation of information (see below), the roles of the educators will substantially change. Basically, no longer will we expect the teacher to be a generalist, with competence in many skills. Many will specialize in measurement and assessment. Others may gravitate toward counseling and advising. Most may remain in some aspect of "teaching"—or more properly—the "facilitating of learning"; they will become the lecturers, the tutors, the discussion leaders, the curriculum writers, the planners, the demonstrators. Others will emphasize the discovery of new knowledge— the researchers. Some, of course, will want to serve as the administrators and managers of the learning systems.

Individualized computer-generated schedules might be developed for every student on a regular daily or weekly basis as he progresses. His small groups, his tutoring, his lectures, his media experiences, his lab work, his field trips, all could be arranged and scheduled as his needs differ and merge with those of other learners. The design and administration and successful operation of this kind of "system" would require an integrated technology of the highest order—combining human and nonhuman resources—involving self-teaching apparatus, media, and computers.

As the responsibilities for learning shift and evolve into new patterns, we will probably see the emergence of a new type of learning specialist, the commercial learning company— basically an extension of today's textbook and educational film industry. If we are to measure learning achievements by the accomplishment of specific performance objectives, then it becomes feasible for individual private contractors to devise instructional programs designed specifically to meet certain kinds of objectives. It would be possible, for example, for RCA, Wadsworth Publishing, IBM, the local school district, and General Learning Corporation all to devise learning programs each of which would be designed to take an individual with X abilities (whatever the speci-

[22]Glaser, in Rubin, *Future of Education*, p. 132.

fied entry-level behavior should be) and raise his ICS spelling score to, say, 325 points. It would then be up to the individual learner (or his school) to sign a performance contract with whichever learning company seemed to offer the best deal.

A variation on this approach might be a tax-supported voucher system. Some states have investigated such a possibility. Instead of requiring attendance in a public school, the state (or any other taxing authority) would present to the learners (or parents) vouchers that could be exchanged for a certain segment of education at a public school, private school, or—conceivably—any accredited alternative commercial learning agency. The concept is the same: the learners would "buy" their learning programs through whatever kind of educational contract would appear most promising.

12.7   Changing Patterns: Flexibility in Time and Space

In the halcyon times of days gone by, the process of education was rather simple and easy to administer. The local board of education built the buildings, hired the teachers, established the curriculum, and decreed that every child should follow the prescribed path. The purest example of this nineteenth-century model may have been the stereotyped frontier town where the townspeople built the schoolhouse with their own hands, hired a schoolmarm from "back east," and forced the children (with threat of a hickory switch) to stay in school until they learned the three R's. The elaborate administrative machinery and educational architecture we have carried three quarters of the way through the twentieth century is basically just a painstaking extension of this traditional model.

Today we are faced with an incredibly complex educational milieu: an increasingly complicated web of personal and worldwide needs and problems to be solved by "education"; a tremendous acceleration in the rate of change of all types; an incomprehensible mutiplication of mankind's storehouse of knowledge (with the total amount of "information" in the world being doubled every eight to ten years); a bewildering insight into the varied learning patterns of different individuals; and—at the same time—a fantastic array of communications and teaching hardware that can revolutionize the teaching/learning process. As we speed irreversibly toward the twenty-first century, any further extension of the "schoolmarm" model seems dangerously inadequate and insecure. The concepts of the "school bulding," the "classroom," "the school year," and the "class period" need to be seriously reconsidered.

Dropping Out and In   Certainly there is less need or justification to maintain the sacred cow of designated learning periods—whether in terms of a fifty-minute subject/class period, a six-hour school day, a nine-month school year, or a twelve-year schooling span. With today's teaching apparatus, computer-based instruction, self-teaching programs, telecommunications channels, and storage and retrieval devices, "schooling" can truly be a continuous process—subject only to the schedules and motivations of the individual learner. Goodlad's look into the future includes the following:

> Teaching and learning will not be marked by a standard day of from nine to three, nor a standard year from September to June, nor a year for a grade of carefully packaged material. Age will be meaningless as a criterion for what one is to learn. Will learning be any less because there are no periods, no Carnegie units—thank God—no ringing of bells, no jostling of pupils from class to class? And what will the school principal and his administrative associates do when it is no longer necessary to schedule teachers so as to produce a balanced diet of subjects? Perhaps we will start doing some really important things.[23]

[23]Goodlad, *Future of Teaching and Learning*, p. 21.

Is there any reason why an individual cannot play back a video cassette at 3:00 a.m.? Spend thirteen hours in a single day to complete a programed textbook in elementary statistics? Take an ICS test in July? Talk to a computer terminal on a Saturday night? Complete the requirements for a B.A. degree at age seventy-eight? Or enroll in a differential calculus "course" at age twelve (if he has met the prerequisites)? On the other hand, is there any need for the learner to be formally "enrolled" in any schooling program at all? Maybe he would be better off by taking a two-week (or two-month or two-year) period for a trip to Australia—or for skiing with his parents—or for lying in a meadow watching daisies grow.

The Where of Learning   As you deliberately design a radical system to challenge the traditions of the school building and the four walls of the classroom, you may determine that learning takes place wherever people and material can come together—the church, the neighborhood center, the office, museums, parks, zoos, stadiums, theaters, the street, a training center, a research laboratory, a factory, a public library, or anywhere else. Certainly the home will be considered more and more of a learning center. Many authors and forecasters have predicted futuristic home communications/learning centers with numerous television and audio channels, computer terminals, automated learning devices, self-teaching materials, telephone access to testing centers, and so forth.

> School, as we now know it—whether egg crate or flexible space—will have been replaced by a diffused learning environment involving homes, parks, public buildings, museums, and an array of guidance and programing centers. It is quite conceivable that each community will have a learning center and that homes will contain electronic consoles connected to this central generating unit. This learning center will provide not only a computer-controlled video tape, microfiche, and record library, but also access to state and national educational television networks.[24]

Suppes adds to Goodlad's predictions by pointing out that the exact combination of elements in the learning environment cannot easily be determined at this point.

> Once we have the possibility, there will be a swing toward the home as the place in which learning takes place, that is, the place where the student stays and engages in learning, even though the exact configuration is simply not easy to see at the present time. . . . The point here is that getting the instrument into the home is not a technological problem; rather the problems are how to use it and how to find out whether people want to use it once it is there.[25]

As you build your innovative system, many new concepts and approaches and designs may affect the formal physical surroundings set aside specifically for schooling purposes. In addition to being an instructional location, a "school" building may become more of an administrative headquarters, communications center, testing operation, and storage and retrieval depot. As Ralph Tyler summarizes it, "the line between schooling and the educational responsibilities of other agencies may shift somewhat, but it is very unlikely that the school will disappear."[26]

We will probably see public libraries and school libraries merge into common resource centers. (In Hawaii the public libraries are operated as part of the state department of educa-

---

[24]*Ibid.*, pp. 21–22.

[25]Patrick Suppes, "The School of the Future: Technological Possibilities," in Rubin, *Future of Education*, p. 150.

[26]Ralph W. Tyler, "The School of the Future: Needed Research and Development," in Rubin, *Future of Education*, p. 168.

tion, and new integrated "community" libraries are being built to meet both school and community needs.) All library services will become truly multi-media operations. As early as 1947, the Carnegie Corporation made a grant to the American Library Association to support a Film Advisory Service "for the purpose of demonstrating that public libraries could serve as centers for the distribution of audiovisual materials as well as books."[27] An increasing number of public libraries have become centers for educational advising and administration of the CLEP examinations, prompting one city librarian to remark, "At last we see the possibility of the library becoming in fact the 'people's university' that Benjamin Franklin envisioned."[28]

In addition to school buildings and a more comprehensive role for libraries, other specialized facilities may evolve to fill out the skeleton of your new proposed educational system—neighborhood discussion centers, commercial testing offices, tutoring centers, and so forth.

12.8   Changing Patterns: The Media Matrix

We now have progressed to the point where we may be ready to deal with the question raised at the onset of this chapter, *"Where do telecommunications best fit into the overall educational picture?"* Even within the context of a rather conventional modification of the present system, several predictions can be made. Trow notes that "television seems most effective in the perceptual area of learning for purposes of general orientation and demonstration."[29] From the standpoint of physical facilities and time al-lotted to media experiences, Saettler conservatively forecasts:

> It also seems likely that as much as 50 percent of the college degree program will be available for credit via television and that school buildings will be more frequently designed for the use of instructional television. Part of the school day may be devoted to the large television class and the rest to small-group discussions and independent study or laboratory sessions.[30]

We will also see a merger of educational resources, media, and materials. With increasing frequency we hear the declaration that much of the "hard-covered software" (textbooks, 16-mm films) go out of date too fast. Learning resources are increasingly written, combined, edited, collated, and distributed where the learning takes place—and as the learners define the materials they need. Many different materials are being utilized: pamphlets, magazines, filmstrips, audio tapes, video cassettes, 8-mm film, paperback books, newspapers, records, reference works, and so forth.

As school and public libraries merge into common community resource operations, so even do the distinctions among commercial television, public television, and school television become blurred. For it matters little where a learner gets his information. What does matter is that the information be presented over an appropriate medium, that it be accurate and up-to-date, and that it be available when and where the learner need it.

In the next few years, the biggest technical advances in schooling hardware probably will be made in three broad areas: (1) television production and distribution breakthroughs; (2) microimage storage and retrieval of information; (3) computer applications for handling learning

---

[27]Paul Saettler, *A History of Instructional Technology* (New York: McGraw-Hill Book Company, 1968), p. 188.

[28]Alfred and Florence Steinberg, "The Do-It-Yourself Way to Earn College Credits," *The Reader's Digest,* October 1975, p. 138.

[29]Trow, *Teacher and Technology,* p. 109.

[30]Saettler, *Instructional Technology,* pp. 248–249.

programs, testing data, and administrative functions.[31] In each of these areas, telecommunications advances—especially in the instantaneous delivery of information over great distances—will play a crucial role.

Another way of examining the potential of telecommunications in schooling applications would be to consider broad categories of learning "problems" and bottlenecks. What kinds of technical limitations are there in the theoretical schooling set-up that we have been discussing? Probably the largest single "problem" is in the handling of information—the storage and cataloging, the retrieval and dissemination of "information" of all descriptions. This need for information handling includes everybody in the system: the learner, looking for accurate detailed data; the teacher, seeking up-to-date pedagogical help; the counselor, needing detailed student assessment; and the administrator, wanting budgetary assistance and cost/effectiveness data. Telecommunications media certainly perform essential roles in these situations.

In addition to the applications listed above—which deal primarily with the *distribution* side of telecommunications—there are many *production* applications that also should be considered. These include both curriculum-structuring applications and audio-visual applications (Section 10.5): producing complete instructional courses and shorter segments, enrichment/reference uses, self-teaching modules, recording field trips, preserving interviews, structuring new courses following instructional design procedures (Section 13.4), an overhead camera, and so forth. In Section 17.4 we discuss a few broad implications and suggestions as to how telecommunications can most effectively be used in a learning situation.

In answering the question *"Where do telecommunications best fit into the overall edu-cational picture?"* many types of answers could be given. One of the best, and shortest, was once given by a student who replied, "Use television—or any other piece of equipment—wherever the hardware can do a particular job better and more efficiently than a human being; and then use human resources for those situations where human contact and interpersonal relationships are essential." In other words, use the technology to humanize the learning process.

The theme of humanizing the learning process through the use of television and related technology has often been repeated. A booklet put out by the Committee for Economic Development stated:

> Properly employed, the new instructional technology will assist the teacher, but it will not displace him any more than computers or the other advanced technologies applied to clerical work have displaced white collar workers. Moreover, instructional technology can contribute to the humanization and personalization of the school.[32]

Machines can help humanize the schooling process by handling those routine instructional tasks that can be most efficiently assigned to machines—thus allowing humans to handle those areas where human relationships are crucial. But technology can also help to humanize education in another way—by facilitating a more humanistic emphasis in subject matter:

> What I mean by a humanistic curriculum is an emphasis on mankind values in the substance of the curriculum and a concern for both the individual and mankind in the environment of teaching and learning. I believe these tasks to be so formidable and their import of such magnitude that I wel-

---

[31]*See especially* Goodlad's *The Future of Teaching and Learning,* an address given before the National Education Association.

[32]Committee for Economic Development, *Innovation in Education: New Directions for the American School* (New York: Committee for Economic Development, 1968), p. 42.

come the computer into the instructional process and charge it with teaching some of the basic skills and concepts which are only the beginning in educating the compassionate, rational man. I submit that the computer can and will do certain instructional tasks better than any human teacher can perform them. The research challenge is to catalog those aspects of instruction that are most appropriate for the machine, on one hand, and for the human teacher, on the other.[33]

The Ultimate Briefcase Carrel   Before leaving the task of devising our new educational system, it may be worth while to explore the possibility of designing the ultimate media distribution/reception device—a portable terminal/display "carrel" that could give the user immediate access to virtually all of the printed and audio-video information stored anywhere in the nation—or in the world.

Imagine a small case, somewhat smaller than a normal briefcase or portable record player. Opened up, it reveals a seven-inch video monitor and a pair of headsets for private viewing and listening. The carrel also includes a photocopier to instantaneously reproduce a hard copy of any visual material wanted. It has a fiber-optic cable that can be connected to any convenient laser terminal (Section 15.4); these fiber-optic laser terminals would be as commonplace as telephone jacks are today (in the home, at the office, in school/communications centers, and even in pay booths). With this laser hook-up, you would then have instantaneous potential access to thousands of channels of information.

There would be several operational modes built into the "briefcase carrel." Through a dial-tone, push-button system, you could have access to the master index catalog (comparable, say, to the Library of Congress index system), which would give a catalog and code number for everything available. Another mode would allow you to punch up any material desired (which would then be sent out from a "National Programing Storage and Retrieval Center" somewhere in the Midwest); this "viewing mode" would give the learner access to every page of every reference book, every lecture ever recorded by the world's greatest lecturers, every instructional program ever devised, every instructional film ever made, and so forth. A third mode could connect the learner directly with any of several individual computer-based instructional programs that the learner is currently working on. There might be a fourth setting, a "testing mode," that would connect the learner with any of the ICS examinations he wished to take.

Futuristic? Certainly. Far-fetched? Somewhat. Impossible? No more so than walking on the moon, or sending back color pictures from Jupiter. Such a distribution/reception device certainly is within the realm of technical possibility. Most of the hardware components have already been worked out—at least theoretically. All that remains to be done is to work out the administrative, financial, political, legal, and pedagogical details (all man-made problems).

Although the potential for significant revisions of our schooling systems exists, the fact is that little has really been implemented on a meaningful scale. After a 1974 comprehensive study of educational practices in sixty-seven elementary schools in the United States, John Goodlad came to the following conclusion: "many of the changes we have believed to be taking place in schooling have not been getting into classrooms; changes widely recommended for the schools over the past 15 years were blunted on school and classroom door."[34] He continued:

Chances are, most teachers seeking to teach inductively, to use a range of instructional media, to individualize instruction, to nongrade or team teach, have never seen any of these things done well, let alone

---

[33]Goodlad, *Future of Teaching and Learning*, p. 16.

[34]Goodlad, *Looking Behind the Classroom Door*, p. 97.

participated in them to the point of getting a "feel" for them or of how to proceed on their own. We simply do not have in this country an array of exemplary models displaying alternative modes of schooling, in spite of assumed local control and diversity.[35]

In the last four sections of this chapter, we have been attempting to design some of the theoretical elements of a totally new and different approach to education. Unrealistically radical and deliberately provocative, this exercise has been undertaken in an effort to stimulate some imaginative thinking about the potentials of reorganizing a few of our traditional educational concepts and approaches. Even if it were immediately possible, would the model we have designed actually be completely desirable? Which of the elements described above would you really like to see incorporated into a new approach to education? Which of the above elements might tend to lead to substantial resentment (due to the possibility of a nationally imposed curriculum or a federal testing program or the threat of teacher unemployment)? Which of the elements might be too revolutionary? Which might prove to be acceptably evolutionary? And which changes are likely to occur whether we want them to or not?

Think about the potential of telecommunications media, the hidden realities of television and related media, the traditional sacred cows of our educational establishments; and think about the importance of taking a "design" approach to instructional technology. Then determine how you would actually respond to the challenge of designing the ideal learning system.

So much for the world of radical theories and idealized models. With these concepts in mind, let us return to today's reality and examine some of the aspects of ITC administration and instructional planning.

---

[35]*Ibid.,* p. 103.

Administration and Instructional Planning

This chapter bridges the gap from the world of theory to the operational aspects of educational telecommunications (Part Four). In terms of the fifteen components outlined in Section 10.4, this chapter will deal with components 1 to 4. The last four chapters of the book will be concerned more practically with the remaining components. This chapter also returns us to the functions organization (introduced in Section 1.7). We want to examine several administrative considerations as well as the first "function" of instructional planning.

13.1   Administration of a Functions Organization

The suggestion was made in the first chapter that television should not be viewed in isolation. Educational telecommunications should be considered as part of an integrated media/instructional system and should not be set aside as a separate "project"—somehow isolated from the mainstream of the instructional/training enterprise. To this end, a functions or processes organizational structure was proposed that would combine all media into an integrated operation and make administrative subdivisions among the various instructional steps or functions—rather than among the differing media (a television department, a book division or library, an audiovisual center, and so forth). The organizational chart (13-1) is a typical representation of the administration of such an organization, broken down into seven units (excluding the broad support categories common to all organizations: budget and finance, accounting services, personnel, public information, secretarial services, and the like).

How exactly are these functional units related? What are the specific concerns and considerations under each broad function? This is what Part Four will be looking at from a practical perspective.

1. *Instructional Planning:* This primary function includes several different substeps. The

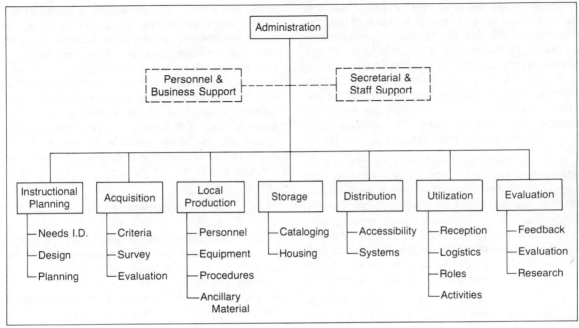

13-1 A Typical Functions Administration of an ITC Organization

specific instructional or training needs must first be identified. An instructional program to meet those needs must then be designed—which may or may not involve extensive use of media and materials. Specific curricular or industrial planning then must involve appropriate members of the instructional staff and supervisory personnel. Decisions must be made as to the exact types of materials that are needed. What will the overall instructional package look like?

2. *Acquisition:* The next step is to acquire—by lease or by purchase—appropriate materials that are needed in the instructional package. Can we find existing films, audio tapes, books and pamphlets, an ITV series, overhead transparencies, or 8-mm film loops that will do the job? If existing materials can be found, it certainly will be cheaper to acquire them than to produce new ones.

3. *Production:* Those materials that cannot be acquired from existing sources must be produced specifically to meet the identified needs. Several factors concerned with local production need to be considered: production personnel (including an on-camera TV teacher?), clearances and copyrights, design and structure of the telelesson, production of ancillary materials, facilities and equipment, production procedures.

4. *Storage:* Once the materials are acquired or produced, where are they to be stored? In what format? Who has immediate access? How are the print and nonprint materials to be cataloged? Are items of various media to be housed together? What role does the traditional library play?

5. *Distribution:* Next, we must be concerned with distribution—moving the materials to wherever they are to be used. What kinds of

accessibility are needed? How fast must learners be able to get hold of various materials? How far must the materials be delivered? How many copies might be needed at any one time? Which materials are to be controlled by the instructor and which by the individual learner? Many such criteria must be considered before the actual different dissemination systems can be examined (open-circuit, libraries, cable TV, resource centers, video cassettes, bookmobiles, and so forth).

6. *Utilization:* There are several elements concerned with the actual implementation and usage of the materials: the reception facilities that are needed; the arrangements for the viewing center or receiving classroom; ideal viewing conditions; the role of the classroom teacher or instructor immediately responsible for the content of the lesson; the specific activities designed for preparation and reinforcement of the telelesson.

7. *Evaluation:* The final function or process is that of evaluation. This includes informal feedback concerning a specific series or lesson in a series, data collected on the effectiveness of the telelesson, formal evaluation of a given project, and major research programs involving a formal design.

There is nothing sacred about these particular seven subdivisions. Some agencies may combine several into a single unit. Other set-ups may divide some of these categories and work with more than seven steps. The important concept is that the distinctions of a book operation and a film center and a television project are discarded—in favor of a total-media approach encompassing all materials but organized by the actual functions or processes in the instructional program.

This kind of approach will enable an elementary teacher to come to one central library/media/resources/learning center and ask, "How can I set up an individualized program for two students who are having trouble with long division?"—and be supplied with a programed work-book, two short video cassettes, a series of flip cards, and a single-concept 8-mm film loop. Or an automotive manufacturer service division might ask the training center, "How can we quickly update our service centers throughout the country regarding modifications in our stratified engine?"—and be supplied with a revised section of the service manual, a one-time closed-circuit TV hook-up, and a follow-up series of three video cassettes. In neither case would the need be answered by saying, "Go to the library, check with AV, and talk to the TV people."

13.2   Administrative Support

Regardless of the type of ITC project, it is imperative that full administrative support be evident on at least three different levels. Any training director or media program chief should be aware of these three levels and the nature of the commitment needed at each level.

Three Levels of Administrative Influence
At the top level, it is important that any extensive commitment to an educational media program have the backing and support of the *controlling policy board*. This may be a local board of education, a board of trustees, or a board of regents; in a business structure it may be the board of directors; for a government agency it would be the controlling legislative body—city council, state legislature, or whatever. At least tacit policy support must be obtained at this level.

It is also important that strong support be evidenced at the *top administrative level*—school superintendent, college president, company president or vice-president. The chief executive of any school or corporation may not be directly involved in the ITC project, but he must be favorably inclined toward the idea of it. In many instances, if the board is in favor of the project, and if the middle-management supervisor in direct charge of the project is enthusiastic, then the top executive will see no reason to stand in the way. On the other hand, it will sometimes be the chief executive who can be counted on for strongest support of a new ITC development.

Success of the project often will be determined by the attitude and support of the *middle-management* officer directly above the media director. This may be the school principal or dean or department chairman or a company branch manager or agency chief. Support at this level can be very beneficial in working around petty obstacles and annoying bureaucratic roadblocks. If there is resistance to the ITC idea at this immediate management level, the project will be subjected to a very rough existence.

Administrative Failures   There have been many failures of ETC endeavors of one type or another. Open-circuit operations have been closed down. Closed-circuit systems have folded. Many state legislatures and local school districts have questioned their involvements with ETV and ITV projects. Several corporations have discontinued ITC-based training programs. In many instances, the problems came about because the particular ITC project was ill-conceived—often with no clear-cut purpose or with little attention paid to real educational needs. Many projects have been started simply because the neighboring school district or city or county or state had established a system or station or network; therefore "we should have one too." In some instances, management channels were not carefully worked out. In other examples, financing was never arranged on a secure footing. In virtually all these cases, one way or another, the ETC breakdown could be traced to some administrative failure—not to some inherent weakness in the medium but, rather, in the way someone tried to use (or misuse) the medium.

The second ETV station in the country, KTHE (Los Angeles), was on the air in 1953. It went off the air permanently ten months later because of a change in funding policies of the foundation that was furnishing its almost total support. Station management had failed to secure

adequate financial guarantees at the top policy level.[1]

In 1975, the FCC refused to automatically renew the licenses for the stations of the Alabama ETV network, due to previous racial imbalances in programing and employment. This reflected some questionable decisions made at the top management levels.[2]

In Hawaii, the statewide network operation was run jointly by the University of Hawaii (which was the licensee for the stations) and the department of education. The main problems were the lack of communication between the higher levels of administration at both institutions and the lack of continuity at the top management levels; the Hawaii department of education had five different superintendents (top administrative level) and six different assistant superintendents for instruction (immediate management level) in the first five years of the existence of the ETV branch. After a legislative auditor's report, the state ETV network was turned over to a newly created Hawaii Public Broadcasting Authority, with more direct control by the state legislature.[3]

In Delaware, the state system fared even more poorly. Created in 1964, the Educational Television Board established a statewide closed-circuit system comparable to South Carolina's. Serving the public schools as well as the university and state college, the ETV board reportedly was more concerned with perfecting its elaborate transmission system than with carefully ascertaining the real educational needs of the state. After an exhaustive report by a state-appointed

[1]Station KTHE, Channel 28, in Los Angeles, was licensed to the University of Southern California, although it was funded almost entirely by the Hancock Foundation. The station went on the air November 29, 1953. When the foundation changed its funding priorities, the station found itself with no other source of support, and it discontinued operations in September 1954.

[2]For a popular account of the Alabama ETV license struggle, see William Bradford Huie, "Why Are They Trying to Force Alabama Educational TV Off the Air?", *TV Guide*, February 8, 1975, pp. 20 ff.

[3]*Audit of the Hawaii Educational Television System* (Honolulu: Office of the Legislative Auditor, 1971).

ad hoc TV committee, which recommended continuing the network on a minimal basis, the General Assembly (including a couple of members who were allegedly upset by some quasi-political remarks made on a school TV program) voted to cut out all funding for the board—and the state-wide system was abandoned in 1970.[4]

Other examples could also be outlined—situations in which there was not proper thought given to the most effective administrative and financial structure possible for a given ETC entity, and consequently the ventures could be considered administrative failures.

13.3   Management: Budgeting and Paying for Solutions

An integral part of any management consideration is, of course, the financing and budgetary elements. Many ITC operations are begun without adequate thought given to the long-range financial problems. The questions of how funds are spent—and with what results—are crucial.

Capital and Operating Budgets   Most ETC programs are established within the authority of an institution that works with both an operating and a capital budget. The operating budget is the basic annual budget (often on a July-June fiscal year) that is needed to keep the entire agency operating on a year-to-year basis. This is the source of recurring anticipated expenses such as salaries, office supplies, travel expenses, utilities, maintenance of equipment, and similar items.

The capital budget, on the other hand, consists of special requests for more-than-ordinary investments: major items of equipment and facilities, such as a new transmission tower, a new color studio, new quadruplex video recorders, a van equipped with facilities for remote produc-

tions, and similar major items. These capital items are those that normally would not have to be considered absolutely mandatory for continued operation during the next year. They are more negotiable than items on the operating budget. The capital budget often requires a special allocation from the taxing authority (board of education, state legislature) or board of directors supporting the ETC operation. Therefore, the capital budget is more susceptible to changes in the economy—recessions, cutbacks, and periods of expansion. On the other hand, the operating budget is normally more susceptible to gradual inflationary pressures—showing a slight increase in the budget request each year.

Many ETC programs fall into the trap of securing a handsome initial capital outlay to start a major project without being able to guarantee an adequate continuing operating budget. This is often true of grant programs. Federal and private foundation grants frequently provide for a decent capital budget for a new building or transmission facility—but often there is no provision for personnel to run the new operation or for replacement parts to keep it running after the first year or so.

Cost/Benefit Ratio   One term that ETC practitioners increasingly must deal with is "cost accounting." Basically, this means a careful accounting for every dollar spent. Where did the money go? As simple as it sounds, this sometimes can be a very complicated process—especially in connection with large operations that have special accounts, reserves, amortization procedures, deferred costs, shared expenses, hidden assets, and the like.

A closely related concept is that of the "cost/benefit ratio" (or the "cost/effectiveness ratio"). Simply expressed, this is the ratio of dollars spent (cost) for value received (benefit). How much did it cost to get so many units of benefit? In dealing with the two variables (costs and benefits), it is usually necessary to hold one variable constant and examine the other. Either: (a) you have a given amount of dollars and you want to

---

[4]E. Sidney Shaw, *The Delaware Educational Television Network: A Review and Evaluation* (Dover, Delaware: Ad Hoc Educational Television Committee of the General Assembly, 1970).

achieve the most benefit possible; or (b) you need to achieve a given benefit and you want to do it with the smallest expenditure possible. An example of the first situation would be a school district that has a special appropriation of $250,000 for its reading program; it hopes to improve the reading level of its ninth graders as much as possible with that amount (thus, there is a given cost, but a variable benefit). An example of the second situation would be a large banking corporation that has to instruct every one of its tellers how to use a new computerized customer statement; this is an objective that it must achieve, and it wants to do it with as little expense as possible (in this case, the cost is variable, but it is necessary to achieve the fixed objective or benefit).

Too many media projects are initiated without adequate thought given to these kinds of cost/benefit questions. What is it that the ITC program really is supposed to do? Too many times an ITC program is initiated on the basis of the second situation illustrated above (maintain the given benefit with fewer dollars) when it should be justified on the basis of the first situation (increased benefits per dollar). In other words, the STV enthusiast will often claim he can help the school save dollars by using television, when he should be stressing instead that he can help raise the educational results by adding television to the system. Each individual ITC case is different, and each one should be thought out and justified carefully.

If *"cost* accounting" is one necessary component in determining the cost/benefit ratio (one must know how to measure costs before being able to compute the ratio), then the corollary component has to be *"benefit* accounting"— being able to measure exactly what educational benefits are derived from a given program. This "accountability for learning" (as it is sometimes referred to) is an intangible and elusive factor to grasp. As we shall see in Sections 13.5 and 13.6, the difficulty of measuring precise educational achievements (and the virtual impossibility of measuring achievement of less precise long-

range goals) is one of the biggest drawbacks to fuller implementation of media in schooling programs. ITC administrators need to be familiar with theoretical models and approaches to these kinds of cost/benefit, problem-solving situations and planning systems.

Problem Solving and PPBS At different times and for different purposes, various schemes have been evolved for logical approaches to problem-solving situations. The scientific method (defining a problem, stating a hypothesis, experimentation or collection of data, analyzing the results and evidence objectively, confirming or refuting the hypothesis, reporting the results) is but one formula for rational problem solving (Section 17.5).

John Dewey summarized a useful model in his "reflective thinking pattern" at the beginning of the century.[5] Dewey's model had five steps:

1. *Statement of the Problem:* Precisely defining the need, limiting the scope of the problem.

2. *Analysis of the Problem:* Attempting to discover the relationship between the problem and its causes and effects.

3. *Presenting Possible Solutions:* Setting down the criteria that a good solution should meet, and then listing possible solutions.

4. *Selecting the Solution:* Weighing each solution against the criteria, and selecting the one that appears most promising.

5. *Implementation and Evaluation:* Putting the solution into practice and evaluating it to see if it works or if it needs to be modified.

During the 1950's and 1960's, this type of thinking led to specific management techniques designed to accomplish greater effectiveness in achieving particular corporate or educational ob-

---

[5]John Dewey, *How We Think* (Boston: D. C. Heath and Company, 1910), p. 72.

jectives as economically as possible. Known ge-
nerically as a program planning and budgeting
system (or PPBS), this technique has several
different models—one of which uses the following
six steps:

1. Determine the precise objective.

2. Compile alternative means of reaching
the objective.

3. Analyze the potential effectiveness
(anticipated percentage of success) of each alternative ("benefit accounting").

4. Analyze all of the costs of each alternative ("cost accounting").

5. Select the alternative with the best
cost/benefit ratio (starting with a given cost or
with a given benefit-objective).

6. Implement and evaluate the results
and modify the program accordingly.

By considering the cost/benefit ratio as one of
the criteria to be analyzed in Dewey's third step,
we can easily see the parallel between the modern management tool of PPBS and the traditional
rational problem-solving pattern.

The important point for ETC administrators is that media projects increasingly must be
subjected to this kind of procedure and cost/
benefit analysis. In many instances, this is altogether fitting and proper. Any kind of instructional
tool certainly must be defensible on the grounds of
cost effectiveness. However, this kind of rigid
PPBS management thinking can also be a serious
detriment in some instances. For example, PPBS
analysis can best be applied to business and
military training situations where we are dealing
with precise and easy-to-define instructional
objectives (for example, what steps to follow in
returning merchandise to our service center or
how to assemble an M-16 rifle blindfolded).

It becomes more difficult to think in precise PPBS terms when dealing with the larger
and less tangible goals of our schooling systems.
How do you measure whether or not today's high
school student will be able to appreciate a given
Shakespeare play when he reaches the age of
thirty? How do you determine, today, whether
tomorrow's adult leaders will be able to effectively
erase all social and economic barriers based upon
race? How do you measure the motivational value
for a fourth grader of watching a color film of the
seashore—as opposed to a series of still pictures?

In most educational situations, telecommunications media can be considered as one type
of alternative tool in reaching a given objective—
one alternative solution in answering an instructional need. ITC approaches must be considered
on a cost/benefit basis along with other alternative
instructional tools; but we must remember that
not every objective can be stated so precisely
as to enable us to measure whether or not a given
tool has immediately achieved that objective.

13.4   Instructional Design

As noted in Section 10.6, the audio-
visual movement started out primarily concerned
with "audio-visual" and "reference" applications
of film, slides, and audio tapes used for sporadic
enrichment purposes. Beginning in the mid-1950's,
the movement became increasingly concerned
with language laboratories, teaching machines,
multimedia presentations, and elemental uses of
computers in teaching—moving into curriculum-
structuring applications. What eventually evolved
was a systems approach—designing a complete
program or systematic course of instruction to
answer needs presented in a particular educational situation. This movement obviously par-
alleled its industrial counterpart—the systems
management approach in the military and business worlds. The procedures were similar: pin-
point needs precisely, define exact objectives or
criteria, organize and amass resources, analyze
the various alternatives, implement a carefully
designed plan of action, and evaluate the results
for possible modification of the program.

In the industrial/management/military scheme, this resulted in PPBS. In educational/media circles, the resulting approach was labeled "instructional design." There have been many specific I.D. models and schemes which, with their various flow charts and lists of steps to follow, all resemble a PPBS outline or even Dewey's reflective problem-solving pattern. One of the clearest and easiest-to-follow I.D. models was designed by Jerrold Kemp in the early 1970's (see illustration 13-2).[6] Kemp discusses eight steps in the I.D. process:

---

[6]See Jerrold E. Kemp, *Instructional Design: A Plan for Unit and Course Development* (Belmont, California: Fearon Publishers, 1971).

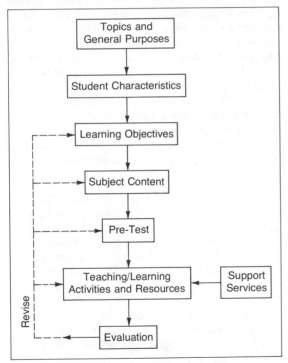

13-2 An Instructional Design Model

From the book Instructional Design *by Jerrold E. Kemp. Copyright © 1971 by Fearon Publishing, Inc. Reprinted by permission of Fearon Publishers, Inc.*

1. *Topics and General Purposes:* This is the practical beginning point for any discussion of a schooling program. What are the broad general goals of the system? What basic subject areas and topics are to be covered?

2. *Student Characteristics:* What are the specific needs of the students? What particular student characteristics will assist them or hinder them in their educational quest?

3. *Learning Objectives:* The careful and precise wording of clear, measurable objectives is a deceptively difficult—but absolutely essential—part of the instructional design process.

4. *Subject Content:* Once the specific objectives have been formulated, we can begin to outline the elements of the subject content that need to be covered in order to achieve the objectives. What facts must be presented? What kind of unit structure will work best?

5. *Pretest:* The pretest of the learners is necessary for two purposes—to determine whether or not the student is ready to undertake the specific course of instruction and to determine whether or not a student may already have achieved some of the objectives.

6. *Teaching/Learning/Activities and Resources:* This is the heart of the instructional situation—the actual implementation; the interaction of the student and the instructional program. If media are to be used, this is where they would be integrated.

7. *Support Services:* These include all the necessary elements that comprise the total schooling situation: personnel, funds, special facilities. Adequate support services must be available before any program can be put into effect.

8. *Evaluation:* Finally, and most importantly, the results of the program must be evaluated. The students' learning must be measured and—based upon the results of the evaluation—the program is adopted, revised, or thrown out.

It is the continual process of evaluation/revision/ implementation/retesting that results in a strong program—the results of which can be validated or guaranteed as to the achievement of stated objectives (Section 17.3).

As one examines the role of television and related media in this total I.D. process, it becomes apparent that media are involved primarily in the sixth step, the teaching/learning activities. (In a broader, nonconventional sense, however, ITC media can also play administrative roles in several of the other steps—formulating objectives, shaping content, assisting in testing.) Whether a schooling system intends to use ITC in a traditional curriculum situation or delve into a total systems development, certain components and elements are common to both approaches. Needs must be determined, objectives must be formulated, the specific roles that television can play must be examined, and planning procedures for the series must be undertaken.

13.5   Determining Educational Needs

Regardless of the geographical scope or level of the ITC operation, the type of ITC administration, and the affluence of the ITC industrial or schooling complex, precise educational needs must be determined before solutions can be formulated. As McBride emphasized:

> Those with prime responsibility must determine for themselves, as well as for the educational system, the nature, the extent, the current status of the educational problem. Parameters must be established if television is to serve effectively. There must be a clear definition of the educational problem to be solved, of the need to be fulfilled through the use of television.[7]

---

[7]Jack McBride, "The Twenty Elements of Instructional Television: A Summary," paper presented at the Seminar on Educational Television Sponsored by the Centre for Educational Television Overseas, Cairo, Egypt, February 6–12, 1966, p. 4.

A Diversity of Needs   One does not have to examine the problem of determining needs from an unnecessarily restrictive perspective. Consider the many legitimate needs ITC can possibly meet. Several years ago, an equipment manufacturer published a series of "application bulletins," listing various needs that have been answered by differing TV set-ups.[8] A random sampling of these applications include the following: accident prevention self-evaluation tapes for a paper mill, shipboard training on an aircraft carrier, dangerous and remote chemistry experiments in slow motion at a major university, connecting regional offices of a large corporation for a nationwide conference, fire-fighting training for a volunteer fire department, police "stakeouts" in high-crime areas, dissections in a medical school, observing whales in their natural habitat as part of an oceanography study, quality control inspections in a sewing machine factory, sales training for a research corporation, and on and on.

There is a wide diversity of various industrial and instructional needs to which television technology could be applied, but it is important that the specific needs are pinpointed and analyzed—that television is not used simply for the sake of using television.

Analysis of the Educational Situation   In a formal school setting, one must begin by analyzing the entire educational situation—starting with an examination of the goals of the school district itself. Often there will be some sort of official statement of the district goals and community commitments. What about the socioeconomic status of the school district? Is it a ghetto area? Middle-class neighborhood? Are vocational programs needed? College preparation programs? What state-aid programs are available? How about federal grants?

---

[8]*Application Bulletins* (Long Island City, New York: Sony Corporation of America, 1971–72).

Another aspect of the analysis of the overall setting would be a look at the teaching staff itself. What are the weaknesses of the teaching staff as a whole? Science? Art? (And what are the strengths of the teachers? Where would you definitely *not* need any additional help?) Where are there vacancies on the teaching staff? Foreign languages? Advanced math? Are there any imbalances in the teaching staff? Age? Gender? Political viewpoints?

Referring back to Kemp's second step in the instructional design model, what about the student characteristics? How much schooling have they had? What quality of schooling up to this point? What are their general academic abilities? Any differences as to age? Gender? What about general aspirations and skills needed? Vocational training? Business courses? Again, what about socioeconomic factors as they relate to individual students? Motivational levels? Results of various IQ, achievement, and aptitude tests may help to paint an overall picture.

Community and Individual Needs   Finally, one must look at both the needs of the community and the needs of the individual student.

Schools have been expected to answer several *community* needs. The National Defense Education Act of 1958 channeled funds into the schools because of the national need to turn out more engineers, scientists, and foreign-language specialists to meet the Soviet space threat. Other community needs include local programs such as driver training and sex education curricula—which are both justified partially on the grounds of the need to protect society (as well as for individual benefits to the students). Some communities have specialized curricular needs depending upon the geography and vocational patterns of the area. Agricultural education may be important—or aerospace training or fishing and maritime-related programs.

Most important, of course, are the needs of the *individual* students themselves. What skills

does Sally need to earn a living? Where is Carlos deficient in his college prep program? What about Mary's interest in psychology? Can Jim work out an independent study program for advanced work in French language and customs? At this juncture, both the students' formal (classroom) needs and their informal (out-of-school, cultural/social) needs must be considered.

Three different classes of student needs are often isolated: *general, liberal arts education* (to equip the student to function creatively and positively in a democratic society); *vocational training* (to equip the student with the skills to earn a living); and *preparation for further education* (college or advanced specialized training).

Only after both the generalized and specialized needs have been analyzed in these various contexts can the schooling system begin to zero in on specific programs that need to be initiated on a pilot basis, or those elements of an existing curriculum that need to be revised, or those programs that are simply outdated. The next step is to draw up specific objectives that need to be accomplished in these various areas.

13.6   Educational Objectives

For centuries, educational institutions have traditionally set forth vague and generalized statements of "purposes"—"to study U. S. history," "to teach plane geometry," and similarly worded imprecise statements. With the movement toward PPBS, systems approaches, and instructional design, the inadequacy of these traditional vague phrases becomes apparent. What content is really to be covered? What specific needs are to be met? How do we know if the learners have achieved whatever they should have from the lesson or unit or course?

Taxonomies and Educational Functions
Some of the earliest work in the classification and specification of objectives came from Benjamin Bloom, David Krathwohl, and their associates.

In a series of handbooks, they codified instructional objectives into three categories or "domains."[9] (See Section 17.3).

The cognitive domain is the easiest to come to grips with. It includes those objectives that deal with the "recall or recognition of knowledge and the development of intellectual abilities and skills." It is concerned with facts, dates, formulas, and the acquisition of concrete information.

The affective domain is probably the most abstract and also the most controversial. It includes objectives that "describe changes in interest, attitudes, and values, and the development of appreciations and adequate adjustment." This domain is the most difficult to define, to formulate precise objectives for, and to measure successfully.

The psychomotor domain is the manipulative or motor-skills area. It encompasses objectives that are concerned with development of physical, manual, and vocal skills. Certain aspects of physical education, foreign languages, typing and business operations, speech and drama, shop work, laboratory techniques, and similar skills areas, would fall into this domain.

In considering the broad functions or purposes of a schooling system, some educators have taken these three domains, then added a few other categories. Some have added specialized purposes (which should possibly be placed in the affective domain) that the schools should be especially aware of—although they do not fall into categorical subject areas: stimulation of creativity, learning to learn, and the habit of analytical thinking. Others have broken down the functions of education into broad classes roughly corresponding to some of the functional needs described above: general liberal arts education for the citizen; socializing/culturalization of the developing child, preparation for further education, and vocational skills and training.

Drawing upon these domains and purposes and functions, it is possible for the instructional designer to compile his own list of broad goals and useful things for schools to concern themselves with. This kind of exercise in categorization is often helpful to the instructional specialist and media designer in that it prepares the educator to come to grips with the kinds of specific, measurable, precise statements of objectives that are necessary for specific learning units.

Specific Learning Objectives   After reviewing the broad goals of the school system and considering the general functions of schooling, we should be ready to develop the precise objectives needed to measure achievement in a given learning situation.[10] Let us take the above example, "to teach plane geometry."

Terminal Objectives   In a complete and sophisticated instructional program, the designer works with both interim and terminal or final objectives. For our purposes, let us work only with the terminal objectives. Therefore, our final statement would probably include the qualifying phrase, "At the end of this unit of instruction, . . ."

Learner-Oriented Perspective   Second, we are concerned with the student's learning—not with the teacher's teaching. Instead of stating that our objective is "to teach," let us state that our objective is that "the learner will learn." Of course, there must be some measurable way of determining whether or not anything was actually learned.

Behavioral/Measurable Factors   Educational psychologists tell us that every achieve-

---

[9]Benjamin S. Bloom, Max D. Englehart, Walker H. Hill, Edward J. Furst, and David R. Krathwohl, *Taxonomy of Educational Objectives, Handbook I: Cognitive Domain* (New York: David McKay Company, 1956). The definitions used in the text are from page 7. See also David R. Krathwohl, Benjamin S. Bloom, and Bertram B. Masia, *Taxonomy of Educational Objectives, Handbook II: Affective Domain* (New York: David McKay Company, 1964). For a further discussion of these taxonomies, see Section 17.3.

[10]A brief handy programed text for mastering the skill of writing educational objectives is Robert F. Mager, *Preparing Instructional Objectives* (Belmont, California: Fearon Publishers, 1962).

ment in learning is measurable in some observable, behavioral manner. The student will be able to execute a movement; he will be able to draw a figure; he will be able to recite ten reasons; he will be able to write an essay; he will be able to score at a given level on an exam. Some designated action verb will be inserted to indicate what the learner is expected to do. Kemp suggests that the educational designer ask himself the question, "What does the student have to do in order to show that he has learned what you want him to learn?"[11] In responding to this question, the correct behavioral, measurable response will probably be indicated. So the phrasing of our specific learning objective for plane geometry now might start out something like this: "At the end of this three-week unit, the learner will be able to . . . [do whatever we deem necessary to indicate he has mastered the content]."

Specific Content Reference   Now we must insert the specific content mastery that we are concerned with. What exactly do we want the students to have learned? Let us assume that this geometry unit was concerned with finding surface areas of straight-sided, two-dimensional figures. Thus, our objective might now read, "At the end of this three-week unit, the learner will be able to compute the areas of triangles, squares, rectangles, and parallelograms."

Performance Standards   As the objective now stands, we have outlined what we are looking for, but we have left the student with quite a few loopholes. How accurate must he be? How many problems must he work? How many mistakes is he allowed? How much time does he have? Can he use any tools or aids? So it is necessary to insert whatever performance standards or criteria are needed. These, of course, vary tremendously from subject to subject; but generally they are concerned with items such as expected stan-

dards, limitations, and allowable aids. Thus, our final wording of this particular objective might be as follows: "At the end of this three-week unit, on the standard exam of twenty problems, the learner will be able to compute—using only a ruler—the areas of triangles, squares, rectangles, and parallelograms, with 90 percent accuracy, completing the test within thirty minutes."

Now we have a specific, measurable, terminal learning objective—in the cognitive domain. This gives us something tangible and firm to work toward. And we will be able to determine whether or not the learner has reached our objective.

These objectives, once formulated, are not set in concrete. There is nothing immutable about having specified an objective and committing it to paper. As a result of trial and evaluation of the learning program, it may become apparent that the objective was unrealistic (only half of the students could accomplish it). Or the objective may be too simple (every student achieves it with only a half-hearted attempt). This is why the concept of continual revision/implementation/retesting/evaluation is so important (see figure 13-2). One of the four steps subject to revision is the formulation of the learning objectives.

There are other realistic limitations on the use of specific, measurable, terminal learning objectives. This concept is not equally applicable to all subjects for all purposes. It is much more difficult, for example, to specify and measure objectives in the affective domain, where we are dealing with intangible results such as appreciation, citizenship, acceptance of an idea, self-images, values, and standards. Some manifestations may be observable. (Does the student voluntarily buy art prints of the masters? Listen to serious music? Read poetry? Does the student participate in student government? Work with needy groups? Socialize with members of minority races?) But attempting to measure these observable behaviors—and to correlate them with specific learning programs—can seldom be as precise and scientific as would be desired.

---

[11]Kemp, *Instructional Design,* p. 23.

There are also limitations relative to the terminal nature of some objectives; not all goals of the school system are to be accomplished in one semester or one year. Many objectives deal with decision-making abilities, analytical skills, attributes of good citizenship, mature appreciation of the arts and literature. "These high-level objectives may not be fully measurable until years later in schooling or until the individual becomes an active member of society and is in his profession or vocation. Therefore, . . . some objectives cannot be completely satisfied during the regular instructional program."[12] Nevertheless, it is important that media specialists and instructional designers, fully aware of this limitation, still try to work with terminal measurable objectives that apparently will be indicative of success in achieving the long-term objectives.

13.7   The Roles and Functions of Television

In the preceding sections, we have been concerned with the model of instructional design as a whole. It is important that the media specialist be aware of this overall pattern of administration and instructional planning. However, except for some specialized applications, media are used primarily for the step of "Teaching/Learning Activities and Resources." This is where the media specialist or ITV director is most heavily involved. To examine how media may most advantageously be applied in the learning situation, one needs to examine the characteristics of a given medium— for example, television—that make it appropriate for one role or another.

Inherent Characteristics   First of all, several broad features of the nature of the medium are obvious. The medium is *ubiquitous;* it is everywhere about us. This is especially true when considering open-circuit school applications of ITV. Parents at home can witness a portion of the

actual instructional situation; institutionalized students can more easily keep pace with their classmates.

Television, when used properly, is inherently a *motivating medium*. It is a change of pace in the classroom—or in the industrial training program.

Television is—or can be (at the discretion of the planners and the classroom teacher)— an *integral part of the learning situation*. It can be used on a regularly scheduled basis, integrated into the curriculum (more so than, say, occasional enrichment films).

From a production standpoint, television can be used to combine and present other audio-visual resources and materials. It can be used to direct the viewer's attention to specific items by *controlling the lens length and shot angle*. It can control and instantaneously *change picture and sound;* it can facilitate *editing* of materials. Film, of course, is a comparable production medium, but television is a more economical method of using basic film techniques. Television encompasses film.

Other characteristics of the medium include the fact that television gives the appearance of *direct eye contact*. An on-camera instructor, looking directly at the camera lens, creates the illusion of establishing eye contact with every individual viewer.

The medium also has the inherent capability to *combine more than one picture*—either by superimposition or by various split-screen effects.

As a means of dissemination, television is a medium of *widespread distribution*—whether one is distributing a demonstration to the classroom next door by means of a simple closed-circuit or presenting a live United Nations debate to schools throughout the entire globe by satellite. "Widespread dissemination" is relative.

In a curriculum context, television is a *means for standardization*—implementing curriculum reforms throughout a large region, pro-

---

[12]*Ibid.*, p. 30.

viding a minimum foundation of educational experiences.

Television is a means of *recording and preserving* the best teaching efforts of a given teacher or curriculum/instructional-design/production/teacher team. Yet, at the same time, television is also a means of *rapid revision* of curricular materials (contrasted with textbooks and films).

Television has the electronic characteristics, as a display device, to enable it to be *viewed in a lighted room*—making it easier to take notes, fill in a workbook, study a printed diagram.

Advantages and Limitations Given these inherent characteristics of the medium, planners and educators can begin to plan specific applications of television. These characteristics lead to certain advantages of the tool, in some situations, and to distinct drawbacks in other situations—depending upon how the medium is used.

Television can be a means of sharing resources—including good teachers. The best teachers can be electronically multiplied to reach numbers of classrooms. Preparation time is another resource that can be shared by means of television. If one classroom teacher has to prepare for twenty-five hours of student activity per week, there obviously is little preparation time. But, if one teacher is being shared by a hundred classrooms (by means of ITV) and is responsible for only thirty minutes of instruction time per week, then the teacher has extensive preparation time for one or two lessons; and the hundred classroom teachers are relieved of the extra preparation for those particular lessons—so they can have more time for preparation in other areas.

The ITV lesson can incorporate a wide variety of still and moving pictures. These can be "live" or carefully prepared and photographed. Close-up shots can be captured that would otherwise be impossible to reveal to an entire classroom. The learner's attention can be focused on a single view. A "front-row seat" can be made available to all students for special demonstrations, guests, interviews, materials, and the like. Many nonclassroom resources can be brought to the students—large demonstrations, dangerous experiments, field trips, inaccessible phenomena, and so forth.

New curricular programs can be introduced over a widespread area—or into a single school—with strong controls and standardization. Subsequent curricular revisions can be easily introduced and administered. Special motivational programs can be designed for students with learning problems, reading disabilities, and the like. New curriculum/instructional patterns can be facilitated: large-group instruction can be used where appropriate, and individualized instruction can be implemented where advantageous. Parents can be involved with open-circuit programs or with cable TV distribution.

Of course, some of the inherent characteristics of the medium—when improperly used—also lead to definite limitations and disadvantages. Television is two-dimensional. It is usually restricted to a relatively small viewing screen. In many schooling installations, finances prohibit a full conversion to color. There may not be enough funds initially for quality program production. Depending upon the instructional/curriculum situation, the programing may not be geared toward the appropriate curriculum needs; there may be scheduling problems with open-circuit broadcasts; poor teaching can be multiplied as efficiently as good teaching; and so forth.

The important concept to be stressed is that—once the entire schooling situation has been analyzed and once the particular instructional needs have been carefully isolated and an instructional program designed—television and related ETC applications must be very carefully structured and delicately designed to be a vital part of the "teaching/learning activities and resources."

13.8   Curriculum Planning and Specific Programs

There are several different ways to approach the actual curriculum planning process. A thorough analysis of the community and schooling situation should be undertaken. Are there community resources that would be an advantage to an educational program? Are there apparent community biases for or against the use of media that would affect the success of a program? A complete analysis of the financial situation of the school district or industrial system should be reviewed. Will a continuing support budget be available? Is the program going to be justified on the grounds of saving money? Or will the program be justified as a solution to a particular problem—which will admittedly cost additional money? (Is the emphasis to be on saving costs or on increasing benefits?)

Consider the implications for planning suggested by the several perspectives presented in Chapters 10 and 11. Will there be a need for in-service training to accompany the program? What existing distribution facilities are there? Is the school concerned with a curriculum-structuring use or audio-visual applications? Is there a strong metropolitan ITV association that can be called upon? Consider the functions organization: Can the existing AV center contribute significantly? Once some of these basic questions are answered and the underlying foundation is laid, then the planners are ready to proceed with the actual construction of the teaching/learning activities.

The Planning Team   The actual planning process for a media series (or for any other teaching/learning activity) involves several layers of educational bureaucracy. The classroom teacher seldom works alone in any involved curriculum undertaking. The same is true for a training director in an industrial situation.

Television is an encourager of the team teaching approach. The television teacher ordinarily plans broad course and individual program content in concert with his peers and superiors. At the primary educational level, this means workshops and planning meetings where fellow classroom teachers from representative schools join the television teacher, curriculum specialists and supervisors and the television producer cooperatively to plan the broad course outline and general program content.[13]

The planning process can be extremely complicated. *Sesame Street,* for instance, involved years of research on a nationwide level. Recent AIT consortium projects involved dozens of educational agencies comprising curriculum committees. MPATI set up curriculum and planning committees covering a six-state area—involving nationally recognized educators. Quite frequently, even district-level programs of a substantial scope involve national consultants, regional planning teams, and the resources of the professional associations (for example, AECT and NAEB).

At the district level, almost anyone may get involved in the planning. The board of education may take a special interest and assign a representative. The superintendent or his assistant superintendent for instruction or curriculum certainly will be giving at least superficial guidance. Individual district curriculum officers will probably be involved in a major undertaking: the director of elementary (or secondary) education/instruction/curriculum; and, most appropriately, the subject area specialist for the content concerned—the staff specialist or subject supervisor.

At the school level, of course, the principal will be involved. Any grade-level or department chairmen will keep on top of the situation. A community advisory group or parents' committee might be established to provide input. On many smaller projects, a committee consisting of the classroom teachers will be directly in charge.

[13]McBride, "Twenty Elements," p. 7.

Additionally, there will be the actual media and production people: possibly audio-visual and library staff, the instructional design team, and the television personnel—TV producer/director and production crew.

Depending upon the nature of the particular project, various personnel at these levels will be involved in numerous committee activities—some involved with content and authenticity, others with teaching and communication techniques, some with research and evaluation, a few with long-range planning, some with financing, others with implementation.

Implementation Considerations There are several implementation considerations that must be examined fairly early in the planning stages. How exactly is this new program to be utilized by the learners? What special arrangements need to be made at the school or classroom level?

How about the size of the learning groups? Will large-group reception of materials be appropriate? Will smaller-than-average size be best (perhaps clusters of seven or eight students working together)? Would individual learning carrels be called for?

What about the necessity for any in-service training of teachers? Will the program call for new procedures the teachers need to master? Will additional teacher-training materials have to be developed? Will in-service, after-school or weekend workshops be established? Maybe team-teaching approaches can be devised. What does this do to traditional scheduling and room assignments?

How will ancillary or supplementary materials be distributed? All these and many other implementation questions need to be considered at this planning stage.

Traditional or Programed Approach? Finally, at this stage—if not before—the curriculum planners and instructional designers must decide to what extent the new program will be validated along the lines of programed instruction (Section 17.3). Or will the program be merely an extension of traditional teaching methods—expose the students to the materials and activities and test them afterwards to see if any "learning" can be detected?

Programed instruction, as a formal concept, takes the idea of instructional design one step further. It is concerned with the validated or guaranteed learning of one discrete unit of instruction. First, the specific, terminal, behavioral learning objectives must be precisely formulated. Second, the exact entry-level qualifications of the learner must be determined; and a pretest to determine if the learner has the requisites for the program must be devised. Third, the objectives are broken down into small manageable learning units; the subject content is divided into individual bits or frames.

Fourth, the instructional pattern is constructed—the development of the materials and learning activities that will facilitate the interaction between learner and program. Next, as part of the actual learning sequence, provision is built into the program for constant feedback from the learner; he is continually reinforced as he responds to the presented material, so that his progress is always being monitored—his mistakes corrected and his correct responses rewarded. Finally, upon completion of the program, the learner is tested to determine the extent to which he has achieved the stated objectives for the program. This evaluation, then, is used to revise the appropriate parts of the program.

With the final test scores, however, the programers have what amounts to "benefit accountability." The programers can—with confidence—predict what the results of the program will consistently be; for example, "Ninety-five percent of the learners who take this program (assuming they have met the entry-level requirements) will be able to achieve a score of 85 on the designated standard exam." Thus, the program is said to be validated at that level. It can be used anywhere, with students who meet the entry-level standards, and the results can be accurately predicted.

This has been but a superficial outline of the process of programed instruction.[14] (A more detailed discussion of the evaluation process of validation is included in Section 17.3.) However, this summary may suffice to indicate the kind of planning and analysis that could go into a modified

programed approach for instructional media. As media programs are designed increasingly along the lines of this type of validated programed concept, the stronger will be the likelihood that those programs will flourish while other projects may continue to be labeled "frills" and eventually be discarded. The key to a successful instructional program—whether or not ETC is involved—lies in the extent and quality of the initial administrative set-up, the analysis of the precise educational needs and learning objectives, the attention paid to instructional design and the proper role of ITC, and the thoroughness of the curriculum planning.

---

[14] For a detailed explanation of programed instruction in programed workbook format, see Sivasailam Thiagarajan, *The Programing Process: A Practical Guide* (Worthington, Ohio: Charles A. Jones Publishing Company, 1971).

four

**Acquisition
and Production**

With this chapter we begin Part Four, an outline of the "practice of educational telecommunications." Although the next four chapters are focused primarily on instructional applications of ETC, in many areas the discussion also applies to public broadcasting: acquiring and leasing of programs, copyright considerations, studio equipment, production procedures, distribution technologies, follow-up and evaluation, and the like. Therefore, this final part can be interpreted as a broad consolidation of elements that apply to both of the preceding two parts—but specifically from the perspective of school-related ITC programs. Each of the last four chapters covers one or more of the operations of the functions organization (see Figure 13-1).

This final part of the book also applies to various industrial, medical, and military applications of ITC. Again, most of the references are to broad schooling applications—but the principles apply to any uses of telecommunications media for training and instructional purposes in any type of situation.

**14.1   Primary Consideration:
Cost/Benefit Ratio**

Once the instructional needs have been clearly identified and the learning program has been designed, we are ready to obtain the needed materials to implement the program. Materials—instructional media presentations—can be obtained in two ways: *acquisition,* by lease or purchase, of existing materials; or *production* of original materials. Where do we start? Which avenue should be explored first? As a general rule, it is usually better to investigate existing materials before deciding to make your own. If an ITC program already exists, why duplicate it by making your own version of the same thing?

The First Rule-of-Thumb   In fact, if there were only one basic truism regarding obtaining of ITC materials, it would be this: *It is almost always less expensive to acquire existing materials than* *it is to produce original materials.* Notice that, in the functions organization (Figure 13-1), "acquisition" precedes "production" as an alternative. It is simply cheaper to duplicate existing materials than to build prototype materials from scratch. It may cost $100 to acquire a copy of an existing half-hour ITV program (including tape costs, rights, and rental fee) while it may cost anywhere from $1,000 to $10,000 to produce a similar program locally.

There obviously are legitimate situations in which, because of unique needs and important local considerations, it is necessary to produce an original local program rather than lease or buy an existing package. This is more likely to be the case with industrial training needs (where a training problem may be unique to one particular corporation) than with schooling needs (where curricular problems are similar from district to district). However, in too many instances a local school district undertakes an original production project merely for the sake of having its own name on the credits or to justify the existence of the production facilities or to perpetuate the myth of local autonomy or to help build the empire and reinforce the ego of the local ITV director.

In debating the question of acquisition versus production, one must consider the cost/benefit ratio. It is fairly easy to examine and compare the cost figures, but it is more difficult to compute the benefit comparisons. It may be an important payoff, for example, that the board of education *can* point with pride to a local production. How is this "benefit" to be priced and evaluated?

Of course, in many instances, materials acquired from an outside source do not quite meet the exact objectives as set forth for the local instructional program. An ITV series leased from across the state boundary may meet only 90 percent of the objectives a local school has set for a particular math program. The school now has to deal with the law of diminishing returns. The cost per unit almost always increases as you get closer to your goal; as you near perfection, each step becomes harder to attain. Therefore, are you willing to spend (maybe) three or four times as much money in order to get (possibly) 5 percent closer to your goal? You may be able to reach 90 percent of your goal for $3,000 by buying an existing series from another school district. You may be able to achieve 94 percent of your goal for $10,000 by producing some of the materials yourself. You may be able to reach 97 percent of your goal for $35,000 by getting into a really elaborate local program. Is it worth it? Where do you draw the line? The cost/benefit considerations cannot be ignored.

Regional and National Cooperative Production In the development of new curricular approaches, more and more resources—manpower, time, financial support—are being pooled on regional and national levels in order to guarantee the production of the best materials available. There is a growing realization that what is good for students in Ohio also would be good for students in Oregon, that what works in New Hampshire also will often work in Hawaii.

It has been estimated that four or five carefully developed national curricular programs (in each major subject area) would meet the needs of more than 90 percent of the students in any given school district. Dwight Allen, head of the School of Education at the University of Massachusetts, once estimated that—in order to achieve a good balance in a typical subject area—50 percent of the curriculum should be nationally developed, 25 percent should be developed at the regional or district level, and 25 percent should be generated at the individual student level.[1]

It is apparent that more materials should probably be produced at the regional and national levels; and the local districts should concentrate exclusively on those programs that need to be tailored solely to meet local and individual needs.

[1]Remarks to State Symposium on Education, Maui, Hawaii, November 15, 1965.

Unfortunately, to the contrary, much of what has been produced in the name of local autonomy— imitating what is also being done by a neighboring school system—has only led to redundancy and duplication of efforts, with the result of a general widespread mediocrity.

With these realizations in mind, recent efforts by the Agency for Instructional Television (AIT), with its numerous consortium projects, and by the Children's Television Workshop (CTW) have provided worthy models for national productions. Regional efforts—pioneered by the Midwest Program on Airborne Television Instruction (MPATI) and by several of the regional networks— attest to a lessening parochial attitude among educators at all levels.

One of the most perplexing questions in the STV field relates simply to the balance of local and regional/national efforts. How do we identify needs and determine how those needs can best be met—considering cost/benefit restrictions? What needs can best be met locally? What needs should be answered on a national level? How do we tell the difference?

14.2   Selection of Materials

After the decision is made to lease ITC materials, the next step is to become familiar with some of the sources of series and individual programs—both for general schooling purposes and for specialized and industrial needs.

Sources of ITC Materials   Several of the major sources of ITV programing have already been mentioned and discussed in Chapter 11. At the national level, the two best-known nonprofit STV sources are probably AIT and Great Plains National Instructional Television Library (GPNITL) (Section 11.7). Many smaller special-interest ITV libraries also exist with programing limited to a particular area such as children's literature (Weston Woods in Connecticut), art (Hermine Freed Video Productions, New York), or vocational counseling (Counselor Films, Inc., Philadelphia).

Several commercial STV producers produce educational materials strictly for schools and market the product just as textbooks are sold. Western Instructional Television (Los Angeles) was the first company to produce ITV programs, on a series basis, for commercial sale to school systems. Many of the commercial old-line audiovisual film distributors (Time-Life, ACI Films, Pyramid, and others) are also distributing their products on videotape and video cassettes for both closed-circuit TV (CCTV) and community antenna TV (CATV) uses.

The 1975 edition of *The Video Bluebook* lists more than 2,500 programs for business, industry, and government, and more than 2,000 "general interest" programs.[2] The 1974 *Program Source Guide* lists 137 different companies and distributors that have programing ("educational" in the broadest sense) available in video-cassette format for noncommercial television uses.[3] Included among the commercial and nonprofit suppliers are numerous universities (some, such as Indiana University and Brigham Young University, have vast production and distribution facilities), ETV stations, and specialized producers.

About half a dozen organizations are concerned with religious programing. Some, such as CETV (Texas) and Video Ministries (Florida), are nondenominational; others, such as United Methodist Communications (New York) and Faith for Today (California), are denominational. About fifteen companies are concerned with medical and related health-care education. Some are broad in their coverage and include a wide variety of topics in their programing (Section 11.7). Others deal with specialized programing: Medic-Media Corporation (Michigan) deals with psychiatry; Yasny Educational Services (California) is concerned exclusively with otology.

A wide variety of companies distribute

[2]*The Video Bluebook* (White Plains, N.Y.: Knowledge Industry Publications, Inc., 1975).

[3]Ken Winslow (ed.), *Videoplay Program Source Guide* (Ridgefield, Connecticut: C. S. Tepfer Publishing Co., Inc., 1974).

business and industrial training programs in ITC formats. The *Guide* lists almost thirty. Some have produced a large number of programs on various subjects: Hewlett-Packard (California) has more than 150 single and series titles available for purchase. Others are interested only in specific limited topics such as computer science education (Data Processing Education Network, Inc., New York) or sales training (Video Educators, Inc., Pennsylvania).

Many other companies deal with very limited specialized topics, such as the International Institute of Coiffure Designers, Ltd. (New York) with its video course in hairstyling. Several "underground video" outfits (Section 10.7) are included, such as the April Video Cooperative (New York) with its catalog of "contemporary themes offered under the headings of community tapes, survival tapes, and environmental tapes." Some listings are quasi-entertainment such as the P. M. Wrestling Film & Video Tape Co. (New York); and, finally, "adult entertainment of a simulated and hard core nature" is available from the X-Video Organization (Ohio).

In addition to these national sources, many regional and state agencies also supply ITC materials. Eastern Educational Network (EEN) and Southern Educational Communications Association (SECA) are two of the regional networks that are heavily involved in this area. Many state departments of education have materials available for school use. Many local stations, Instructional TV Fixed Service (ITFS) operations, and even closed-circuit systems have packaged materials that would be applicable for other systems for their adaptation and use. With these local and uncataloged sources, it requires more effort to contact the producer or distributor, find out exactly what is available, and what kinds of rental or purchase arrangements can be worked out; but the extra effort is often worth it.

Criteria for Selection   Before any materials are actually acquired, several factors and criteria must be considered. Listed below are some of the questions and considerations that a school district should normally examine. Other criteria might suggest themselves for a particular application.

First of all, the broad goals and purposes of the schooling operation must be examined. Will a given ITV series be compatible with the overall direction of the school's philosophy? The specific, behavioral learning objectives, of course, should probably be the primary determining criterion. Will any proposed series really lead to satisfactory achievement of these specific objectives?

There should be a thorough examining and review of materials that are available. Preview arrangements can be made with virtually every library and distributor so that representative programs or the actual lessons can be ordered and evaluated by the school system. Normally, several persons are involved with the previewing of materials—content supervisors, classroom teachers, district personnel, media specialists, and, possibly (depending upon the subject and age level), even students and/or parents.

Several other criteria must be considered during this process. Is the content applicable? Is the pacing at the right level for these particular students? Is there any jarring geographical incompatibility? Does the production quality affect the presentation negatively? Is there any other evidence that the programs might not work in this particular learning situation?

Also, what evidence is there that the programs *should* work? Have the telelessons been validated to any extent? Is there any empirical evidence that they have been successful in various situations? Have the programs been widely used by different school systems? What about the reputation of the library or distributor? Are the verbal assurances of the distributor adequate to convince you of the program's merit?

Finally, what about the costs? Are they in line with the budget of the school system and the media operation? Do the costs balance out in the cost/benefit considerations? Individual half-hour programs may cost anywhere from nothing to well over $100 to rent—depending on many factors (including costs of the original production, scope of distribution, size of the anticipated audience, and other factors).

Once these types of questions have been answered, the school system is in a position to make the most responsible decisions regarding leasing or purchasing of existing materials. However, it may well be that some situations and objectives cannot be met with materials from outside of the schooling system—because of specialized local needs (local history, state government, local vocational training programs, regional language problems, and the like). For such situations, the school will have to undertake its own local production.

14.3   The Telelesson

After making the decision to move into ITC production, an instructional system (schooling or industrial) is often tempted to start thinking immediately about studio and equipment and staff. How much space will we need? What kind of cameras do we buy? What staff positions do we fill? Actually, the more appropriate starting point would be consideration of the product the system will be turning out—the telelesson itself. One must start by thinking in terms of the purpose of the productions, the extent of pretesting or validation, the number and length of lessons, and the production format.

Purpose and Validation   First of all, we must examine the exact role that television is expected to play in a given instructional setting. What is to be the purpose of the ITV telelessons? Total teaching? Major resource? Supplementary?

What will television's role be in the *total* course of study? Will it serve, along with conventional classes, as a device for *enriching* already adequate, on-going instruction? Will its role be that of a *cooperative teacher,* assuming *part* of the instructional

burden, with classroom teachers doing the rest? If so, which part? How often per week and how long per day? Will televised instruction offer *total instruction?* Why?[4]

After determining the explicit purpose of the lesson or lessons, telelesson planning must be done in terms of those specific objectives—not in terms of "Wouldn't film be nice here?" "How many graphics should we include?" "Let's do a stand-up lecture." Think first in terms of what needs to be accomplished for the learners; then think about how to accomplish it.

An important consideration at this point should be the question of validation of the television materials (Section 17.3). Are the telelessons going to be pretested at all? Do the precise objectives lend themselves to accurate measurement and validation? Are the objectives important enough that financial resources should be made available to conduct the necessary pretesting and revision of the ITC materials to guarantee that the objectives will be met? If the objectives are important (and if the specific terminal learning objectives can be measured), then the budget should include provisions for trial production of the telelessons, testing with a representative group of students, revision and new production, and further testing—until the designers are assured that the telelessons will enable the learners to achieve the desired objectives a satisfactory percentage of the time.

Design: Number and Length of Lessons It is necessary to determine the number and length of the telelessons based upon the educational objectives—*not upon scheduling convenience.* Again, think in terms of what needs to be accomplished for the learners, not in terms of time slots to be filled.

Maybe one or two programs will suffice to accomplish a given purpose. Perhaps only one

---

[4]Lawrence F. Costello and George N. Gordon, *Teach with Television: A Guide to Instructional Television* (New York: Hastings House, Publishers, 1961), p. 71.

supplementary program needs to be added to an existing classroom presentation. At this stage of planning, instructional designers certainly will be aware of the distinction between curriculum-structuring applications and audio-visual/reference uses. Maybe a heavy commitment to sixty-four or more telelessons will be needed. (Some of the early MPATI series were designed with 128 programs per series—to be utilized four times a week for a one-year course. Some Samoa "series" were even longer.)

Perhaps a five-minute demonstration is all that is needed. Maybe a full two-hour presentation should be planned. Keep in mind that attention spans at lower grade levels normally dictate shorter telelessons. Maybe fifteen to twenty minutes should be considered the maximum at the primary levels. Thirty minutes would normally be the limit at upper elementary levels. Secondary viewing periods could reach a maximum of forty to fifty minutes. College-level and adult training periods could last for an hour or longer. At best, these are just rough guidelines. "The best length for a lesson will depend upon content, the quality of the students' attention, their degree of motivation, their interest in the subject at hand, the skill of the teacher and production staff, and the quality of the television system."[5]

Lesson Format Before buying equipment or hiring personnel, we should think about the types of programs that may be produced—in terms of production formats. Depending upon the job to be done, will simple single-classroom equipment be adequate or must elaborate studio facilities be installed? Will film equipment be necessary? Will staff artists need to be hired? How much will be needed in the way of scenery and prop construction and storage? Expensive electronic editing equipment? Color facilities?

Preliminary consideration should be given to basic formats, some of which are suggested below:

Straight Lecture: The "talking face." Is this all that is necessary?

Illustrated Lecture: Adding a heavy portion of visuals, films, slides to the lecture format.

Demonstration: Working with experiments, models, handling actual objects, animals, and manipulating things.

Voice-over Visualization: No visible talking face. The illustrated lecture with no identified lecturer, just an off-camera narrator.

Interview: Often effective to add another dimension to a one-person program.

Panel/Discussion: Can be effective or can be deadly.

Drama: Many objectives in the affective domain lend themselves to dramatic vignettes, role-playing situations.

Filmed Field Trip: On-location recorded footage of a particular place or event or activity.

Documentary: A more extensive film or studio treatment of a topic involving a mixture of guests, graphics, on-location inserts, and the like.

Or any other combination of the above. This list is not intended to be an all-inclusive catatalog, but it should give the planner and media specialist a starting point to think about some of the various approaches for planning. It would be extravagant to plan for an elaborate studio and extensive film facilities if fairly simple "talking face" programs are all that is needed. On the other hand, it would be shortsighted to build a small studio if there were the possibility that more elaborate productions might be called for in the near future.

14.4   Producing Personnel

In virtually all ITC projects, one person usually winds up ultimately in charge of what is produced. In most of these operations, this one

---

[5]*Ibid.*, p. 73.

person is the designated "TV teacher," the one who actually appears on camera.

Necessity of the Television Teacher   The TV teacher usually fulfills two primary roles: (1) that of producer and (2) that of on-camera "talent"/ host/performer/presenter.

As producer, the TV teacher is responsible for the content of the program. He or she is the content expert who ultimately pulls the program together as researcher, writer, and authority who makes the final decisions. In many situations, however, this role is overshadowed by a platoon of district supervisors, subject area specialists, teams, and committees to the extent that the TV teacher really loses much of his or her identity and authority as a "producer" as such.

As the on-camera talent, the TV teacher has much more of an identifiable image. His or her face is the one that carries the message to the multitudes. However, this role is also occasionally played down. Especially at the college and industrial levels, ITC operations are turning increasingly to teaching teams or rotating hosts— a series of content authorities, with their own areas of expertise, who take turns as the primary talking face on a given series.

Occasionally at the lower schooling levels, ITC projects are doing away with the continuing TV teacher altogether. This is especially true with the major projects that have a substantial amount of research involved and adequate production budgets, for example, *Sesame Street, The Electric Company,* and the various AIT consortium projects. By using a variety of sophisticated production formats and by combining the talents of a number of on-camera personalities, it is possible to accomplish more than a single TV teacher could do. Even at less ambitious levels, it is possible—using voice-over narration and extensively visualized programs—to substitute other arrangements for the TV teacher. The question to be asked is: "Is there a better way to accomplish the objectives for this program than to rely on a conventional TV teacher?" Often, the

TV teacher simply represents the cheapest and quickest way to get the job done—not necessarily the best way.

Sometimes the TV teacher is a compromise—an attempt to lessen the fears of those who feel that they may eventually be replaced by mechanical teaching apparatuses. As long as there is a talking face on the tube, there appears to be less threat of automation completely taking over. At least one college TV instructor has admitted that he would have preferred pedagogically to put his telecourse together without appearing on camera at all. His visuals, film clips, charts, and a voice-over explanation were all that was needed—especially with his complete set of ancillary materials: workbook, lists of readings, diagrams, and so forth. However, he was told that if he did not appear in person in the series many of his colleagues would be certain to express considerable ire and insecurity about the dehumanization (and threat of unemployment) represented by television. So his "talking face" was prominently integrated into the telelessons.

Characteristics of the TV Teacher   Assuming, then, that the on-camera television teacher will be a fixture in many ITV series to come, the question arises, "What makes for a good TV teacher?" What are the qualities we look for? Who will work best in front of the camera? Many authorities and experts have come up with their own lists of qualities and characteristics of a good TV teacher. A 1964 list by Diamond is as complete, yet concise, as any.[6] He states that a TV teacher should have the following traits: (1) a thorough knowledge of the subject, (2) classroom teaching experience, (3) the ability to communicate, (4) creativity, (5) well-organized work habits, (6) the ability to work with others and to take criticism, (7) a sense of humor, and (8) the ability to improvise.

---

[6]Robert M. Diamond, "The Television Teacher," *in* Robert M. Diamond (ed.), *A Guide to Instructional Television* (New York: McGraw-Hill Book Company, 1964), p. 248.

Most of these traits are self-explanatory and hard to argue with. The first three, however, are especially important. The TV teacher must be thoroughly knowledgeable about the subject matter. Some experiments have been tried using actors in place of an on-camera teacher. The results generally have been disappointing. "Skilled teaching, on or off television, is a subtle art, and so is acting. Both require certain superficial 'gifts' related to the development of skills needed to affect communication of *both* cognitions and emotions. Here the similarity ends."[7]

Secondly, actual classroom teaching experience—and lots of it—is mandatory. Skilled teachers know how a class will react to a certain statement, how much time to leave for consideration of a rhetorical question, when to slow up the pacing. There is no substitute for years of working in a classroom with prodding, probing, questioning, and confused students to condition a TV teacher for working alone in front of a camera with no immediate feedback. However, experiments have been tried using a small select group of students in the studio to provide live feedback to the TV teacher.

> Almost without exception this method was discontinued immediately. Two things tend to happen: first, with students in the studio, the teacher loses eye contact with the students in the viewing audience, and second, the entire logic behind a program is often lost when the teacher attempts to pace his presentation to the students in the studio . . . the ability of the television teacher to complete a body of information without interruption.[8]

The third quality is that elusive natural "spark" that distinguishes the good communicator—one who is at ease and spontaneous, able to excite and stimulate the audience in a particular situation. "This television teacher must be an experienced communicator, one who can present ideas and concepts and formulae and theorems effectively, vividly, forcefully."[9]

Of course, these are the same desirable traits that one would look for in any good classroom teacher, are they not? It might be said that most *good* classroom teachers would also be good candidates for TV teaching. However, there is always that indefinable chemistry that takes place when one steps in front of a studio camera. Some fluent and excellent classroom teachers suddenly freeze up and are unable to communicate to a lens. Others, more reticent and introversive in the classroom, are able to relax and open up in a spontaneous manner when chatting informally with a TV camera. There is no sure-fire guaranteed method of determining who will and who will not be an effective on-camera teacher until one steps in front of a camera and tries it.

Production Staff   The size and organization of the production staff needed to pull any ITC endeavor together vary tremendously. The size of a production team may range from one person (Chuck Jameson operating the microscope camera in his biology class, Section 11.1) to hundreds of persons working on varied and independent elements of a complex production (*Sesame Street,* with specialized animators, set designers and builders, film crews, research experts, actors, and so forth).

The typical studio television production involves personnel to carry out the following types of functions: operating cameras, obtaining good audio reproduction, arranging lights and scenic elements, working with the performers, making graphics and other visual components, mixing pictorial inputs (including film and slides from a

---

[7]George N. Gordon, *Classroom Television: New Frontiers in ITV* (New York: Hastings House, Publishers, 1970), p. 166.

[8]Diamond, "The Television Teacher," p. 248.

[9]Jack McBride, "The Twenty Elements of Instructional Television: A Summary," paper presented at the Seminar on Educational Television Sponsored by the Centre for Educational Television Overseas, Cairo, Egypt, February 6–12, 1966, p. 5.

film chain), handling recording and other engineering responsibilities, and several other specialized tasks. A "normal" TV studio crew can vary from five to twenty persons.

Coordinating all this will be the television director—often referred to as the producer/director (or P/D) in many PTV/ITV operations. He or she directs all the camera and audio operations, gives commands to the on-camera talent, orders which pictures and sounds are wanted when, and generally is in charge of actually putting the program together live or on tape. Depending upon the type of production, he or she should have been involved—along with the TV teacher, curriculum experts, writers, researchers, and the like—in the planning process for the ITV lesson from the beginning. The P/D is responsible for the bulk of the actual production preparations. Figure 14-1 represents the producer/director working with the on-camera instructor.

Some producer/directors have gathered considerable experience in commercial broadcasting; others may have started out as educators; many have moved into ETC production directly from their university broadcasting training. The important consideration is that the P/D's be *trained and disciplined as communicators* and be skilled in the use of the specific production techniques of telecommunications media. For if P/D's are good *communicators*—if they really know how to analyze the audience and define specific needs, how to put together a meaningful message with appropriate visuals and pacing and motivation, how to use the medium they are working with, how to anticipate communication blocks or "noise," and how to gauge feedback—then they are, by definition, performing the same type of role as the teacher. Communication is education; education is communication (Section 1.5). As a *communicator,* the P/D should truly be considered an equal partner of the TV teacher—sharing the responsibility for constructing and delivering the message. (Of course, depending upon the experience of each partner and the specific situation, sometimes the subject matter

specialist—teacher—should prevail and sometimes the communications specialist—television P/D—should be the dominant partner.)

The remainder of the production crew would consist of the following types of personnel: assistant or associate director, audio engineers, camera persons, technical director or "switcher" (who actually punches the buttons to get the right picture on the air), the floor director or stage manager (who is in charge of coordinating everything on the studio floor), various floor assistants (cable pullers, graphics flippers, grips, and the like), lighting and staging supervisors and assistants, film chain operators (handling motion picture film, slides, and the like), video engineers (who control the camera electronics), recording engineer, and similar positions—depending upon the complexity of the production. Many other positions may have been involved before the actual studio production: researchers, writers, graphics artists (illustration 14-1), production assistants, film crew, set designers and builders, maintenance engineers, and so forth.

Regardless of the scope and level of the production, engineering and technical personnel are indispensable categories of staff. "Heading into even modest production without proper technical help . . . is unwise, uneconomical, and self-defeating. At the hands of amateurs, minor mechanical problems will turn into complex electronic ones. Little difficulties will snowball and waste hours of staff time."[10] A program can get by without good art work. A project will survive with unskilled or voluntary camera persons. A studio can function without the associate director. But nothing will be accomplished if the equipment is not working.

There are several methods of obtaining and using staff positions. Not every production function requires a full-time permanent staff member. Perhaps only a few key people really need to be kept on a full-time payroll. Many positions may be filled with temporary or part-time

[10]Gordon, *Classroom Television*, p. 156.

14-1 Producing Personnel: Producer/Director Discussing Camera Angles with On-Camera Instructor (left) and Graphics Artists Preparing ETV Materials (right) (Courtesy: RCA and KUON-TV, University of Nebraska)

persons. Many of the larger operations hire crews on an hourly or daily basis as they are needed for specific productions. Other types of specialists—content authorities, researchers, writers, artists, film makers—may be hired as "consultants" on short-term contracts (to get a certain job done, not to work a given number of days). Sometimes a complete production is contracted or "farmed out" to a commercial production center to get the program recorded. Many school-related ETC operations can use student help as production crews and assistants—from the college level down to the lower grades. Students may be paid at minimum wages; they may be part of a work-study program; they may be earning academic credit for related laboratory experience and internships; or they may simply be unpaid volunteers. A variety of arrangements have been worked out in various production centers.

14.5 Production/Origination Equipment

In a book of this scope, we cannot pretend to present a thorough catalog of various items of television production equipment.[11] All we can do is outline a few considerations and basic categories. Detailed, up-to-date specifications and information on specific items of equipment are available from equipment manufacturers; and the latest information about costs

---

[11]For more details on production equipment and production procedures, see any of the following standard texts: Edward Stasheff and Rudy Bretz, *The Television Program: Its Direction and Production*, 4th ed. (New York: Hill and Wang, 1968); Gerald Millerson, *The Technique of Television Production*, 6th ed. (New York: Hastings House, Publishers, 1968); Herbert Zettl, *Television Production Handbook*, 3rd ed. (Belmont, California: Wadsworth Publishing Company, 1976).

and comparative systems are available from local and regional distributors and installers.

Origination Categories All ETC originations could be divided into three broad categories: *"classroom" originations, location recordings,* and *studio productions*. Each of these three classifications has its own characteristics and production considerations; and each encompasses a wide spectrum of equipment possibilities.

Classroom Originations This category refers not only to formal classrooms as such but also to any instructional or functional setting where television equipment is installed on a more-or-less permanent basis—but without substantially altering the basic purpose of the room. It may be a traditional classroom, an auditorium, a laboratory, a counseling office, a conference room, a training center, a hospital operating room, an administrative office, a library, or similar existing space. Equipment installations in these rooms are typically for limited audio-visual/hardware uses—image magnification, overhead cameras, administrative messages, observation purposes, and the like. Two typical classroom installations are illustrated in figure 14-2. Without the television equipment it would be impossible for every learner/viewer to peer through the microscope,

eavesdrop on the interview of the emotionally disturbed child, stare over the surgeon's shoulder, sit in the district manager's office, or otherwise enjoy a "front row seat."

Equipment for this kind of installation is normally relatively inexpensive. Industrial cameras costing a few hundred dollars are sufficient in many cases. Where high resolution and exceptional picture quality are required, much more expensive cameras might be needed. Audio hook-ups are minimal—if any are required; and switching and control equipment usually are kept to a minimum.

Location Recordings This category encompasses all television recordings that originate at some location other than where permanent TV facilities are installed. Although it is possible to broadcast live from such locations—using a microwave hook-up or by leasing AT&T lines— normally ITC materials originated on location would be recorded for later utilization. Such location recordings might include activities like a field trip to the local zoo, an interview at City Hall, analysis of a specific procedure on the assembly line, a guest lecturer, a student performance for self-analysis, safety procedures in the warehouse, a handicraft festival in the local park.

Film cameras—both 16-mm and 8-mm—

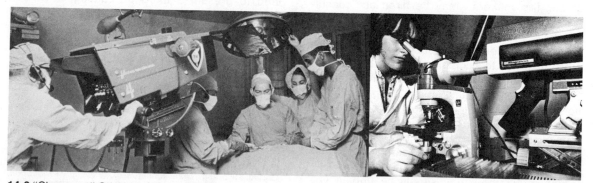

14-2 "Classroom" Originating Equipment: Operating Room Installation (left) and Camera-Microscope Combination (right) (Courtesy: University of California Medical Center, San Francisco, and Philips Audio Video Systems Corp.)

14-3 Location-Recording Equipment: Combination Camera/Recorder/Monitor on a Movable Cart (left) and Hand-Held Camera and Backpack Video Recorder (right) (Courtesy: Patrick Loughboro and Magnavox Corporation)

might be the least expensive appropriate recording medium for many of these location uses. Portable television cameras and recorders are becoming increasingly feasible each year; by 1975, the simplest monochrome backpack camera and recorder combination was available for well under $2,000 (illustration 14-3). Network-quality "minicams" are becoming the standard for broadcast journalism remote recordings and live coverage. The next step is to incorporate a couple of cameras, a simple switcher, and a video recorder into a van to record multicamera productions on location. In a more elaborate remote recording truck, one can easily spend hundreds of thousands of dollars on the best cameras, elaborate switching equipment, high-quality audio gear, top-of-the-line quadruplex recorders and editing equipment. Addition of a portable control room in another section of the semitrailer results in a studio on wheels.

Studio Productions   Most people immediately think first of the TV studio when television production is mentioned. The studio gives the producer complete control over production conditions, where everything (including artwork preparation; special effects, lighting, setting, and audio resources; editing facilities; and engineering maintenance and testing equipment) can be centralized and regulated. Two different types of studio productions are illustrated in figure 14-4.

Studio facilities range in cost from about $50,000 for simple conversion of existing classroom or instructional space to a minimally equipped basic studio (assuming no major building modifications are needed for electrical/lighting, air conditioning, and sound requirements) to a multi-million-dollar capital outlay for new facilities.

Origination and Production Hardware
Regardless of the type of production, certain basic equipment is needed to pick up sound and pictures, process them, and record them (or transmit live programing). The following discussion treats briefly some of these items of hardware.

Audio equipment consists basically of microphones, mounts, cables, patch bays and control boards, other input sources, and recorders. Microphones range in cost from less than $100 to several thousand dollars for sophisticated high-quality microphones. They may be mounted on booms, stands, desks, held in the hand, or hung around the neck; the latter lavalier microphones are probably the most flexible and serviceable for many ITC applications. Most studios include an audio patch bay in which the various microphones can be connected to inputs on the control board, which is used to mix the audio from the microphones and other sources. Other input sources include items such as audio-tape recorders, or sound from another video recording, records, cassettes and cartridges, audio from a film sound track, and the like. The resulting program audio can then be transmitted live, recorded on an audio recorder, or—in most instances—recorded along with the program video on the video recorder.

Cameras range in complexity from sim-ple monochrome industrial vidicon models with no viewfinders (for less than $500) to top-quality, high-resolution professional color models (costing close to $100,000). Lens arrangements also vary from a simple fixed lens to a complicated zoom lens arrangement. Cameras can be hand carried (usually for location work), mounted on a basic tripod (when no camera movement is needed, as in a "classroom" origination), or mounted on any one of a variety of pedestals, cranes, dollies, or booms (which allow more flexibility in movement—and take up considerable studio floor space).

A sophisticated studio installation usually includes a synchronizing generator to drive all cameras with the same electronic pulse, camera control equipment to adjust all of the electronic variables on the cameras, possibly a time-base corrector to upgrade the video signals, and considerable testing equipment, oscilloscopes, special monitors, and other support equipment.

Another specialized camera set-up is the film chain or telecine operation. This includes the TV camera, plus one or more possible sources of projected material—16-mm film projector (with a specially adapted shutter system to synchronize with the television electronic scanning system), 8-mm film projector, slide projectors, and maybe even an opaque projector. These different inputs are controlled by a multiplexer, an optical device of prisms or mirrors that can feed one input at a time to the TV camera.

The switcher is the device that enables the director to switch from one camera to another. Essentially, it is a system of pushbuttons (one for each camera) that enables the operator to select the camera he wants on the air at any given moment. Of course, the switcher can get quite complicated. Most of them have fader arms to enable the gradual fading in or out of cameras and the superimposition of two cameras. Special effects amplifiers can create many different fancy wipes, transitions, and inserts. Other devices and special keys can allow elaborate special effects and combinations of two or more composite pictures.

14-4 Studio Productions: A Local Series Produced by KUON-TV, Nebraska (top), and a Closed-Circuit ITV Course Produced by the University of California, Santa Barbara (bottom) (Courtesy: KUON-TV, University of Nebraska, and University of California, Santa Barbara)

The switcher is located in the control room or booth, along with other necessary gear and monitors. Associated with the control room are the audio control booth, the master control room (for the engineering and recording operations), lighting control, and related operations. The larger and more complex the studio installation, the more likely that each of these operations has its separate isolated space.

As the program is being put together in the studio and control room, it is (usually) fed to the video recorder in the master control area. It may be recorded on either a helical-scan slant-track recorder or on a quadruplex recorder (Section 4.5). The most sophisticated color "quad-head" machines with editing capabilities cost $100,000 or more. The simplest monochrome ½-inch slant-track recorders start at about $1,000; higher-quality slant-track recorders (color, with editing facilities, meeting broadcast standards) can cost more than $20,000. After being recorded, program materials may need to be further edited, reassembled, inserted with other materials, and so forth. This postproduction editing can be handled with relatively simple editors (starting at about $4,000) or can involve computer-controlled editing complexes costing as much as several hundred thousand dollars.

In addition to all this actual origination equipment, consideration should also be given to other studio and production hardware. Lighting facilities certainly are a big factor here. Settings and props may be a major item. There must be some sort of art facilities for making charts, title cards, maps, and other visuals. Studio display devices include easels, crawls, drums, and similar stands for mounting and holding these graphics. Special instructional furniture, lab tables, display units may need to be furnished. Consideration should be given to other studio projection devices—slide projector and rear screen, overhead projection, and the like.

This has been only an introduction to origination and production equipment. It is beyond the scope and intent of this book to present an in-depth treatise on production procedures or a catalog of production equipment, but the above brief discussion should serve to summarize some of the factors and elements that need to be considered before thinking about getting involved with ETC production facilities.

14.6   Rights and Copyrights

Before moving into actual planning and production procedures, the ETC producer needs to consider questions of clearances, rights, and copyrights—possibly the most vexing of the problems involved in the use of television. The fundamental rationale behind copyrights is simple enough: If someone creates something worthwhile, he or she should be able to control its use by others and to realize some financial rewards for the efforts expended. The creation might be a poem, picture, map, photograph, story, motion picture, speech, or television program.

Compensation for TV Teachers   One of the first issues to be faced is that of compensation for the TV teacher. How is the on-camera instructor to be rewarded for his or her effort in preparing an instructional program? Unlike the researcher, or writer, or producer/director (all of whom are typically paid for their work on the program as part of their ongoing duties), the TV teacher is in a unique position. First of all, the on-camera teacher usually returns to the classroom once the program or series is completed. (Very few long-term "professional" TV teachers are found throughout the country.) Secondly, the TV teacher has a continuing interest in the ITC material produced. He or she will probably be directly involved with its implementation and continued use, and will be concerned when the material becomes outdated and needs to be redone—or discarded.

Most TV teachers receive released time for producing their ITC materials, and get no other form of remuneration. A minority of TV teachers get extra money—either in the form of extra compensation while working on the program(s) or by being placed in a higher salary category. The thorny question, which is seldom satisfactorily resolved, is to what extent the on-camera teacher should receive any financial benefits comparable to residuals in the commercial TV and film world. Does the teacher receive any additional pay if the series is used for the next four years? Does the teacher receive any of the money if the school district sells the series to another user?

A related question, of course, deals with copyright of the telelesson and its contents. If the copyright is in the name of the teacher, then he or she obviously has considerable control over its future use. If the school system holds the copyright, then the TV teacher may have little to say about future considerations. In most instances, the school holds any copyright—on the grounds that the materials were produced with school facilities and staff and that the TV teacher was paid for doing his or her assigned job (putting together the TV programs). The argument goes that the TV teacher deserves no additional compensation, any more than the producer/director (or curriculum specialist or writer) does—as a creative contributor to the program. This is an understandable viewpoint from the school's perspective.

However, the whole question is still irritatingly vague from the teacher's point of view. Both the National Education Association (on behalf of the elementary and secondary teachers) and the American Association of University Professors (on behalf of higher education) have conducted studies and have issued policy statements, but little has really been resolved. Koenig summed up the situation in the late 1960's—and his summary is still valid:

> The television teacher's compensation varies widely from one institution to another. Ordinarily the teacher does not receive additional pay for appearing on educational television. Working conditions vary as widely as compensation practices. Although many schools provide release time for tele-

vision teaching, there is no specific standard for determining how much release time the teacher will receive. Finally, teachers are not legally protected from the improper use of their programs at another date; if a program is out of date the teacher has no legal right to stop its distribution.[12]

The best advice that can be offered is simply that, before entering into an ITC production situation, the TV teacher should have a written contract that clearly spells out agreements such as the following:

1. Compensation, including any "residual" payments for reuse of the materials

2. Teaching load and released time

3. Ownership and copyright of recorded and supplementary materials

4. Reserved rights for revision, editing, and withdrawal of materials

5. Approval for circulation of materials beyond the producing institution

Using Copyrighted Materials in Production
One side of the coin, discussed above, is protection of the rights of the TV teacher or creator of materials. The other side of the coin is the extent to which the TV teacher and producer/director can use materials created by others in putting together their local telelesson. As one is preparing ITV recordings, there are only two defenses for using someone else's materials without permission: (1) public domain and (2) fair use.

Works in the public domain are not covered by copyright. Several kinds of material may be considered in this category. First, the material

---

[12]Allen E. Koenig, "Rights for Television Teachers," *in* Allen E. Koenig and Ruane B. Hill (eds.), *The Farther Vision: Educational Television Today* (Madison: University of Wisconsin Press, 1967), p. 255. *See also* Thomas F. Baldwin and Donald G. Wylie, "ITV Rights: Model Policy Statements," *The NAEB Journal,* May-June 1966, pp. 30–36.

may never have been copyrighted at all. Small town newspapers, for example, are often printed without copyright notice. Second, the copyright on the material may have expired. Under the current copyright act, enacted in October 1976, copyrights expire 50 years after the death of the author of a literary work. Thus, it is likely that any work over 100 years old (from the first copyright date) is in the public domain. The situation is especially complicated as the 1976 copyright act extended copyrights due to expire to a maximum of 75 years. The best advice, if in doubt, is to check with the copyright holder. However, anything *originally* published by the federal government (with very few specialized exceptions) is in the public domain and can be freely used. Finally, any phrase or expression that has been in common usage for a long time is considered to be in the public domain. Despite what may circulate to the contrary, it is not possible to copyright the alphabet or "Mary Had a Little Lamb."

The category of fair use is a more difficult concept to pin down. Generally, it may be safe to use copyrighted material—without permission— if the use is incidental and you are not violating the rights or jeopardizing the reasonable financial returns of the creator or copyright holder. But how far can one go with the defense of fair use? The answer is not at all clear. The 1976 copyright act specifies four criteria for determining fair use.

First, the *purpose and character of the use* help to determine whether a use is fair. One may use brief excerpts from copyrighted material in a critical review of the work. Such use of copyrighted material in scholarly works, teaching situations, and research papers, has been allowed by the courts when similar commercial ventures would have been found to be infringements.

The second criterion is the *relation of the use to the nature of the work*. Having students act out a play in the classroom so that it can be discussed by the class is fair use but having the same students perform the play before the entire school is not. Yet, one would probably be safe in having the same students recite copyrighted poems at

a school assembly. Courts have always been strict about common law copyrights—which apply to works that have not been published and which are not otherwise in the public domain. For example, as strange as it may seem, photographic slides an ITV teacher may find in the Learning Resources Center may not be as safe to use without permission as are pictures from a national magazine.

The third criterion is the *amount of the portion used in relation to the work as a whole*. If one wanted to quote one line from a thirty-line poem, it would probably be a fair use. However, if the TV teacher were to use a copyrighted eight-word jingle in its entirety, he probably would not be protected by the fair use doctrine.

The final criterion, and the one that will determine to a large extent the amount of the judgment for a copyright infringement, is the *effect of the use upon the potential market for, or value of, the copyrighted work*. The BBC series *Civilisation* is available on film to audio-visual centers. Obviously, if a school copies it off the air when it is broadcast and then uses these "pirated" videotapes (rather than rent or purchase the films through an authorized distributor), the loss to the producer and distributor is direct and ascertainable. A TV newscast, however, is a different matter. It has little tangible value after it has been telecast; therefore, schools probably can justifiably record newscasts off the air without permission as long as the recordings are used for academic purposes.

The Golden Rule Test　All of which brings us to the best test one can apply to any use of copyrighted material. In deciding whether it is safe to use someone else's copyrighted material in a local ITC production, two basic questions need to be kept in mind: (1) Did someone other than the federal government create this work within the past 100 or so years? (2) If so, does the creator have a valid reason to *care* if the material is used? In other words, if the situation were reversed, and you owned the copyright to the material, would *you* care if someone used the material without your permission? Would you consider it a fair use if you were the "used" rather than the "user"?[13]

However, even in those cases where you feel that it is legal to use someone else's copyrighted material or original creation without permission, the copyright holder should be informed of your intentions, if you are planning to use more than brief excerpts, and the source should be acknowledged, in any case. Professional courtesy dictates that copyright holders be consulted and asked how they might want acknowledgments handled.

14.7 Production Procedures

Regardless of the level or complexity of ITC production, there are similar stages or procedures that any television program goes through in the process of turning out a completed product.

Preproduction Planning and Writing Well before the actual program begins, there is considerable work to be done. The administrator in charge of a new ITV project often fails to appreciate the many hours necessary for television program preparation and overestimates the ability of a burdened staff (and TV teacher) to pull together a decent telelesson without adequate time for planning and lesson preparation.

In preparing a script outline or run-down sheet (few routine ITV telelessons are produced from a complete word-for-word script), the TV teacher (or producer/writer) typically spends anywhere from five to forty hours to get one lesson ready. Considerable work is involved in research and updating old lecture materials, planning

graphics, filling out sketchy areas of the lesson, securing clearances for copyrighted materials, arranging for guests, securing props and paraphernalia, preparing and duplicating the script outline, and so forth.

Even for an unscripted "classroom" origination or location recording, considerable planning is necessary: schedules and personnel must be coordinated, some planning must be done to ensure that what is needed will be what is recorded, equipment must be arranged for, and so forth. It takes considerable forethought to make sure that everything and everybody is in the right place at the right time ready to do what is expected of it/him/her.

In a studio ITV lesson, once the script is prepared the producer/director and the production staff must attend to many preproduction details. The script must be carefully reviewed and analyzed for production requirements. Arrangements must be made for shooting any film footage or slides that are needed. Graphics must be prepared. Settings must be planned and constructed. Props must be checked out. Paperwork and forms must be filled out: to schedule the studio, to reserve equipment, to order unusual technical gear, to make staging requests, to turn in a lighting plot, to schedule a crew, to order special visual effects, and so forth. Conferences must be planned with the TV teacher and the producer/director. The P/D should meet ahead of time with the key members of his crew—the associate director, the floor manager, the technical director, and others.

Rehearsals may be necessary for larger productions. A really complex program may require a dry-run rehearsal (possibly in a rehearsal hall or empty classroom) where facilities are not needed; this is to familiarize the talent with major movements and rough blocking of the action. Most productions start out with a technical or walk-through rehearsal; the director works with the talent and the camera operators and floor crew in blocking out major movement on the studio floor, going over transitions and rehearsing the rough spots. No attempt is made to have a

[13]For a further discussion of the use of copyrighted materials in ITC productions, *see* Fred S. Siebert, Donald G. Wylie, and Thomas F. Baldwin, "Using Copyrighted Material for ITV," *The NAEB Journal*, May-June 1965, pp. 44–47; "Copyright Law Revision: An AECT Position Paper," *Audiovisual Instruction*, November 1973; Fred S. Siebert, *Copyrights, Clearances and Rights of Teachers in the New Educational Media* (Washington, D.C.: American Council on Education, 1964).

continuous uninterrupted rehearsal at this stage. Finally, there should be—ideally—a dress rehearsal; going through the entire program without stopping, making notes of things to be corrected, and then working out the problems after the rehearsal.

In reality, production and time limitations do not often allow the P/D the luxury of having all the rehearsal time he needs. So, with script hastily marked and fingernails chewed close, he bounds into the control booth—followed by his associate director and others—and tells the floor director through the intercommunication system to order "Quiet on the floor," orders the recording engineer to "Roll tape," tells the audio operator to "Fade in music," orders the technical director to "Fade in camera two," and nervously looks down at his notes to see what he has already forgotten to do.

Production/Recording   It might be helpful to think in terms of three different levels of production/recording of ITC materials. Each of these levels has its own requirements for production procedures and recording techniques.

The simplest level we might call *immediate utilization,* which involves one-time playback of some material—probably recorded under the "classroom origination" category. Audio-visual uses such as self-confrontation or microteaching are included here. Preproduction planning is minimal at this level. (In our extended print analogy, this might be comparable to uses of the chalkboard.)

The next level might be labeled *storage or semirepetitive uses* of material—where we anticipate holding on to some material for a limited time (say a year or two) for playback on the demand of a teacher, for some exchange within a district, and for similar limited uses. Some lesson preparation and production planning are necessary here. (Our print comparison might be with the spirit duplicator or ditto machine.) Both of these first two levels use only the slant-track recorder.

The third level, *repetitive programing,* uses either a high-quality 1-inch slant-track recorder or quadruplex recording. Thorough preproduction planning and rehearsals are necessary at this stage. This is (primarily) for studio productions for which extensive distribution is anticipated. The master tapes probably are retained and duplicate copies are made for large-scale distribution and playback. (Our print analogy would be to a high-quality printing press.)

Postproduction Concerns   Even after the studio production is finished, the process is not complete. First, there may be production considerations still to be wrapped up. For instance, some postproduction editing might be needed: to add titles or closing credits, to insert program material recorded earlier (or later), to clean up mistakes, to polish difficult semianimated sequences that were not possible to assemble in "real time" during the studio recording, to edit together portions of a documentary-style program recorded at different times, and for similar purposes.

Increasingly, we find ETC operations, at all levels, making more use of electronic news gathering (ENG) equipment and techniques—miniaturized cameras, portable recorders, and postproduction editing procedures. There is less emphasis on "real-time" studio productions as television producers are moving into the field for more location productions, simulating the traditional film techniques of single-camera recording and editing.

Also, some testing and revision of the recorded program material may be needed—especially if the program is to be validated to any extent as part of a programed instruction sequence of learning activities. This testing and revision process may go through several versions of the production. Even in a nonvalidated traditional ITV telelesson, there is continual feedback from the teachers in the field, portions of the content become outdated, the TV teacher sees ways that the production could be improved, and eventually the telelesson should be completely revamped. In a sense, very few ITV lessons can ever be considered completely finished—never to be revised.

14.8   Ancillary Materials

In addition to the television presentations, ancillary or supplementary materials are also produced in connection with most ITC programs. Most of these materials are produced for the classroom teacher or receiving instructor. However, some are produced for the direct use of the learners.

Classroom Teacher's Guide   The most common ancillary item is the classroom teacher's guide, a handbook to be used by the classroom instructor in utilizing the individual programs in a particular series. The guide may be mimeographed, dittoed, or printed—complete with heavy cover and illustrations. It may either be reproduced "in house," or be contracted out to a local printer. Different examples and varieties of guides are illustrated in figure 14-5.

Typically, the handbook contains some general information about the series—its purpose and design, usually a general introduction or foreword including a few platitudes and words of encouragement from the TV teacher to the classroom teacher. The real value of the guide, however, is in its specific information about each lesson. Normally, the handbook contains items such as the following about each separate program:

1. *Objectives:* The explicit purposes or learning objectives for the individual telelesson.

2. *Vocabulary:* A list of all new terms or words that are introduced in the TV lesson.

14-5 Various Classroom Teacher's Guides (Courtesy: D. G. Wylie)

3. *Materials Needed:* A list of all materials and tools that will be needed by both the teacher and the students during the actual telecast—rulers, maps, pencil and paper, art materials, science supplies, and so forth.

4. *Lesson Summary:* A synopsis or outline of the content of the telelesson—including any verbatim material the classroom teacher should know (such as words to a song, a math formula, or important dates).

5. *Suggestions for Preparation:* Ideas and activities for the classroom teacher to use in preparing the students for the telelesson.

6. *Follow-up Activities:* Suggestions for reinforcement and follow-up classroom activities, including quizzes, homework assignments, readings, projects, and the like.

7. *Reference Materials:* A bibliographic list of readings, films, and other sources of background information for the classroom teacher.

In addition to the all-important classroom teacher's guide, other materials might also be produced and distributed to the schools for teacher and classroom use: station or CCTV-system annual schedules that list all series being offered with a brief description of each one; charts, overhead transparencies, slides, or audio materials, to be used in the classroom to accompany a given ITV series; written supplements giving additional background information on a particular program or series; testing materials to be used with the students.

Student Supplementary Materials   Many kinds of ancillary materials might also be produced and distributed directly to students for their use and consumption. Syllabuses and reading lists are common examples. Occasionally, charts and diagrams are reproduced to be handed out to students. Workbooks in which the students figure out problems might be included. Some-

times "structured note-taking" worksheets are used, in which the basic outline of the program is reproduced and students fill in blanks and take notes during the course of the telelesson. Short excerpts and readings might be distributed to the students. Self-testing materials to be used by the students are a possibility. Commercially produced free pamphlets might be available to be distributed to students in some specialized courses. Self-teaching "programed" materials might be designed. Even audio-visual materials in nonprint forms may be utilized.

One interesting example of the latter category would be 45-rpm records. The Hawaii Educational Television Network produced a junior high school series, *Seasons of Change,* dealing with adolescent problems of self-identity and social awareness. To help reinforce the affective message of the TV series, original pop music was commissioned—setting various themes in a soft-rock format that the student could identify with. Two of the more popular songs were then pressed on a 45-rpm disc, and the record was given free to every seventh-grader in the state (at a cost of twenty-five cents per student—less than the cost of some duplicated print materials).

The idea of producing and distributing ancillary materials—whether for teacher or for direct student use—is an example of the concept introduced in Section 1.7. Television, or any specific telecommunications medium, cannot be viewed in isolation. All media and methods and approaches must be examined and used as part of an integrated program to get a specific instructional job completed.

**Storage
and Distribution**

After the message is produced, it must be made accessible to the learners. If it is a live presentation, it must be transmitted instantaneously through some sort of electronic delivery system. If it is recorded for later usage, then additional distribution alternatives are possible. If it is to be recorded, we must also consider the problems of storage, cataloging, and retrieval processes. Generally, in this chapter, we will be concerned more with produced ITC (both curriculum-structuring and enrichment programing) than with audio-visual/hardware uses; applications in the latter category—single-classroom and overhead camera techniques—usually do not involve extensive distribution.

In discussing distribution of telecommunication messages, we are considering *all* types of electronic information—dots and bits of electric signals—for this is what really distinguishes telecommunications media from other nonelectronic media (film, charts, 35-mm slides, overhead transparencies, and so forth). Telecommunications distribution capabilities include anything that can be reduced to electronic signals—high-quality studio TV productions, computer instructions, data bits, film, microteaching demonstrations, voice communications, still pictures, Teletype, facsimile, and slow-scan pictures from Mars.

15.1   Storage and Retrieval Operations

In virtually all television recording situations, the primary medium of video recording is videotape. Depending upon the purpose and intended utilization of the material, any of several recording formats may be suitable.

Recording Formats   In Section 14.7, we outlined three different levels or categories of production/recording. *Repetitive programing* implies ambitious studio productions combined with an anticipated need for extensive distribution. This would probably justify quadruplex video recording if a quad machine is available and feasible. Otherwise, a high-quality "top-of-the-line" 1-inch slant-track recorder would generally suffice. With auxiliary equipment (such as a time-base corrector), the 1-inch or ¾-inch helical-scan recorder can result in a recording that will meet broadcast technical standards.

The *storage or semirepetitive* category of recording refers to those materials that will receive some limited amount of distribution within a school or some restricted interschool exchange. The material need not meet top-quality studio production standards; it probably will not need to be stored and kept for an indefinite period. A 1-inch slant-track video recorder is generally adequate. Perhaps even a ½-inch recording would do the job. In some instances, materials might be recorded directly onto a ¾-inch video cassette (the U-matic format).

The simplest level, the *immediate utilization* category, requires no more than the ½-inch or even the ¼-inch recorder. This material will be looked at a couple of times and then erased. Little permanent quality is required. Many of these uses fall into the audio-visual/hardware classification.

Storage and Distribution Formats   Once ITC materials are recorded on videotape, they may be transferred to some other appropriate format for storage and distribution purposes. Frequently, materials are dubbed down (duplicated onto a lesser-quality format) for economy of space and distribution convenience. For example, a 2-inch quadruplex master tape is used to make 1-inch slant-track copies for use by schools, or a high-quality 1-inch master recording is dubbed down to a ¾-inch cassette or ½-inch cartridge format for distribution and playback. Sometimes, to save money and storage space, materials recorded and distributed on a 2-inch or 1-inch format are dubbed down to a ½-inch tape for reference and archives purposes. Of course, the video material may be stored in the original recording format—which is the case with most of the programing produced by ITC operations.

Film   Film, itself, is both a recording medium and a storage format that should not be overlooked in any classification system. Many television programs incorporate 16-mm film segments into the production process—for location footage, interviews, animation, time-lapse photography, and similar purposes. Film also could be considered as a storage and distribution medium. An entire videotaped program can be transferred to 16-mm film for storage and ultimate distribution purposes. Although more expensive, this would be appropriate, say, for school systems that are adequately equipped with film projectors but have little in the way of television reception equipment.

Video Cassettes   Until 1970, all video recording, storage, and distribution were handled on conventional reel-to-reel machines. Although almost everyone *could* learn and master the loading and threading procedures, it nevertheless was a bother, and many mistakes were made in threading the tape. Then, in the early 1970's, the ¾-inch U-matic video cassette was brought to the marketplace (after several years of ballyhoo and false starts). Two later models are shown in figure 15-1. The simplicity of dropping the self-contained cassette into the slot and pushing the "Play" button was, in itself, a minirevolution in the distribution and playback field. It was now possible for any elementary-school student to have direct "hands on" control over the material he or she wanted to see—providing not only a psychological boost but representing a substantial savings in playback convenience. The ¾-inch cassette was soon followed by the ½-inch cartridge video player which provided some of the same convenience and ease of operation; and in the mid-1970's a ½-inch cassette system (designed initially for home use) began to be used by schools and businesses.[1]

---

[1]As with audio cassettes and cartridges, the term "cassette" refers to a self-contained unit with two reels—a feed reel and a take-up reel; and "cartridge" refers to a self-contained single reel that is automatically threaded into its player.

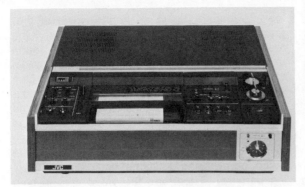 

15-1 Video-Cassette Recorders, ¾-inch U-type Format, with Off-the-Air Tuners: JVC Model CR-6100U with Built-in Clock/Timer (left) and Panasonic Model NV-2125 with Remote Control Unit (right) (Courtesy: JVC Industries, Inc., and Panasonic/Newcraft, Inc.)

Innovative and Hybrid Systems   During the past few years, several unique film and hybrid storage and retrieval systems have been announced and partially developed. In the late 1960's, CBS announced and demonstrated its Electronic Video Recording (EVR) system, a marriage of film and electronics that resulted in a nonperforated, dual-track film, ⅜-inch wide, in a self-contained cartridge. It held much promise, but one of its biggest drawbacks was the inability of a producer to record directly on the medium. Almost simultaneously, RCA announced its original SelectaVision system using holography (three-dimensional laser recording) to emboss signals on clear plastic tape. By 1976, neither system had been developed to take its place in the competitive marketplace. Kodak has meanwhile pushed ahead with its own combination of super 8-mm film and the self-contained Supermatic cartridge film video player (illustration 15-2).

Video Discs   The most promising recent developments, however, have been in the area of video discs. By the mid-1970's, two main sys-

15-2 Kodak Supermatic Film Video Player, Featuring Automatic Threading, Instant Review, Stop Motion, and Cartridge Loading (Courtesy: Eastman Kodak Company)

tems were competing for dominance in the U. S. market—the RCA "mechanical-capacitive" format and the Philips/MCA "laser-optical" format.[2] Prototype models of both systems are shown in figure 15-3. Winslow sums up the potential advantages of the video-disc system:

> There is widespread agreement on the projection of the *base cost* of an hour's information on the videodisc at about 1¢ a minute—not 40¢ or $4.
>
> The videodisc player itself could be only one-third the retail price of similarly functioning players in the film and tape technologies.
>
> From a single, paper-thin-roll-it-up-and-send-it-through-the-mail videodisc you could selectively call up any one of the over 100,000 single picture frames stored on one side and display it indefinitely at the press of a button.
>
> You could mix stills and full motion randomly, manually, and on a pre-programmed basis.
>
> You could go forward and backward at will, jumping from the first to the last part of a program in only several seconds.
>
> Audio fidelity could be better than that currently provided by a good-quality LP audio-disc or tape, and there could be four channels if you wanted them.
>
> In certain of the approaches being discussed [e.g., the Philips/MCA format], there is no mechanical contact made with the disc itself. This means that all these advantages would never degrade through use—the disc would never wear out!![3]

Considering the sophistication of these processes and the tremendous capacity for storing information (up to an hour's worth of programing on one side of a floppy thin disc), the reader is wise to avoid the natural temptation to compare the video disc with the standard audio LP record; they are completely different. One similarity should be noted, however. It is impossible to record directly on the disc. Therefore, the television material must be recorded on videotape and then transferred to the disc. This transfer—to create the master disc—can be expensive (although pressing additional copies is relatively inexpensive). Therefore, the video disc is best suited to mass distribution applications.

Cataloging and Shelving All stored ITC materials need to be labeled and cataloged in some manner. The larger the operation, the more imperative a good cataloging system becomes. In a large library system, it may be decided to catalog all nonprint materials with the same classification system as is used for books—either the Dewey decimal system or the Library of Congress system. In smaller operations, the video and audio materials may simply be numbered consecutively, with a corresponding printed catalog that identifies each program. Or materials may be grouped together by subject areas, with some sort of loose-leaf cataloging system.

Consideration also must be given to the physical storage and shelving arrangements. Appropriate temperature and humidity conditions must be maintained. Videotapes and cassettes must be stored standing—not flat (see illustration 15-4). Are all like materials to be shelved together—books in one place, audio cassettes and tapes in another section, records in another room, video tapes in their own corner, films in a vault, and so forth? Or will shelving be by subject area—with mixed media resources scattered throughout the instructional materials center? What materials need to be kept in the library? What should be kept in the classroom? What needs to be centralized at the district office? Which materials can be checked out for at-home viewing?

---

[2]For a brief summary of the two principal video-disc systems, see Ken Winslow, "A Videodisc in Your Future," *Educational and Industrial Television,* May 1975, pp. 21–22.
[3]*Ibid.*

15-3 Video-Disc Formats: RCA SelectaVision VideoDisc Player (left) and Philips/MCA Disco-Vision Player (right) (Courtesy: RCA and MCA Disco-Vision, Inc.)

15-4 Videotape Storage Facilities, Illustrating Vertical Storage and Easy Accessibility (Courtesy: MPATI)

Many considerations such as these should be examined at the onset of any ambitious media/instructional program. Careful analysis of recording formats, storage formats, cataloging system, and shelving requirements will alleviate potential headaches as the program progresses.

15.2   Levels of Accessibility

Many schooling and industrial ITC projects have been designed and implemented, with elaborate delivery facilities, before giving any careful analytical thought to the actual distribution requirements and criteria the system really has to meet. Perhaps one of the most useful methods is to plan in terms of levels of accessibility that are needed.

Scheduled Playback   This first category or level of accessibility is probably the least expensive—and most inflexible—type of distribution arrangement. This "scheduled" designation refers to a predetermined timetable when specified materials are to be played back or made available to the learners. Normally, this type of scheduled playback is established on a regular daily or weekly basis. The traditional open-circuit station broadcast pattern is the most common example; the third-grade series *Let's Wonder About Science* is broadcast every Monday and

Wednesday at 10:30 a.m. (with repeats at 11:15 a.m. on Tuesday and Thursday)—and that is that, take it or leave it. Comparable business or military uses are many, for example, the 8:30 a.m. Monday morning recorded manager's message to the branch offices, the regularly scheduled Thursday afternoon closed-circuit showings of video programs for the "Wilderness Survival" training course, the monthly nationwide closed-circuit hook-ups for the company sales staff.

Not only is this the most economical means of sharing ITC materials on a widespread basis but it also implies the least freedom of choice for the classroom teacher or student or recruit or branch office salesman. As a general rule of thumb, the more widespread the distribution, the less the individual receiver has to say about the selection of the materials.

Comparable scheduled playback uses of nontelecommunications resources would include items such as bookmobiles (which visit the small rural school every Tuesday morning), assemblies (with a guest speaker every fourth Friday during third period), and weekly industrial production conferences (scheduled to coincide with the plant manager's inspection tour).

Request Scheduling   This next category reflects, to a much greater extent, the particular needs and desires of the classroom teacher and/or individual learner. This enables the receiver to request certain materials—which, of course, will be delivered at some time in the near future. This takes considerable planning and forethought as to specific needs, but it does give the classroom teacher more flexibility. This request scheduling could involve video materials (cassettes or tapes) from the school coordinator, programs from the district office, materials from the state ITV operation, or even leasing a special program from one of the national ITV libraries.

The advance time necessary to ensure delivery of the materials when they are needed may be anywhere from twenty-four hours to two or three weeks. Although this involves more plan-

ning and work on the part of the receiver, it does represent an improvement over the scheduled playback, which permits relatively little input from the individual receiver. Of course, this request scheduling also represents more cost. It is almost always more expensive (on a per-unit basis) to deliver a single copy of any item or program to an individual user than it is to make the program available to a wide number of people at the same time.

There are numerous examples of non-ITC materials available on a similar request scheduling basis. Films from an audio-visual center are probably the most common example; the teacher orders them well ahead of time and, usually, they arrive by the time they are needed. Special books that are not available in the local library may be ordered from the state library or even from the Library of Congress. Perhaps some expensive laboratory equipment can be scheduled through the district science supervisor for a demonstration on a certain day. A series of training films can be scheduled for use in a data processing seminar for the corporation's accounting executives.

Immediate Access   This category, of course, implies the utmost in flexibility and user convenience and applies to any use for which the intended receiver—classroom teacher, student, or salesman—can gain immediate access to the particular materials he is interested in. (The term "immediate" is relative, of course; it may take five or ten minutes to actually get hold of the materials.)

Some ITC examples in this category include remote-controlled random-access systems that permit the student to sit at a learning carrel and call up any information desired (Section 4.5), a video-cassette check-out arrangement in the school library, a collection of video materials housed in the instructional materials center, and so forth.

There are many non-ITC examples of comparable materials and resources used in the same immediate access manner: reference works in the library, maps in the classroom, globes and other models and realia in the resource center, experimental apparatus in the laboratory, and the like.

This immediate access category implies not only the most expensive means of making video resources available but it also, of necessity, is the most limited in terms of quantity and choice of materials available. It is simply impossible (at this stage) to make everything available to everyone on an immediate basis. (However, as we build toward the ultimate distribution web discussed in Chapter 12, this ideal comes closer; it is just a matter of connecting enough reception outlets with enough materials centers using enough advanced technology channels.)

15.3 Criteria for a Distribution System

In this discussion of distribution arrangements, we have been following a somewhat modified problem-solving approach (Section 13.3). Deliberately, we have not yet talked about specific distribution methods (this would be a discussion of alternatives or solutions) because we have not yet fully considered all the factors relating to defining our needs and establishing criteria that must be considered: What kind of *producing/recording system* is needed as far as quality of materials is concerned (repetitive programing, immediate utilization)? What kind of *storage format* might be most appropriate (videotape, video disc, film)? What kind of *accessibility* would be needed (scheduled playback, request scheduling)? These and other criteria must be examined before we can discuss solutions.

Range of Distribution Another important criterion to be considered, before looking at specific solutions or distribution alternatives, is the extent or the range of the distribution anticipated. How much territory must be covered? We could consider the eight geographical levels of scope and structure as outlined in Chapter 11.

If we are concerned solely with *classroom* distribution, for instance, dissemination arrangements are very simple. Probably nothing more is needed than the playback mode of the video recorder or simply a video player. Maybe video discs? At the *school* level, we probably should consider distribution systems that enable us to tie a whole school together for simultaneous playback or transmission throughout a building—possibly a wired system of some sort. Would there be need for an immediate access system?

At the *district* level, we might have several different kinds of needs. Do we want to tie the whole district together to allow for simultaneous distribution to all school buildings? This is what we will need if we rely heavily upon scheduled playback services. If we want to design a system to incorporate some provision for request scheduling, does this necessarily have to be plugged into the electronic delivery system, or can we economically incorporate some sort of physical delivery system?

The *metropolitan ITV association,* by definition, is primarily concerned with open-circuit station services. However, this metropolitan group also might want to get into other kinds of services—request scheduling, for example. What system would be needed at this level?

Are we to be concerned with *state* level operations? Is there to be a state-operated network? Need it be a multichannel system? What kinds of immediate-delivery and request-delivery needs are likely to be anticipated? What kinds of *regional* hook-ups might we want to tie into? Are we concerned with immediate delivery, scheduled service, or some kind of request-library service?

At the *national* level, we might be involved with any of the three classes of accessibility. What local provisions might be needed to take advantage of any particular national service? How about *international* possibilities? Eventual satellite hook-ups? For what purposes? How many channels? What materials are available?

Miscellaneous Criteria and Considerations
The answers to the above questions will begin to outline the shape of the distribution system that is needed to fit a particular situation. There are still additional criteria and miscellaneous factors to be considered, however.

Technical Standards   What kind of technical standards are needed? It would be rather inefficient, for example, to install a delivery system capable of good-quality high-resolution transmissions if most of the content to be delivered is low-quality technically. On the other hand, for a medical or specialized industrial application it may be necessary to have a very high-quality delivery system that can transmit exceptionally high-resolution video information. For instance, most transmission systems use an *RF (radio frequency)* system, utilizing standard broadcast channels such as the regular station transmitter and home receiver. However, it is possible for a closed-circuit hook-up to use a direct *video* distribution system, which sends only one picture through the complete system to a single-channel monitor, resulting in a higher-definition picture (Section 16.1).

Recording Capability   In discussing various ITC operations, one should make the distinction between the production/recording system and the distribution system. Is it necessary, for example, to have recording machines as part of the distribution system? If the delivery system is solely for the transmission of acquired materials, then no recording equipment is needed in the system; maybe even film or video discs would be the backbone of the delivery system. However, if recording capability is desired as part of the delivery system (for instance, to allow classroom teachers to record off the air or to tape materials for later use), then we are dealing with a different set of criteria for the system.

Public Coverage   Maybe an important criterion is that the public be involved in aspects of the overall ITC program. Perhaps the school district is expanding its adult education program and wants to reach people at home. Perhaps the school ITV project is dealing with some sensitive affective area of the curriculum (for example, family life education or sex education), and it wants to involve parents in previewing the materials (in the late evening) before they are telecast into the classrooms (the next day). Or perhaps a particular program relies heavily upon teacher in-service education materials that most conveniently can be broadcast for teachers in the early morning hours. All these purposes would involve at-home viewing arrangements—either open-circuit broadcasting or use of a community cable TV channel. On the other hand, there may be some sensitive information involved in some ITC projects that definitely should *not* be broadcast to the general public. How are these materials to be kept confidential if an open-circuit station is used? (Some medical projects have done just that by sending scrambled signals over a PTV station—with a decoder located at the hospitals involved in the project.)

Multichannel Capability   An important consideration for many projects is the number of channels that may be needed at any one time. Will it be necessary to be able to transmit four or five channels of instructional materials simultaneously (in which case a broadcast station operation is not much help)? Or will it be possible to get by with only one channel at a time?

Time Accessibility   This is another criterion related to total system capacity. Is it necessary to be able to use the system twenty-four hours a day? Is it necessary to own and have complete control over the system? If a delivery system is needed only on a part-time basis, maybe arrangements can be worked out to use some already-existing system—an open-circuit station or community cable channel or state closed-circuit network. Maybe it would be possible to use some existing system during the hours when it is not otherwise in use—utilizing semiautomated recording equipment to record material during special late-night transmissions (Section 15.5), similar to the National Public Radio (NPR) "netcue" system (Section 6.5).

> A variety of tools are now available for local recording and playback; some are suitable for capturing transmission during otherwise off hours. It seems appropriate to begin testing the potential use of such an approach to evaluate its utility. With such devices it is even conceivable that commercial broadcast stations might cooperate with the transmission of classroom materials during the wee hours.[4]

Costs　Depending upon the project, several other criteria, considerations, and factors must be examined and weighed before looking at specific solutions or distribution alternatives. One important criterion in any project, of course, is that of costs. How much money is available to do the job that everyone wants done? If the funds are not there, which criteria receive top priority, and which ones do we sacrifice? Financial resources ultimately must be a very important determining criterion.

15.4　Alternative Methods of Distribution

After having considered all of the criteria that go into determining what a distribution system should do, we are now ready to look at the specific alternative methods of distribution—the solutions to the problem. Basically, there are three broad classifications of distribution systems: over-the-air systems, through-the-wire systems, and physical-delivery systems.

Over-the-Air: Open-Circuit and "Short-Circuit"　This broad classification includes all systems of transmission that send signals through the air, utilizing various broadcast frequencies of the radio spectrum. For practical purposes, it is often divided into two separate categories—regular broadcast station operations, and "closed" open-circuit systems generally not receivable by the public.

Open-Circuit Stations　This category is the most familiar to most of the public; it, in fact, represents the heritage and the backbone of today's educational telecommunications operations. Included are radio stations (AM and FM) and television stations (VHF and UHF). Although the bulk of ETV programing is carried by noncommercial public TV stations, the commercial television stations also should be considered as an alternative distribution method. There are more than 700 commercial TV stations in the United States, and many carry instructional programing for local schools and perform other services broadly related to the overall ITC picture.

Network operations also should be included in this category. Although the school or the viewer does not receive the network directly (it picks up the network signal from a local station transmitter), the network operations dominate much of the ETC picture. State networks (Section 5.5), regional public TV networks (Section 5.6), and the national network (PBS)—both in their origination/production activities and with their transmission services—represent the warp and woof of the fabric of open-circuit ETC.

Another important element to many viewers is the repeater or satellite station and the translator station, which help to bring a primary station's signal to otherwise inaccessible or isolated viewers. The repeater station is a regularly licensed station (often VHF) that simply picks up the signal from the primary station and relays it to the community it serves. The translator station is a low-powered UHF repeater station that works on the same principle. These relay stations play an important part in the ETV picture because the noncommercial licensee is dedicated to serving all of the people of a given area, however remote some may be from population centers. For example, the Hawaii ETV network operates its primary station (KHET, Channel 11 in Honolulu), one repeater station on the Island of Maui, and eight smaller translators repeating the signals throughout the remaining valleys and bays on the other islands. Commercial stations, on the other hand, need to reach the largest concentrations of population at the most reasonable cost; it is not commercially feasible to spend thousands of dollars for a translator to reach an isolated valley that has only two hundred potential consumers.

Although they do not play an important role at present, another possible alternative is open-circuit airborne transmissions. It is technically possible to transmit direct-to-home or direct-to-school programing from either an airplane, such as Midwest Program on Airborne Television Instruction (MPATI) (Section 11.6) or from a satellite (see below).

"Short-Circuit" Operations The other category of over-the-air delivery systems has sometimes been referred to as "short-circuit" systems or, more properly, "closed" open-circuit stations. Technically speaking they are open-circuit operations; but they are not broadcasting stations in the sense that they can be picked up by the general public.

The most common example of this category for hundreds of schools throughout the country is the Instructional Television Fixed Service (ITFS) stations (Section 4.5). Limited in power, but capable of covering an area up to forty miles in diameter, this class of point-to-point telecasting is "in effect, a multi-channel, owner-controlled, closed-circuit, inexpensive system capable of covering a large geographic area."[5]

One of its main advantages is that it is possible for the school district or college using the system to transmit on four different channels.

Microwave transmission, which operates in the same upper ranges of the UHF spectrum as ITFS, is even more specialized and limited in its applications. Used for relaying telecommunications signals on a point-to-point basis (including network relay stations), its most prominent ITC applications would be specialized statewide networks such as Texas Educational Microwave Program (TEMP) and Indiana Higher Education Telecommunication System (IHETS) (Section 5.5).

Satellite transmissions are also often included in this grouping. Although some satellite plans eventually call for direct satellite-to-home transmission (putting such transmission in the "open-circuit station" category), most envisioned satellite services are for relaying signals to national or regional network centers or directly to ground-based open-circuit stations. Thus, at this point, satellites are seen basically as auxiliary delivery services to existing stations. However, the Public Service Satellite Consortium, incorporated in 1975 (Section 11.8), estimates that at least a seven-channel domestic ETV satellite is needed to serve both PBS and non-PBS activities.

One other "closed" open-circuit distribution system would be the use of FM subcarriers. It is possible to use part of the standard FM radio signal for auxiliary transmission services, such as stereo broadcasting or broadcasting in two different languages. One of the more imaginative uses, however, is to use part of the FM signal for the transmission of "slow-scan" television pictures. The Flint, Michigan, educational FM station, WFBE, first demonstrated the use of the subcarrier to transmit visual information in 1971. Since only a small amount of electronic video information can be handled on the limited subcarrier frequency, it is obviously impossible to transmit thirty frames every second (which is standard for moving video pictures); but it is possible to transmit enough information to build up a still picture (one frame) every ten seconds. The

[5]Francis J. Ryan, "The Case for ITFS," *Educational and Industrial Television,* March 1972, p. 30.

instructional advantage of being able to transmit a still video picture every ten seconds, along with a standard FM audio broadcast, has many intriguing educational applications.

Through-the-Wire: CCTV and Cable TV This broad classification includes all distribution systems that physically connect the sender and receiver with cable. Again, for convenience, this grouping is sometimes thought of in two different categories—regular closed-circuit (CCTV) systems and community antenna TV (CATV) or cable TV.

Closed-Circuit TV Systems   A distinction is sometimes made between open-circuit broadcasting and closed-circuit transmission as if these two means of dissemination were irrevocably opposed to each other in terms of philosophy and purposes and quality. This is not necessarily so. Actually, the only distinction that may be valid is that closed-circuit systems are *generally* more economical for reaching small numbers of viewers in specialized locations (for example, for formal instructional programing on a limited basis); and open-circuit broadcasting is more effective for reaching large numbers of viewers at home or in a widespread area (for example, for general, evening viewing).

The term "closed-circuit" includes a wider variety and scope of applications than any other type of telecommunications transmission. At the simplest level, any hook-up between a single camera and a receiver is technically a closed-circuit system. At the school level, several rooms may be connected together for a CCTV system. District-level CCTV systems or a campus-wide university CCTV program can be a fairly ambitious operation; the Hagerstown system is probably the best prototype model (Section 3.9). At the state level, South Carolina (Section 11.5) best illustrates what can be done with an elaborate multichannel closed-circuit system. Even national hook-ups are possible for network previewing purposes, temporary conferences, and similar transcontinental specialized applications.

Depending upon how one defines closed-circuit applications, there are probably two or three thousand "major" CCTV systems in the nation.

A specialized use of CCTV technology is the random-access retrieval system (Section 4.5), which enables a viewer—seated at his own learning carrel in an instructional materials center—to order whatever individual piece of video information he wants at that particular moment. While this represents considerable flexibility for the learner, it is an expensive system to install and operate. Advances in some of the physical-delivery systems (video cassettes and video discs) provide the same kind of flexibility at substantial cost savings (see below).

CATV: Cable Television   Community antenna TV (CATV) started innocently in the early 1950's merely as an extension of the viewers' receiving apparatus. In communities where home television reception was poor or marginal, local entrepreneurs could erect an antenna on some nearby good reception site (say, a hilltop), and send the TV signals—by cable—down to the homes in the poor reception area (say, in the valley). Subscribers were willing to pay a little each month in order to get good reception and they were happy; the CATV operators collected their monthly revenues and they were happy; stations got the benefits of extended signals and a larger potential audience so they were happy.

But, as CATV began to get more ambitious, problems arose—along with the new potentials. First of all, cable television (as it came to be called) started importing distant TV signals from far-away stations; this made local stations unhappy. Then, some cable TV operators began to originate programing, distributing some of their own productions directly into the homes of their subscribers; this made all stations nervous. Then the FCC decided that the cable industry needed some regulation; this made everyone somewhat disgruntled.

For several years the predictions have waxed unrestrained: Cable TV would lead to unlimited distribution channels; the "wired city" was just around the corner; consumer/viewers would be able to conduct all of their business, by means of two-way interactive systems, through the cable; pay cable (at last subscription TV would get off the ground) would spell the end of the "free" commercial broadcasting system; cable TV would be the panacea to solve all school distribution problems . . . et cetera.

By 1975, more than 10 million homes *were* hooked up to more than 3,200 CATV systems—almost 15 percent of all TV viewers were connected to cable;[6] distribution technologies *had been* developed that enabled at least eighty channels to be carried in a single coaxial cable; and, by 1975, thousands of individual schools *were* making some use of cable distribution availabilities. Most larger cable systems are willing to provide some hook-up services for schools free or for a nominal charge. Others have made a channel available to schools for their own use (the FCC at one point required larger cable systems to make channels available to schools, to municipal governments, and to the public at large). Some cable systems have cooperated with schools to make production services available. However, the pattern has yet to be set. The potential has yet to be tapped.

Cable today is mainly an extension of over-the-air broadcasting. It can pull in a signal that has previously been unavailable, or it can make an existing signal sharper and brighter. Cable *could* become a comprehensive communications medium in its own right, offering a wealth of entertainment and information services, but its performance has yet to match its potential.[7]

Lasers and Fiber Optics   Related to the through-the-wire systems are new technological

developments that may someday make coaxial cable look quite crude by comparison. Laser beams can be modulated with electronic signals and an immense amount of information can be carried on one beam. A single laser beam the thickness of a regular lead pencil can carry all of the electronic signals generated by all of the radio and television stations currently operating in the United States! However, laser light is affected by adverse weather conditions and cannot be used to send information very far through the atmosphere (as compared with open-circuit radio transmissions).

However, laser light can be sent through tubes or pipes containing fiber optics—bundles of threads of very pure and carefully fabricated fiber glasses. Such a fiber optic channel can conduct light very efficiently over long distances. Using lasers as the light sources, tens of thousands of channels can be transmitted this way.

The growing volume of data communications, video business conferences, "cashless society" credit transactions, etc., are helping to support technology that will both make interconnection less costly and provide alternatives to open broadcasting. In perhaps ten to fifteen years, we should begin to see bundles of fiber optics supplement, or even replace, much of the current electronic distribution system.[8]

Physical Delivery Systems   This final classification is the most limited in some ways, yet the development in recording and storage formats make this a viable and exciting alternative. A physical-delivery system refers to any method of getting the program from the point of production to the point of utilization by physically handling or delivering the program—in short, any system which does *not* use either over-the-air signals or through-the-wire dissemination. Basi-

[6]*Broadcasting and Cable Television: Policies for Diversity and Change* (New York: Committee for Economic Development, 1975), p. 64.
[7]*Ibid.,* p. 60.

[8]Tressel *et al., Future of Educational Telecommunication,* p. 57.

cally, this category covers the ITC material that goes from the recording medium to the storage format and then directly to the user.

Traditional systems that fall into this category include the following: sending video-tapes through the mail; handling videotapes or films, through a school or district audio-visual center, by regular school delivery; leasing programs from a national library and playing the tapes directly in a classroom; and so forth.

The newer storage formats, however, greatly extend the possibilities for individualizing instruction. With the video cassette or video disc, it is possible to set up systems in which the learner is directly responsible for picking up and playing back the materials. He or she can check out the video cassette or disc from a library or instructional materials center, take it to a viewing room, insert it in the player, and have complete direct control over the retrieval and utilization process. The learner can, for instance, interrupt the playback process to take notes or take a break, review part of the telelesson, or skip portions of the material altogether. The pupil is in complete control of the process. In short, this gives the system even more flexibility than the random-access system involving a wired carrel, but it is much less expensive.

As the storage formats become even less expensive and easier to use, the potential of this type of physical delivery of materials and direct user-controlled systems is likely to increase. In addition to the simplicity and convenience of operation—and potential cost savings—there is the added psychological advantage that the learner has in feeling that he has some control over the learning situation. He is not at the mercy of some unseen master scheduler/technician pulling the strings from the bowels of the system; instead, he—the learner—is as much in charge of the system as if he were looking up print references in the library.

Aside from the possibilities for individualizing instruction, the physical-delivery category also offers other advantages in flexibility and

scheduling and tailoring materials for instructional situations. For small-group classroom or large-group presentations, the video-cassette and video-disc formats facilitate teacher-controlled presentations in that the instructor can alter the pace of the ITV presentation, interrupt the tele-lesson to make a point, go back and review a portion of the program, or postpone the presentation until a more appropriate time. So, both for individualized instruction and for teacher-controlled situations, the physical-delivery distribution category offers more flexibility and direct control than either the over-the-air or through-the-wire systems.

15.5   Selecting the Best Alternatives

Within the problem-solving context, once you have established the criteria that the solution must meet, and once you have listed the various alternative solutions that might be applicable, the next step is selection of the best solution or solutions. This is done by matching each solution against the established criteria—and seeing which of the proposed alternative solutions appears to best meet the criteria. Regarding the question of distribution methods, for example, we could now compare the various distribution alternatives (solutions) with the list of things that we would want a given distribution system to do (criteria).

Designing a Distribution Grid   One practical approach is to design a chart or grid with the criteria and the solutions listed as the vertical and horizontal axes. For example, in figure 15-5 we have listed the various criteria for a given fictional system down the vertical axis (the left-hand column), and we have listed several alternative solutions on the horizontal axis (across the top of the grid).

In this hypothetical instance, the first criterion listed is "cost limitations"; cost would be a prime consideration in any situation. The proposed system has "recording capability" as the

	ETV Station	Commercial Station	ITFS Station	Cable/CATV	District CCTV	Random Access	Cassettes	Microwave	Mail/A-V	Video Discs			
Cost Limitations													
Recording Capability													
Wide Distribution													
In-Service Needed													
Request Scheduling													
Classroom Distribution													
Public Information													
3 Channels at Once													
Much Ancillary Print													

15-5   Problem-Solving Distribution Grid

List all necessary criteria in the left-hand column. List all feasible distribution alternatives along the top. Check those distribution methods that can meet each criterion. Then select the distribution method (or methods) that can best be used (or combined) to form the distribution system.

next criterion; it evidently will be necessary to have recording capabilities at the reception end of this system. Next, "wide distribution" is included; let us assume this is a system being designed for a geographically widespread school district. "In-service" is another important factor; it will be necessary to incorporate quite a few teacher education materials.

"Request scheduling" is listed as a criterion; this probably implies that no immediate-access capacity is anticipated. "Classroom distribution" indicates that many individual classrooms will be tied into the system—as opposed to a few viewing centers for large-group instruction. Another criterion, receiving less priority, is "public information"; the school district feels that it would be important to be able to include parents and community programing at times. The "three channels at once" criterion also is of relatively low

priority; this would seem to imply that the school district would like to be able to have occasional access to a multichannel capability, but evidently it is not of utmost importance for the design of the system. A final criterion is that of "much ancillary print" material; whatever electronic system is adopted, the district is evidently going to integrate heavy amounts of print materials into the instructional design—so the system will have to take that into consideration.

Once the various criteria have been identified and listed, the school district then lists the various alternative methods that might be feasible; these are listed across the top of the grid (ETV station, commercial station, and so forth). The next step, then, is simply to compare each possible alternative with each criterion; and to place a check mark—or use some sort of numerical coding or ranking system—to indicate how well each solution would answer each particular criterion. For example, the ETV station, the commercial station, and the cable TV system all might meet the "cost" criterion rather inexpensively; these are existing systems that might make time available to the school district free or at a low cost. The cassettes and physical-mail-AV distribution categories might not be too expensive. The ITFS, district-wide CCTV, random-access, microwave, and video-disc systems all might be fairly expensive to install and/or operate for a small district operation.

A similar analysis would be made for each of the criteria, with appropriate checking or ranking used to fill in the grid. Then it is possible to look at the completed grid and get a fairly good picture of how the various alternatives compare. For instance, in this particular situation—with these particular criteria (and every projected system would, of course, list different criteria in

a different priority order)—it may turn out that the best-looking potential solution would be a mixture of open-circuit ETV or commercial station, combined with a physical-delivery system, probably video cassettes.

Combination Solutions In fact, in most situations a combination of delivery methods will be involved. Except at the simplest classroom or school level, seldom does a program get from the production center to the learner without going through several different distribution combinations. Often a combination of over-the-air and through-the-wire media is employed; or physical delivery is involved somewhere along the line; or perhaps two divisions of the same broad category will be used—such as a cable TV hook-up to a district's own CCTV system.

Figure 15-6 illustrates the various distribution links that might be involved in getting a

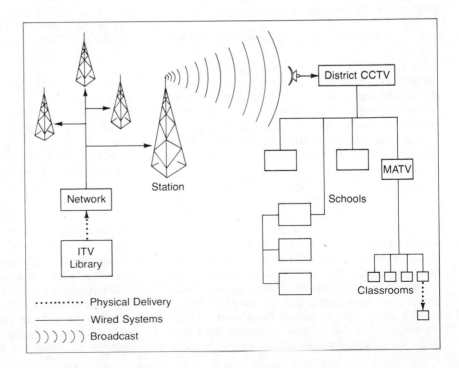

15-6 Typical Distribution Combination

hypothetical program from an ITV library to a classroom. The library mails a 2-inch quadruplex videotape to the regional network (physical delivery); the tape is then sent through its leased network lines to the various stations (through-the-wire and, possibly, microwave); the stations then broadcast the program (over-the-air); it is received at the head end of a school district's CCTV system and distributed to the individual school buildings (through-the-wire); at the individual school it is sent through the school's master antenna television (MATV) system (Section 16.1) to individual classrooms (through-the wire); however, the school's ITV coordinator makes a ½-inch video recording of the program, and it is played back later on a different video recorder in the science room (physical delivery).

This kind of distribution path is not at all uncommon. The larger the system or the more centralized the production operation or the higher the quality of the materials, then the more complicated the delivery mix is likely to be.

CADAVRS Many complex and sophisticated adjuncts to distribution systems have been developed over the years. Various remote-controlled, computer-assisted, and automated devices have been used as integral parts of various systems. One such example is the CADAVRS (computer-assisted-dial-access-video-retrieval system), developed in the late 1960's. This approach is a video counterpart to the NPR "netcue" system (Section 6.5).

The heart of the system is an encoder, tied into a regular transmitter at any normal PTV station. This encoder is capable of generating special coded audio tones—combinations of beeps and blips of various pitches (not unlike the audio tones of a push-button telephone). The receiving part of the system is a decoder, connected to a video recorder at a given school, capable of responding to one particular coded audio combination of, say, three different digits—for example "3-2-6." Thus, whenever the transmitter sends out the coded audio tones for 3-2-6,

the one decoder locked into that number will be activated and will turn on that one video recorder at that given school.

The system is designed primarily for late-night off-hours transmission of specially requested video materials. For example, there might be twenty high schools tied into a single PTV station through a CADAVRS system. On any given day, Central High School might have requests for two films or videotapes needed by different teachers for showing the next day—one ten-minute physics demonstration and a seventeen-minute film on population control. These requests are telephoned into the CADAVRS office at the station. A half-hour videotape is threaded onto the school's slant-track video recorder (which is turned off, but left in the "record" mode), that is plugged into the CADAVRS decoder. Late that night, hours after the normal sign-off for the station, a CADAVRS-shift engineer is still transmitting special requests to sixteen of the participating schools. At 3:07 a.m., he gets to the Central High School request for the physics demonstration, threads up the appropriate tape, punches 3-2-6 on his encoder, which transmits the "turn-on" signal to Central High School, waits thirty seconds (for the high school machine to warm up), and then transmits the physics tape over the air; it is recorded at the school. When the program is finished, the engineer punches up the appropriate code to shut down the school recorder. At 4:58 a.m., he gets to the request for the film on population control, and goes through the same procedure again for "3-2-6."

All over the city, all through the night, the same scene is being repeated—with high school recorders starting up in darkened unmanned audio-visual rooms, one at a time, to run mysteriously for a few minutes, and then quietly shut themselves off as if by some unseen remote-controlled electronic hand. The broadcasting operation is carried out for the smallest audience imaginable—an occasional startled high school custodian, a handful of bleary-eyed at-home insomniacs who will watch anything on the tube at 3:07 a.m., and the one machine for which the program is intended. What the system has done is to transform a scheduled playback distribution system into a request-scheduling mode. Many other imaginative combinations of distribution hardware may be possible for specialized applications.

15.6 Nonelectronic Information and Materials

In addition to the various telecommunications materials distributed through the several electronic channels discussed above, there also are the nonelectronic materials that need to be distributed as ancillary materials for the total program. How do we get the classroom teacher's guides, testing materials, student workbooks, and related materials delivered to the users (Section 14.8)?

In the case of physical-delivery ETC systems, the same distribution arrangements would probably be used for the supplemental materials as for the program material itself—hand delivery, U.S. mails, audio-visual truck, commercial delivery services, and the like. For over-the-air or through-the-wire systems, some sort of physical delivery of ancillary materials must be arranged; except for facsimile transmission, these materials are not normally reduced to electronic information for instantaneous transmission. So the teacher's guides and student workbooks are sent through the mails, or delivered by auto, or otherwise disseminated through some sort of slower means of overland express.

Of course, most of the materials must be delivered in advance of the actual ITC programs, and arrangements must be made for this early distribution. The teacher's guide, for instance, should be in the hands of the classroom teachers well before the beginning of the televised series. There also are annual program schedules that should be delivered well in advance of the start of the school year—so that school administrators and media coordinators and teachers can see what is being scheduled at what times. Other general public information must also be disseminated through nonelectronic means.

Occasionally—for reasons of instructional urgency or for testing and validation purposes—it may be necessary to start telecasting an ITV series before the series has been completed. A semester-long series of thirty telelessons may only have twenty programs recorded at the beginning of the semester. So the first programs must be distributed while the last ten productions are still in the planning stages. This means that, of necessity, the teacher's guide probably will not be completed either; yet the classroom teacher should have the complete guide before the beginning of the series. In this situation, some sort of compromise must be attempted. Perhaps a special overview and outline of the thirty programs can be prepared, and then guide pages are sent out, mimeographed or dittoed, in a piecemeal fasion—five or six programs at a time—before the transmission of the ITC materials.

One other special problem should be considered—handling and distributing the ancillary materials at the school level. The guides and student materials must get into the hands of the intended users—the teachers and students. It is fairly easy to get the electronic information distributed—the TV receiver is turned on at the right time and there is the material; but getting the guides delivered, ironically, can be more of a problem. Someone at the district level may temporarily intercept the guides when they are received from the ITV library. They may be sent to the wrong office—where nobody knows what to do with them. Once they arrive at the school level, some clerk or secretary may hold onto the guides mistakenly—not having received instructions as to their further distribution. There have been occasions when the print materials have been dutifully delivered to the school library (where else should print materials go?) where the guides have been cataloged and shelved for routine reference uses.

Generally, there should be some designated school coordinator or media liaison person
to whom all such ancillary materials should be
directed. This teacher or officer or staff person
would then be the one to make sure they are
properly delivered—and ready for utilization.

Utilization:
Reception and
Implementation

The most exciting television productions and the most sophisticated distribution system are all wasted if the ITC materials are not utilized properly in the learning situation. Whether we are using a conventional classroom, an auditorium, a plush corporation viewing room, or a converted barracks, we must be concerned with the way the materials are viewed and integrated into the learning situation—be they curriculum-structuring materials, enrichment programing, or audio-visual/ hardware applications of television. There are certain common elements in all of these situations: the reception equipment; the logistical arrangements of physical elements and personnel; the viewing conditions; the role of the classroom teacher or reception instructor; and the utilization procedures before, during, and after the ITC presentation.

This chapter will be concerned both with practical/physical considerations and with theoretical aspects of integrating ITC materials into the learning situation. As such, the chapter will be intermixing the abstract and the concrete, the generalization and the specific. Although the classroom teacher is the most important single element in most traditional utilization situations (Section 16.4), we should begin by looking first at some hardware considerations.

16.1   Reception Facilities

The first concern will probably be with the physical facilities needed for the reception and viewing of the ITC presentation, starting with the receiving system that will bring the signal to the TV set. If we are dealing with a relatively basic school closed-circuit system or a simple one-room hook-up of a video-cassette (or video-disc) player and one receiver, then the receiving system is an uncomplicated wiring together of source and reception/viewing point. If, however, we are concerned with the reception and internal distribution of a signal from a distant station—especially a weak signal from a UHF or Instructional TV

Fixed Service (ITFS) transmitter—then the school needs a more complicated receiving system.

Master Antenna System   Many schools that rely primarily upon over-the-air delivery have a master antenna television (MATV) system (see figure 15-6). Basically, this may be thought of as a limited closed-circuit set-up—one without any studio or camera input; the only source of an ITC signal is the broadcast transmission plucked out of the air.

The MATV system is composed of several elements—starting with the antenna. In the case of a reception system designed for UHF or ITFS distribution, the antenna is a highly directional antenna selected for a single channel or certain group of channels—it may even be a parabolic dish (for ITFS or higher UHF channels), similar to a microwave antenna. Placement of the antenna is crucial for weak or distant television signals; the antenna should be tried in several locations—on different buildings at varying heights—before the installation is made permanent.

Normally with a UHF station, and definitely with an ITFS transmission, the received signal is converted down to an unused VHF channel—which can then be distributed throughout the school building(s) and viewed on regular receivers (see figure 16-1). This down-converter is usually placed near the antenna, at the head end of the MATV system.

Assuming that the signal is to be distributed to a number of classrooms or receiving stations, amplifiers must be installed at regular intervals. If the MATV system is working with a particularly weak or distant signal, it may require a preamplifier placed at the head end of the system, even before the down-converter.

Through a system of coaxial cables—with adequate amplification—the signal can then be fed into a number of receiving locations. In each room, a terminal box or tap-off is located on the wall with a connection for the antenna lead from a standard TV receiver.

It is possible to use this basic internal MATV cabling system as the nucleus for a complete CCTV hook-up. An origination point ("studio") can be cabled into the head end of the system, and the MATV/CCTV coaxial cables and amplifiers can carry any locally originated materials—from cameras, video recorder, film chain, or whatever. In fact, the original coaxial cable can be used as a two-way transmission system; installation of terminal conduit boxes provides for originating a signal—as well as serving as the terminus for the antenna. With this type of dual-capacity system, a camera can be connected to the terminal in any room—which then becomes the origination point.[1]

Criteria for Selection of a TV Receiver   The first consideration in selecting a television display device is whether a conventional receiver is to be used or whether a video monitor (or "jeeped" set) is needed. The conventional receiver is adequate for most typical installations where an ordinary RF-modulated signal is distributed—wherever a television signal is modulated on a specific TV channel (or radio frequency) and distributed through the cable, conveying the same video and audio information as a broadcast signal on a given channel. On the other hand, if a higher quality signal is desired, then the raw direct video signal can be transmitted through the cable. This direct video system (which does not include audio) takes up the entire cable capability; no RF modulation is used. Therefore, a special video monitor is necessary—without certain RF circuits, without any channel selection capacity, without audio components, but with much higher resolution. An ordinary TV receiver can be "jeeped" to bypass these RF circuits and can be used for a direct video feed; however, it would not have the higher resolution and technical quality of a video monitor.[2]

---

[1] For a fuller discussion of master antenna systems, see Philip Lewis, *Educational Television Guidebook* (New York: McGraw-Hill Book Company, 1961), pp. 61–63 and 124–127.
[2] For a complete description of RF modulation and direct video signals, see *ibid.*, pp. 111–113.

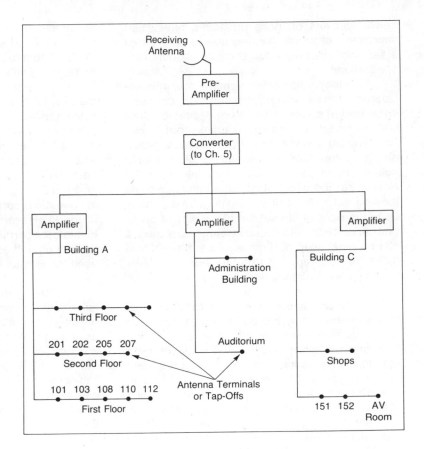

16-1 Diagram of a Typical School Master Antenna System

A second question is whether or not color receivers are needed. Even though the initial ITC project involves only black-and-white materials, if color applications are anticipated at some future date then it might be more practical to invest in the color receivers at the outset. Consideration should also be given to the fact that, as inexpensive color production equipment is becoming more available, fewer monochrome materials are being produced for distribution and acquisition.

Another important early consideration is the size of the receiver. For most normal classroom or conventional-sized viewing room purposes, a 21-inch to 24-inch receiver (or receivers) is probably adequate. If a system is intended for small-group utilization, perhaps a 17-inch or 19-inch receiver is sufficient. If individual learning carrels will be used, then one might want to think in terms of 9-inch receivers. Of course, many installations might want to consider large-screen viewing apparatus. Whereas large-screen television projectors have not been completely satisfactory in the past, brightness and contrast of newer versions of these projectors have been improved sufficiently that they may be considered a feasible alternative for some auditorium-type applications.

Many other criteria should also be considered in the selection of the television receivers. What kind of safety features are desirable? Special electrical grounding? Tamperproof back? Shatterproof shield? Would the added expense of a glare-free tube be justified? Might not lockable doors on the front of the tube be desirable? How about the accessibility of controls? Should they be out in front of the set for convenience? Or in the back of the set out of the way of eager and curious hands? Special institutional receivers can be purchased with these and additional features; but perhaps an ordinary home receiver would suffice for many applications.

Mounts and Stands   In most instances, receivers are simply placed on special stands or carts that enable the TV sets to be easily rolled from one location to another. In the use of these stands, convenience and safety features need to be considered. The carts should be tall enough to allow students to see the screen over the heads of their classmates (Section 16.3), but not so high that the stands tip over when they are moved around the room or from one viewing room to another. (See illustration 16-2.) The set should be bolted or otherwise permanently mounted to the stand, which should have large casters and a wide base to make it as tipproof as possible.

Many systems have receivers permanently mounted. Special mounts can be fastened to the wall or to the ceiling so that receivers can be hung or supported in auditoriums or viewing rooms—locations from which the sets probably will never be moved (figure 16-2).

Audio Accessories   For many different types of installations, certain kinds of receiver accessories may be desired. The receiver should have a special jack for plugging in one or more external speakers, which are commonly used pieces of equipment. The speaker might be used in a viewing room, for example, whose bad acoustics demand that additional speakers be placed in strategic locations.

Depending upon the learning situation, the teacher may not want the entire class to participate in viewing the telelesson. Therefore, the teacher may set up a separate viewing corner for a small-group situation—using headsets. The external speaker jack can be attached to a connection box that may have a number of headsets plugged into it. At times, the teacher may plug in an audio recorder and record the audio portion of the telelesson; this can then be played back to reinforce certain parts of the lesson. Other configurations and ancillary audio equipment can be used for various kinds of special purposes.

Using the Receiver   A TV set, this familiar and commonplace piece of equipment, can become a formidable object in adverse situations, assuming awesome proportions when it demands a minor adjustment in the midst of a telelesson. Perhaps this is because the classroom instructor instinctively realizes that thirty pairs of eyes are eagerly waiting for the teacher to make a mistake and turn the horizontal hold adjustment to stop the vertical roll.

If for no other reason than to bolster one's self-confidence, the receiving teacher should be thoroughly checked out with the receiver at the earliest opportunity. Become familiar with all the knobs and adjustments. What does each accessible control do? If knobs are twisted while the set is off, how are the distortions in sound and picture corrected? Practice misadjusting and then correcting the picture. Determine what problems can be quickly corrected and what problems call for more expert hands. Familiarity breeds confidence.

One hint that many teachers have used to advantage is appointment of one of the students as a TV set marshall to be in charge of turning on and adjusting the receiver. Another helpful hint: Establish the habit of checking out

16-2 Two Representative Receiver Supports: Wall-Mounted Receiver (left) and Typical Cart-Mounted Set (right) (Courtesy: Patrick Loughboro and RCA)

the receiver first thing in the morning, by turning it on and making sure everything is in working order. One final hint: Shortly before its scheduled use, turn on the set and let it warm up, adjust the picture, and then turn down the volume and brightness—now the receiver is warmed up and immediately ready to use on schedule.

Maintenance A final consideration regarding reception facilities is maintenance of the system and all its components—specifically the TV receivers. For larger installations, it might be feasible to have an engineer assigned to handle repairs on the receivers. In a high school or university situation, it may be possible to have students from an electronics or engineering course assigned to handle routine problems. A second alternative would be to have a service contract with a local dealer or with the installer who contracted for the sales and installation of the receivers; for a set annual fee, he would handle all routine preventative maintenance (periodic

check-ups) as well as emergency calls. A third arrangement would be simply a per-call arrangement with a local dealer; each service call would be paid for separately. Depending upon the needs of the specific school or industrial situation, either of the last two alternatives might be most cost-efficient for smaller systems.

As an added cushion—to guarantee that very little instructional time will be lost due to a broken receiver—many larger systems make sure they have a certain number of reserve sets. These receivers—perhaps one additional set for every ten in use—are kept available for instant service, ready to be wheeled into operation as soon as a problem occurs. One final word of advice from many experienced ITC practitioners: Make sure the receiver really does need servicing before the repairman is called in. Many a classroom teacher, building coordinator, or school principal has stood red-faced while a repairman resignedly pushed a plug into a socket or flipped a power switch.

16.2   Logistical Arrangements

In many ITC-based learning situations, there are relatively few logistical problems. In a well-equipped corporation AV viewing room, the TV receivers and associated paraphernalia are always available and ready for use. In many single-classroom ITV applications, the classroom teacher has few logistical problems: the overhead camera and the TV set are permanently assigned to the room—always ready for immediate utilization. In other situations, two or three teachers may share a single receiver, but they are able to work out scheduling problems among themselves with a minimum of conflicts. In another school, video-cassette players may be available in the library for students to schedule individually; the ITC materials and equipment are handled like any other library or audio-visual items.

In many school ITC situations, however, there are many different elements and personnel to be scheduled, moved, coordinated, and meshed together. A school utilization specialist or building ITV coordinator inherits a logistical assignment that resembles a juggling act.

Four Physical Elements   There are basically four different physical ITC utilization elements to be coordinated—brought together at the same time: the viewing room, TV receiver(s), students, and teacher.

Viewing Room   In some schools, only certain areas are designated as television viewing rooms—perhaps just an AV room, or the auditorium, or part of the library or learning resource center. In many schools, a limited number of regular classrooms are equipped with MATV terminal boxes or antenna tap-offs; when the STV program was initially planned, no one foresaw any need for all rooms to be equipped. In many college and industrial and military training programs, of course, there may be just one or two media rooms where all such mediated instruction is scheduled to take place. As long as the demand for the fa-

cilities is not continual all day long, there is usually no major trouble in scheduling learning groups into the facility as needed. However, at the school level, where a number of teachers are repeatedly making demands for use of a television viewing facility throughout the school day, this scheduling problem can be a major concern.

Receivers   In many schooling situations, the administration has been able to equip numerous rooms with antenna tap-offs—although the school could not afford to furnish every room with a receiver. In such situations, there will be a shortage of receivers, with perhaps three or four teachers sharing one TV set, which is moved from room to room. This pattern can be worked out logistically, but it takes careful scheduling and juggling of the receivers.

Students   The learners, of course, must be in an equipped room, with a receiver, at the right time. When using off-the-air programs, utilizing them at the time they are broadcast, this can become an especially crucial scheduling problem. If the school system can record the programs off the air and make arrangements to play them back at times more convenient to the class and the classroom teacher, the scheduling problem is alleviated somewhat—but, of course, this requires more of an investment in video recorders, videotape stock, and technician time.

Teacher   For virtually all elementary and secondary ITV usage, the classroom teacher also needs to be scheduled as one of the four physical elements. (In some college, industrial, and adult education programs, the ITC programs are designed to be used without the necessity of an actual instructor being present at the time of viewing by the students.) In most schooling situations, the classroom teacher would, of course, be scheduled along with the students; they would be considered together as a single unit. However, in many instances—where teaching specialists are used or where students shift from teacher to teacher for different subjects (from junior high school on up)—the logistics are compounded. Even at the lower elementary levels, team teaching arrangements are often advantageous. For example, three fourth-grade teachers may decide to capitalize on their respective strengths—so that Ms. Smith handles the language arts for all three groups, Mr. Jones teaches science to all fourth-grade students, and Ms. Jackson takes all three groups for music and art. We can begin to appreciate how much this compounds the problem.

In this case, let us imagine the scheduling and logistics details when Mr. Jones (who wants to utilize the fourth-grade science series broadcast over WZZZ) takes his class and Ms. Smith's students down to the AV room (which is the only room large enough to hold a double class) for the 10:00 a.m. broadcast. When the station repeats the program at 1:30 p.m., he makes arrangements to teach Ms. Jackson's students (while she handles his group for art), but he has to use Mr. Johnson's room (whose class is in physical education) because it has an antenna tap-off, borrowing a TV receiver from Joyce Loomis (fifth-grade teacher) who is not using it during that hour.

Usage Patterns   At some point, of course, the teachers involved may feel that it is not worth it; the hassle is too much—the disruption to the students' concentration is too great. Whatever the advantages of the particular ITV series are, the discontinuity to the learning situation may offset the benefits. This implies that there would be some discernible break-off point at which it can be said that the benefits no longer outweigh the drawbacks. Perhaps it would be helpful to think in terms of priorities of desirability in the logistical arrangements.

Receiver-Equipped Classroom   The most desirable arrangement, obviously, would be to have a TV receiver permanently placed in every classroom where the teacher might want to use ITV. (Ideally, there should be two receivers in

every classroom-sized viewing room—to guarantee the best possible viewing angle for every student and to provide a redundant back-up system in case of malfunction.) This would certainly make for the greatest accessibility and convenience for the students, the teachers, and the STV coordinator. However, it is prohibitively expensive in most situations; it means tying up a receiver all day long when it might be used only once or twice a day.

Moving Receivers   The next desirable alternative would be to have every classroom equipped with an antenna terminal connecting it to the MATV system. Then receivers can be scheduled and moved into the classrooms on portable stands as needed. One receiver may be able to serve three or four rooms in this manner. It causes some wear and tear on the equipment, but it is less expensive than buying three times as many TV sets.

Moving Students   Generally less desirable than moving receivers is the moving of students. If the facilities are limited, however, and not enough rooms are equipped with either receivers or antenna connections, then it will be necessary to move the students to the equipped viewing rooms. The drawback to this arrangement is the interruption in the learning situation. Frequently the students must be moved out of a viewing room as soon as the ITV presentation is over, with resulting loss in motivation and interest as the students are shuffled back to their classroom. The four or five minutes immediately following the telelesson can be the most valuable period for capitalizing on the students' TV-inspired interest—answering their immediate questions, getting started on related projects, and so forth. If this time is lost—and a myriad distractions are introduced as the students are herded back to their own room—the immediate follow-up to the telelesson is severely hampered.

Large-Group Viewing   Even more distracting is the moving of the class—along with several other classes—into a large-group situation to view the program in the auditorium, cafeteria, or other big viewing room. Not only is the learning situation interrupted but also the added distractions of the large-group situation present additional supervisory problems. In many instances, of course, when the situation is carefully designed and thoroughly controlled, the large-group viewing arrangement can work out quite satisfactorily. As a general rule, however, it is resorted to as a last-ditch makeshift expedient—which makes it the least attractive of the various usage patterns.

16.3   Viewing Conditions

What about the actual viewing conditions in the classroom or other viewing center? Where should the TV set be placed in the room? How near and how far away should the viewers be seated? At what height should the set be placed? How many students can view a single set? What about lighting conditions? These are some of the practical concerns of the classroom teacher or reception instructor.

Placement of the Receivers   Several considerations should determine where in the room the set, or sets, should be placed: glare and lighting conditions, maximum viewing area, and size and age of students, among other items. Ideally, as mentioned above, there should be two receivers in the average classroom; however, economically, this is seldom feasible—so most of the discussion below is based upon having one receiver available for a normal-sized classroom situation.

Glare   The principal source of glare on the face of a TV tube is likely to come from windows. Therefore, the easiest way to avoid this problem is to place the set so that the back of the receiver is against the windows or in a corner primarily facing away from the windows. The more the face of the receiver is turned away from

the windows, the less likely that any reflection can bounce from the face of the tube to any of the viewers.

Maximum Viewing Area   As we shall see below, the maximum viewing area should be based on a pie-shaped quadrant, about a quarter of a circle. Therefore, in order to minimize wasted floor space, it is most economical to place the receiver in the corner of the classroom—ideally a corner with an outside window wall along one side of the receiver. This placement minimizes the glare and also cuts down on distraction from any foot traffic near a corridor door. (See figure 16-3.) If two sets are used, they normally are placed in the two front corners of the classroom.

Height of the Receiver   Generally, the center of the set should be placed from four to six feet above the floor. Some practitioners offer the common-sense observation that the receiver should be placed at about eye level of the standing

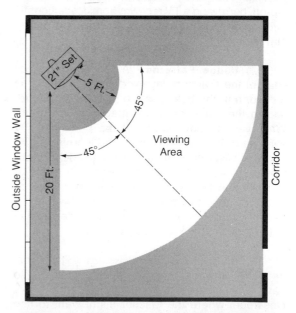

16-3 Floor Plan of Representative TV Viewing Classroom

teacher. Kindergarteners sitting on the floor obviously do not need a set placed at the same level as full-grown students seated in chairs. The closer together students are packed, the higher the receiver needs to be (so that students can see over the tops of heads in front of them).

Seating Arrangements Advice on the placement of the receiver and the seating arrangement for students comes from a variety of sources, many based upon practical experience and a few based upon actual experimentation and research.[3] The following suggestions are based upon a general consensus drawn from these sources.

Horizontal Viewing Angle As indicated in figure 16-3, viewers generally should be seated no more than 45 degrees from a line drawn perpendicular to the face of the TV tube. The narrower the spread the better. Some authorities say that, in order to avoid the possibility of any distortion at all, students should be seated no more than 30 degrees from this perpendicular.

Minimum Viewing Distance Research indicates that viewers should sit no closer than two times the actual width of the TV tube (not to be confused with the diagonal tube measurement). Thus, for a 21-inch tube (with a width of about 18 inches), this would mean a minimum distance of about three feet. Others say it should be more like four to five feet to avoid any discomfort. One determining factor would be the height of the receiver and the vertical viewing angle for the students nearest the set; this upward tilt should not be more than 30 degrees.

Maximum Viewing Distance The maximum viewing distance is determined to a great extent by the nature of the televised materials.

Some research indicates that the maximum distance should be about twelve times the width of the TV tube. This would be about eighteen or nineteen feet for a 21-inch receiver. Others, counting on more carefully constructed TV materials, would stretch this to a rough rule of thumb of one foot per inch of diagonal screen size. Thus, about twenty to twenty-one feet could be considered the maximum distance for a 21-inch receiver.

Maximum Number of Viewers "The *number* of students who can comfortably watch an ITV program in a classroom will vary according to the size of the seats, their age, and the distance a teacher needs to seat the pupils one from the other—owing either to disciplinary or hygienic considerations."[4] It is possible to cram up to fifty students in the viewing area for a 21-inch receiver. However, a more reasonable number—assuming they are seated at tablet-arm chairs comfortably spaced—would be about twenty-five to thirty students. The table (16-4) shows some compara-

16-4 Viewing Distances and Seating Capacities for TV Viewing

Size of TV tube[a]	Minimum viewing distance[b]	Maximum viewing distance	Maximum number of viewers[c]	Optimum number of viewers[d]
17"	3'9"	14'9"	34	17
19"	3'10"	15'2"	38	20
21"	4'10"	19'0"	54	26
23"	4'11"	19'4"	56	27
24"	5'5"	21'5"	72	33

[a]Diagonal screen size.
[b]Based upon seat row spacing of 5'2".
[c]Based upon crowded 3'0" chair spacing.
[d]Based upon desk and chair spacing of 5'2".
Source: Dave Chapman, Inc., Design for ETV: Planning for Schools with Television (New York: Educational Facilities Laboratories, 1960), pp. 32–35.

[3]Two of the better, well-documented sources on viewing arrangements are Dave Chapman, Inc., *Design for ETV: Planning for Schools with Television* (New York: Educational Facilities Laboratories, 1960); and Alan S. Neal, "Viewing Conditions for Classroom TV: An Objective Study," *Audio-visual Education*, September 1968, pp. 707–709.

[4]George N. Gordon, *Classroom Television: New Frontiers in ITV* (New York: Hastings House, Publishers, 1970), p. 191.

tive data on minimum and maximum viewing distances and on maximum and optimum seating capacities for various-sized TV receivers.

Other Viewing Conditions   In addition to seating arrangements, the classroom teacher should also be aware of other special environmental factors regarding the TV viewing situation. Acoustical problems may be bad upon occasion. Electronic sound can sometimes be distorted or garbled. Be considerate of others in adjoining rooms, and do not fight bad acoustics by turning up the volume. Double-check the fine tuning on the receiver; try a slightly different placement of the set; or use auxiliary speakers.

The most common mistake relative to lighting in a TV viewing situation is the tendency to cut down on the lighting too much. *Television should never be viewed in a dark room.* The picture brightness on the face of the tube is so intense that viewing in a darkened room greatly accentuates the contrast between the TV tube and the rest of what the eye perceives. Viewing TV in a normally lighted room will not cause eye strain; viewing TV in a dark room *may* cause eye strain over a sustained period of viewing. Leave the lights up—at least partially. It is easier on the eyes; it is better for taking notes or referring to other materials; and it is better for disciplinary reasons.

Generally, the only other environmental factor that might be affected by TV viewing is ventilation. If window drapes must be pulled (to cut out some glare on the receiver—in which case the room lights should be left on) and/or if the door must be closed to cut down on extraneous noise (or to keep from bothering adjoining rooms), then ventilation could be a problem. However, if the lighting and acoustical problems are properly taken care of, the room should not have to be sealed off for TV viewing purposes.

The first three sections of this chapter have dealt primarily with some of the physical aspects of ITC utilization. We now turn to a more theoretical (and occasionally practical) look at

the classroom teacher. Finally—in the last three sections—we will examine the actual integration of the ITC materials into the learning experience.

16.4   The Reception Instructor

Many ITC practitioners, distracted by the glamor of television hardware and preoccupied with the production and distribution of worthwhile ITV programing, fail to recognize and appreciate the single most important component in the total instructional media mix—the receiving instructor or the classroom teacher. He or she is ultimately in charge of the educational situation and is directly responsible for any learning that takes place.

Key Element: the Classroom Teacher With few exceptions, the classroom teacher is really the keystone upon which the success of the entire learning situation depends—whether or not media are involved. Many writers have emphasized the damage that an inept receiving teacher can cause. One ITV producer commented,

> You can transmit ITV to your heart's content all school day long, but if . . . the classroom teacher [is] hostile or bored or ignorant of how best to use television, you might just as well address your program—regardless of its merit—to the monitor in your own studio.[5]

Gordon states the situation in stronger terms: "In other words, a classroom teacher (or surrogate) may—intentionally or unintentionally—sabotage any TV lesson in a number of ways. . . . They may be overt or devious, well meaning, or malicious."[6] The corollary positive statement is also true: A good, enthusiastic classroom teacher can take a mediocre telelesson and turn it into a worthwhile learning experience.

---

[5]Judith Murphy and Ronald Gross, *Learning by Television* (New York: Fund for the Advancement of Education, 1966), pp. 64–65.

[6]Gordon, *Classroom Television,* p. 182.

The key ingredient is probably the classroom teacher's attitude; in most instances this will be more important than specific abilities the instructor may or may not have. Classroom teachers, first of all, should have a good internalized attitude and positive self-image. They need to realize that the classroom teacher is the key indispensable element in the learning situation and that—despite any initial irrational fears—they are not about to be replaced by a mechanical device of any nature (unless they deserve to be replaced). If the teacher is a truly dedicated and concerned professional, he or she will respect the fact that the learning situation exists to further the education of the learners—not the convenience of the teachers. Therefore, instructional telecommunications—if carefully designed and well integrated into the learning experience—will be a constructive factor.

This positive attitude must then be transmitted to the students. The classroom teacher should explain the whole concept of ITC; the explanation should cover why television is being used; what to expect from it; what it can and cannot do; why one should not compare ITV with the entertainment values of a commercial variety-entertainment program. (This last point is crucial; most students will instinctively associate "television" with prime-time commercial programing, and it is important that they be made to realize the role of television and related media as educational tools in a formal schooling context.) Second, the teacher should instill a positive attitude toward the particular ITV program and—especially—toward the on-camera instructor. "Indeed, the attitude of the teacher toward the program is one of the crucial elements in its acceptance by the group."[7]

Role of the Classroom Instructor Many times classroom teachers themselves downplay the importance of their revised role in the ITC milieu. They see their role as lecturer diminished, but fail to recognize the extent to which other roles have increased in importance (Section 12.6).

Diagnostician Seldom have classroom teachers had sufficient time to diagnose learning problems of individual students, but, as less time is needed for specific lesson/lecture preparation, the teacher can devote more energy to analyzing individual needs and tailoring educational experiences for students.

Planner The classroom reception teacher becomes engaged in a team teaching relationship not only with the on-camera teacher (and this is a very important team teaching situation) but with others at the local school level: the principal, the ETV coordinator (Section 11.2), the librarian, and fellow teachers. This involves a tremendous amount of planning, and the classroom teacher really can be considered the team leader—as far as planning for his or her particular students is concerned.

Discussion leader The classroom teacher, although possibly not as involved in making large-group presentations, will become increasingly active as a discussion leader. In most instances, the reception teacher will be working more with small groups. In many ways, these activities assume more importance than the large-group mediated presentations.

Tutor In a similar fashion, many teachers find that they can devote more time to working individually with students—as the on-camera teacher assumes more responsibility for working with the total group. This is a very rewarding exchange of emphases for many instructors.

Evaluator Also, the classroom teacher plays a more important role as evaluator of the entire learning experience. Not only must the teacher be concerned with participating in evalua-

[7]Charles F. Hunter, "Training Teachers for Television Utilization," *in* Allen E. Koenig and Ruane B. Hill (eds.), *The Farther Vision: Educational Television Today* (Madison: University of Wisconsin Press, 1967), p. 300.

tion and, indirectly, revision of the television materials but also he or she frequently is able to spend more time working on evaluation of individual students—which brings us back to the role of diagnostician.

Many other specific roles could be examined—curriculum designer, counselor, content specialist, researcher. Obviously, not every teacher will embrace all these roles to the same degree. Depending upon individual inclinations, various teachers will take advantage of the opportunities afforded by the introduction of ITV into the classroom to do different things, emphasize different roles, and grow in different directions.

Using the Teacher's Guide   For many school uses of ETC there is no classroom teacher's guide or any other type of ancillary materials involved—most audio-visual/hardware applications and many non-curriculum-structuring enrichment/ reference uses (image magnification, self-evaluation, analysis of a TV drama, illustrative materials, and so forth). However, in most major, sequential, curriculum-structuring applications of ITC, the single most important aid to the reception teacher is probably the classroom teacher's guide. The teacher should study the guide thoroughly before starting to use the series; determine the scope and purpose of the entire series; decide on his or her own direction and what needs to be accomplished with the telelessons. What should the students know before starting in with the first telecast?

The individual guide pages for each lesson should be carefully scrutinized far in advance of the scheduled telecast. What vocabulary terms need to be introduced? What needs to be done to prepare the students for today's lesson? What reinforcement activities should be planned? What additional materials will the class need? Costello and Gordon stress the importance of the determination by the classroom teacher of the unique contribution she or he can make:

She would do well to ask herself, after studying the teacher's guide, "What can *I* offer, from *my* special training and background that is *not* to be offered on television? What, if anything, has the lesson missed? What extra emphasis or drills do the children in *my* classroom need that other pupils viewing the program do *not* need? Is this lesson too complicated? How can I simplify it? Is it too simple? How can I enrich it?"[8]

Training for TV Utilization   The classroom teacher certainly should be aware of the opportunities for receiving training in the proper utilization of television in the classroom. Specialized training of this nature can be considered in two realms: *preservice education* (those courses and segments of curricula offered by teachers' colleges and schools of education or communication—to be taken before teachers are certificated and start their actual professional careers), and *in-service education* (those activities and workshops designed to upgrade and update a teacher's skills—while he or she is a practicing professional in the field).

Preservice Education   In a plea for more adequate preservice media training programs for prospective teachers, Charles Hunter summed up some of the important areas that should be included in such a curriculum:[9]

1. A full discussion of the advantages of television as a teaching tool.

2. Proper attitudes toward ITC and toward the on-camera teacher.

3. Basic production training and knowledge of studio procedures—so that the class-

[8]Lawrence Costello and George N. Gordon, *Teach with Television: A Guide to Instructional TV* (New York: Hastings House, Publishers, 1961), p. 131.
[9]Hunter, "Training Teachers for Television," pp. 300–306.

room teacher can better appreciate the total ITC endeavor.

4. The results of the vast amounts of research conducted in the ITV area.

5. More intensive work in educational psychology and learning theory as applied to telecommunications media.

6. Emphasis on creativity and imagination in the utilization process.

7. Viewing conditions in the learning situation.

8. Special ITV-related problems such as equipment malfunction, scheduling complications, and related aggravations.

Hunter also pointed out the advantages of establishing such programs on an interdisciplinary basis, involving departments of education, radio–television, and audio-visual instruction.

In-Service Education   Many opportunities arise for various kinds of in-service media programs while a teacher is in the field. Local schools and districts often establish do-it-yourself in-service workshops. Metropolitan ITV associations (Section 11.4) often initiate programs. State departments of education and state ETV networks have been instrumental in holding workshops and utilization training programs. Colleges and universities, of course, offer many related opportunities and course work. Federally funded (U. S. Office of Education) programs and institutes have been established across the country.

These activities and programs take many different forms: after-school meetings and seminars; district and state meetings and workshops during one-day teacher institutes or "professional days" of various descriptions; weekend sessions; summer workshops and institutes; and, of course, open-circuit broadcast in-service programing. Some of these programs are credit courses that can be applied toward an advanced degree or for some sort of incentive program; others are noncredit workshops or short sessions. Some are required by the school or district; others are entirely at the teacher's initiative.

Many of these in-service opportunities deal with ITC utilization in general, preparing classroom teachers for their first experience as receiving instructors. Other in-service activities are concerned with training teachers for the implementation of a particular STV series. The conscientious teachers will be aware of these opportunities and undertake those in-service training activities that can help them to do the best possible job in the classroom.

Copyrights and the Classroom Teacher Some considerations regarding copyright problems in relation to ITV production were discussed in Section 14.6. However, the receiving teacher and school administration should be aware of other copyright problems.

It is relatively easy to say what you may and may not do with rented videotapes, whether played back on a video recorder in the classroom or distributed through a school's CCTV system. You *may* play back a recorded tape to any-sized class or combined classes, usually repeating it as many times as you wish—within a limited time period. You *may not* make a video copy of the program, without permission, and use the copy in place of the rented tape in the future. (Similarly, you may not make a video recording of a film without permission, even if a copy of the film is owned outright by your school.)

Recording off-the-air broadcasts is a bit more complicated. It is easiest to start with a couple of negative examples. Any televised special that has been made into a film or videotape for renting to schools cannot be recorded off the air and used in the classroom. For example, you cannot make a copy of a program such as a Jacques Cousteau special and use that videotape thereafter in your classroom. Instructional programs broadcast by the local PTV station may also be off-limits if the ITV/STV service leased the programs from someone else. On the other hand, PBS announced in late 1975 that it had worked out arrangements with the Agency for Instructional Television, Great Plains National Instructional Television Library, and other distributors for a uniform policy on "convenience recordings." This policy would allow schools to record most ITV programs off the air and then to use any such recording for a period of one week before it would have to be erased. Locally produced STV programs probably may be taped for indefinitely delayed use. Usually a policy statement detailing what may and may not be taped off the air has been prepared by the STV service (station, district, state, or metropolitan ITV association).

There is still the question of what can be done with quickly dated broadcast programs such as news telecasts, timely specials and documentaries, and on-the-spot "actuality" broadcasts—those programs that are likely to have a relatively short span of usability or applicability. The courts generally have affirmed the right of teachers to *record* such broadcasts and to *use* them in the classroom on a "delayed broadcast" basis. The right to continue to use the programs much later is less certain. If possible, the programs should be cleared for classroom use by writing to the station or network that produced the program. If the classroom teacher feels that the use of the program is a fair use after applying the golden rule test (Section 14.6), he or she possibly *may* use the program as desired.

16.5   Utilization: the Readiness Period

Since earliest audio-visual days, the actual implementation of the mediated learning material (or the utilization session) has traditionally been broken down into three periods: (1) preparation, or activities before the telecast or media presentation; (2) the presentation itself; and (3) postlesson, or the follow-up activities after the telecast or film. We have labeled these periods the "three R's of utilization"—readiness, reception, and reinforcement.

The first step for the readiness period should be a review of *goals and purposes*. The classroom teacher must be sure of exactly what is to be accomplished during this media utilization session. He or she will definitely take into consideration the lesson objectives as listed in the classroom teacher's guide; but the receiving teacher may want to modify these objectives for a particular class. He or she may want to achieve some other, related, purpose with this particular lesson. Whether working from a formal lesson plan or not, the classroom teacher certainly should have objectives clearly in mind before starting any specific planning for a particular lesson.

Next, the classroom teacher will want to consider *general preparation* for the viewing session. He or she will work to instill positive attitudes (as discussed above) toward the concept of television instruction, as well as toward the on-camera television teacher. The receiving instructor will work on general motivation, creating a constructive classroom atmosphere—including all aspects of the physical viewing conditions. He or she will want to stimulate and reinforce interest in this particular subject matter—the importance of the content, the thrill of discovery of concepts and principles, and the excitement of today's topic.

Finally, the classroom teacher will work on *specific preparation* for today's lesson. What needs to be done to get the pupils ready to gain as much as they can from the telelesson? What must be done in order to achieve the learning objectives planned for this presentation? There are many varied activities which might be fitting for this specific preparation—depending upon the nature of this particular session. It may be appropriate to review the last lesson; what issues were raised that will be explored further? Perhaps some previously assigned student projects should be continued—group projects, individual research. Maybe some reports that were assigned last week should be given before today's lesson. What questions can be asked—perhaps written on the chalkboard—that might be answered in today's lesson? How can other activities, course work, reports and assignments be tied in with today's telecast (or video-disc playback)?

In many instances, specific items of information or background data will need to be given to the students before they can get the most out of today's telelesson. A most obvious area would be new terms and vocabulary words that should be introduced before the telecast. Other information possibly should be reviewed in specific preparation—key dates, important formulas, a summary of main points or concepts. Perhaps new personages or locations should be introduced to the students before the lesson. Much of this information can be presented on the chalkboard or by use of an overhead projector or perhaps may be prepared and handed out in dittoed or photocopied format.

One other aspect of preparation might be *parent preparation and communication*. In many specific subject areas (for example, new math, personal hygiene, or family life education), it may be desirable to involve parents as much as possible with the content of the televised portion of the curriculum. This is one of the strengths and justifications for the use of open-circuit channels for in-school materials (Section 15.3). Several sex education ITV programs, for instance, have utilized a broadcast schedule whereby parents, at home, could preview each individual lesson—telecast in the late evening—before the lesson was to be used in the classroom. This gave the parents a chance to discuss the content with their children ahead of time and/or discuss the lesson after the children had seen it; or, if they wished, the parents could exercise the option of having their children excused from one or more specific telecasts.[10] Parents can be involved in (prepared for) the broadcast lessons through

several channels—the school might want to send letters to the parents, each child might write a note to his or her parents, the local newspaper can be used as a medium of information, and—certainly—the local parent-teacher (PTA) group can be utilized as a forum for discussion and information.

Most of the considerations and suggestions outlined above—under preparation or the readiness period—are not necessarily unique to a mediated presentation. These same principles and activities apply to any kind of learning situation. In fact, these are some of the common techniques and ideas used not only in *educational* settings but in any kind of *communication* setting. One of the shortest and simplest precepts for making a speech is often given in beginning public speaking courses: "Tell 'em what you're going to tell 'em. Tell 'em. Tell 'em what you told 'em"—introduction/body/conclusion. This is the basis, of course, for the three periods of the utilization session: prepare 'em; present the lesson; follow-up and reinforce it. So, most good teachers—following the basic principles of good teaching/communicating—will adhere to the points outlined in this and the next two sections.

16.6   Utilization: The Reception Period

Many neophyte classroom/receiving teachers have made the initial mistake of assuming that during the actual telecast the classroom teacher was relatively free. In fact, some detractors of ITV have often claimed that television was just being used so that the teacher could "sneak back to the faculty lounge for a cigarette break while the boob tube took care of the kids." Actually, nothing could be farther from the truth. The presentation period itself can be a very hectic time for the classroom teacher—especially at the lower elementary levels. He or she has two definite roles to play—two different jobs to do.

The Teacher as Participant   First of all, in many instances the teacher must be a model

---

[10]*See* Donald N. Wood, "Trial and Ordeal of an ETV Sex Series in Hawaii," *Educational Television,* July 1969, pp. 14 ff.

viewer—setting an example for the students to follow. Depending upon the subject area, the grade level, and the attention span of the students, this can be the most important role during this period.

The receiving teacher should set the tone for the viewers, showing interest, paying attention to the screen, radiating encouragement for both the on-camera instructor and for the viewers—as illustrated in figure 16-5. The classroom teacher also should be setting an example for the pupils to follow in responding to and participating in the telecast. This is especially important in subjects such as music and foreign languages where the on-camera instructor is constantly asking the viewers to sing along or repeat a phrase, or in any other subject where the on-camera instructor is trying to get the viewers actively participating and responding—by a nod of the head, a show of hands, laughter, writing in a workbook, or by any

other overt reaction. The classroom teacher must be an active participant. This is especially true in well-disciplined classes where the pupils might be initially unsure to what extent they should be responding and making noise.

Upon occasion, it may also be appropriate for the classroom teacher to participate as instructor. Depending upon the grade level and type of presentation, it might make sense for the classroom instructor to serve as a clarifier/interpreter. Perhaps he or she can work at the chalkboard during the lesson—jotting down key phrases, making an outline, in essence helping to take notes for the class. At times, even oral comments might be appropriate—underscoring an important point, urging students to remember a key fact, relating what the TV instructor said to something covered in the preparation period. This type of participation is demonstrated in figure 16-6. If the telelesson is being played back

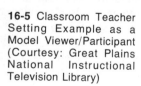

16-5 Classroom Teacher Setting Example as a Model Viewer/Participant (Courtesy: Great Plains National Instructional Television Library)

16-6 Classroom Teacher Leading Classroom Responses during Actual Telelesson Reception Period (Courtesy: Great Plains National Instructional Television Library)

on a video recorder in the classroom, then the teacher has control over the presentation and can stop the tape to make or clarify a point. This role can be overdone, of course, and the classroom teacher can become more of an interrupter than an interpreter; nevertheless, the possibility is there for constructive participation.

The Teacher as Observer The classroom teacher must also constantly be an observer of the students—gaining every bit of feedback possible from the pupils. The teacher can be noting when the class becomes perplexed, what parts of the telelesson need to be explained, where there are contradictory statements, where individual students are having trouble. In fact, this period can be a rare opportunity for a teacher to objectively observe individual learners reacting to a presentation—while the teacher is not actually ego-involved with the presentation. It is one of the best opportunities to gather neutral feedback regarding individual learning patterns.

Based upon this observation and data

gathering, the teacher can make notes on individual students ("Joe needs help with the formulas"; "Sarah is puzzled by the abstractions"; "Both Mildred and Bobby Jean need to review the grammar summary") as well as planning changes and modifications in the plans for the follow-up activities ("Review the first three steps again"; "Explain the apparent contradiction regarding the westward movement").

16.7 Utilization: The Reinforcement Period

Most ITC practitioners, and most classroom teachers, would generally agree that of the three utilization periods the reinforcement or follow-up period is probably the most crucial. The heaviest emphasis is usually placed on reinforcement; the most time is usually spent on reinforcement; and classroom teachers direct most of their creative energies toward reinforcement.

The starting point, as with the readiness period, is with the learning objectives that the classroom teacher wants to achieve. They may be taken directly from the classroom teacher's guide or they may be specifically tailored to meet the purposes of this particular classroom; but it is important that the classroom teacher keeps these objectives clearly in mind as activities for the follow-up period are planned.

The starting point for designing the re-inforcement session should be the classroom teacher's guide, where the teacher will find a multitude of different ideas and projects—usually more than can be handled in any one classroom period. These ideas should be reviewed, then tailored and blended with projects and activities that the classroom teacher creates independently of the guide.

Review versus Application Activities
One distinction that the classroom teacher may want to keep in mind is the difference between those activities that simply go over the material presented in the telelesson (review) and those activities that build upon the telelesson (application activities). Many follow-up sessions combine the two in varying proportions—depending upon the purposes of the classroom teacher. To a limited extent, this classification of follow-up activities can be roughly correlated with the divisions among the three educational/learning domains. (Section 13.6).

Review activities *generally* are better suited for programs in the *cognitive* domain. They are designed to review and summarize facts, dates, names, concepts, formulas, theories, and other bits and pieces of information. Some of the techniques used are lecture/review by the teacher, student questions, oral quiz, small-group reviews, written tests, and similar approaches. Practice and exercise activities in the *psycho-motor* domain would probably also fall into this category.

Application activities *usually* are well adapted to *affective* domain material (as well as the cognitive area). The classroom teacher assumes the pupils have gotten the basic material

they need from the telelesson, and that they are now ready to apply the content in some sort of creative/interpretive/developmental project. They may be assigned to work on projects, write scripts, do independent research, give an oral report, create some artistic work, give a speech, write an original story, or any other kind of creative endeavor that asks the students to apply or build on what they have gotten from the TV presentation.

Size and Scope of Activities   The class-room teacher may want to categorize follow-up reinforcement concerns and think in terms of the size and scope of activities. Although mechanical in approach, it does help to remind the teacher to consider a variety of different configurations.

Total-group activities generally involve the teacher as the focus of a review approach. The format may be a question-and-answer session, an oral quiz, a follow-up lecture (which can be deadly). Inasmuch as the class has just been giving its attention to a (televised) large-group presentation, normally this kind of total-group experience should not be followed for very long after the telelesson—depending, of course, upon the age level, subject matter, purpose of the teacher, and so forth.

Group activities would include com-mittee assignments, panel discussions, game shows, dramatizations, group projects, and sim-ilar endeavors. Generally, these activities would be more suitable for application purposes.

Paired activities, or "dyads," could be designed for a number of different purposes—both review and application projects. These would include research reports, reviewing with flash cards, script writing, small projects, and the like.

Individual activities, of course, take many forms—writing, independent research, reading, creative projects, and so on. Again, the purpose and the function of the individual assign-ment can be tailored to meet any of several dif-ferent kinds of objectives.

Other Reinforcement Considerations

There are many different ways of cataloging and organizing various follow-up/learning activities. One attractive utilization handbook printed by the Virginia state department of education outlined follow-up activities and considerations under seven headings—discussions, dramatizations and oral presentations, projects and demonstrations, aesthetic and creative projects and demonstrations, other educational media, related reading, and additional resources. Under these headings about ninety separate kinds of activities are listed.[11] In 1969, MPATI commissioned a small hard-cover book devoted almost entirely to utilization and follow-up activities and suggestions.[12]

The classroom teacher must keep in mind various kinds of constraints and limitations imposed upon the utilization session. The length of the follow-up period, for example, will quite possibly be determined by the school's bell schedule. Occasionally a ten-minute follow-up period might be all that is needed, but the class is kept together for another twenty minutes. In another situation, perhaps forty minutes are needed, but only ten minutes are allowable. In many instances, when students are moved from their own classrooms to a viewing room—or when lunch and recess schedules make an immediate follow-up session impossible—the teacher must contend with a delayed follow-up period, perhaps later in the day or even the next day. This greatly compounds the problems of motivation and sustaining interest.

Occasionally, the ITV presentation can be the stimulus to other learning arrangements and opportunities. A major ITV series being used by several teachers may lead, for instance, to a multidiscipline approach to several subject areas—following some sort of thematic treatment suggested by the common ITV presentations. Or several teachers at the same grade level may take advantage of the class rescheduling necessitated by juggling groups to fit the ITV schedule and decide to experiment further with a team teaching approach to several subject areas.

In many school situations, the introduction of a major ITV program has provided a valuable service simply in that it has brought together a number of teachers and administrators, on a planning basis, and forced them to take a long hard look at the entire schooling situation. The original need may have been only to figure out an approach to the television situation; but the result often has been a major reevaluation of the school's goals and purposes, curriculum, physical plant, teaching staff, and related issues.

Finally, one other aspect of the utilization session that should be introduced here is the role of the classroom teacher in evaluating the television presentations and providing feedback to the STV service. This gets us into a most important concern—and Chapter 17.

---

[11]James H. Gay-Lord, *Instructional Television: A Utilization Guide for Teachers and Administrators* (Richmond, Virginia: Virginia State Department of Education, 1968), pp. 25–31.

[12]*See* Mary Howard Smith, *Using Television in the Classroom* (New York: McGraw-Hill Book Company, 1961).

Evaluation and Research

As we have been looking at "The Practice of ETC" (Part Four), we have been essentially following an instructional design approach (13.4). The final step in any such structure (or functions organization) is invariably evaluation. This final chapter will deal with this activity and the related concepts of feedback, formal research, and program validation. When TV is used systematically in instruction, it should be *evaluated* and *validated;* these are part of the process of instructional design—they are aspects of instructional technology (Section 12.4). In addition to reasonable evaluation and validation, scholars and practitioners are always asking other questions about STV and ITV in general: Can it teach? Can you put the transmitters in airplanes, or in satellites, to cover greater distances? Is color better than black-and-white in television lessons? How big should title cards be? Should on-camera TV teachers look directly into the lens? Some of these questions are worth asking—some are frivolous.

17.1 Definitions of Terms

Like many other items discussed in this book, the differences between research, evaluation, and validation cannot be cleanly made. The processes are not that distinct. In our definitions it will be obvious that there is considerable overlap.

Definition of Research Combining a couple of different dictionary definitions, with a bit of original phrasing, we can define "research": *The systematic process of objective study and investigation of a problem, using the scientific method and inductive reasoning, in order to establish facts or truths.*

Less formally we can define research, evaluation, and validation by the sorts of questions that we tend to ask when we are engaged in each of these. In *research,* questions to be investigated fall into three basic categories:

1. *Historical Research:* Why did this person do what he or she did? What happened?

Why did this thing occur? What relationship did *this* have to *that?* How similar is what happened in the past to this which is happening now? What were the causes of this particular effect?

2. *Descriptive or Normative Research:* How many of this and that are there in the real world? In real life is this related to that? What is the actual situation (or norm)? For this and that which vary in real life, what are the most frequently found amounts of each (or the average amounts), and what is the average amount of variability we can expect to find in samples of this or that taken from real life?

3. *Experimental Research:* Does this cause that? Trying first this, and then that, with all similar variables controlled so they do not confuse us, what will tend to be related to that, and how is it related? In this series of comparisons (experiments), can we state any principles or conclusions that add to our broad understanding of a phenomenon? As a result of this experiment can any new laws or principles be stated?

Definition of Evaluation   Whereas formal research is usually concerned with answering either theoretical or practical questions on a generalized basis, *evaluation* usually is concerned with appraisal of a specific activity, program, or some other single identifiable thing. We can define "evaluation": *The process of analysis and examination of a specific project, activity, or event, in order to determine future action*.

The concept of determination of a future course of action is always important as the end result of any evaluation process. Whether we are concerned with the effectiveness of a new fertilizer or deciding what to do about a billion-dollar federal investment in a welfare program, there are only three possible courses of action in evaluating anything: (1) Continue the status quo (project, activity, program) without change; (2) modify the solution (program, course of action); or (3) terminate the project or activity.

Thus, in evaluation activities, we tend to ask questions such as: In this project, should we continue doing things as we have been doing them? Are we achieving our goals in this activity? Should we modify this project or activity, and, if so, how? Should we terminate this project? What changes do we need to make to improve this program? Can we come closer to our objective if we do thus and so?

Definition of Validation   The concept of validation is concerned with guaranteeing that a given course of action (that is, an instructional program) will always have consistent results— if all variables are controlled. "Validation" can be defined: *The process of determining if predetermined objectives are consistently achieved when a specific instructional unit, a lesson or course, is presented to a target group of students (who meet certain entry-level standards)*.

In validation activities we ask such questions as: Did students perform as well on our final (criterion) tests as we said they would? If this instructional program were implemented with a comparable group of students, would we achieve the same results? Can we give the lesson again exactly as it was just given, and therefore legitimately expect the same results? Were students able to complete the lessons in the time allowed?

These three terms should take on fuller meaning as we consider each one in turn—starting first with "evaluation" as the final step in an STV program.

17.2   Evaluation of School Television

Evaluation might be considered a type of research activity in which the effectiveness and acceptability of using TV in a specific instructional setting is examined in order to decide whether to continue the use. Evaluation can also include formal and informal procedures for providing feedback to those conducting an STV program or preparing STV lessons. All these types of evaluation will be considered in this section.

Evaluation of an STV Project   It is common for those conducting a pilot project at any level to prepare a report on the successes and (if those conducting the project are honest with themselves and others) the failures of the project. In fact, a project cannot serve as a model for others if no one knows about it; and the pitfalls uncovered by the project will not help others if they are not reported.

Project evaluations exist for every organizational level that has utilized STV. There are reports on single classroom applications, projects serving a single school, district-wide projects, state operations, and nationally produced programs. Many of these reports have received wide circulation; some have been cataloged by the Educational Research Information Center (ERIC) (Section 17.5); and many were never circulated outside the school district.

Among the better-known project evaluations are those reviewing the progress of the Hagerstown project[1] (Section 3.9) and the Chicago Junior College project[2] (Section 3.9). These are encouraging, positive reports. On the other hand, a project evaluation will occasionally result in the termination of an ITC program (Section 13.2).

One of the earliest guides in planning the evaluation of a project was published by Midwest Program on Airborne Television Instruction (MPATI).[3] Abridged and adapted, the MPATI *Local Evaluations Handbook* (Draft) suggests the following for a district or school that wishes to conduct an evaluation of a specific STV project:

1. Organize a local study group or committee responsible for planning and directing the evaluation efforts. If at all possible, the group should be formed some weeks before the beginning of the school term. Composition of the committee or group certainly should include classroom teachers who will be utilizing television in their classrooms; and it may also include parents or other citizens with interest in the school.

2. At the earliest opportunity, convene the committee and devote the time required to orient the group to the objectives of the STV project, the conduct of the project, and the general objectives of the evaluation effort.

3. Next, select questions that the evaluation effort will attempt to answer. The kinds of questions that might be listed include: What are the effects of the ITV lessons on the educational interests of the students? Is there greater student use of the library facilities and services? What are ways the library is/could be supporting the ITV lessons? Do students undertake a greater number of special projects when TV is used? What equipment problems occur when TV is used? Are the sets adequate in number, kind, and condition? Have teachers encountered difficulties in using TV materials? How can problems be alleviated? Should the project be continued? Expanded? Cut back or eliminated? These questions are only suggestions—the committee could spend at least a little time in "brainstorming," that is, in thinking of and recording all of the questions, outlandish and sensible, that might relate to the STV program. Once the group has exhausted its store of questions (and has recorded them), it can then proceed back through the list— this time with a critical eye—and modify or delete questions which, by agreement, are impossible to study or not worthy of careful study.

4. At this point, the committee should have agreed upon a common group of questions and these questions will then form the basis for the evaluation that is to be conducted. The next task is to develop evaluation procedures suited to each question. While examining each question

[1]*Closed-Circuit Television Teaching in Washington County, 1958–1959;* and *Washington County Closed-Circuit Television Report, 1956–1961* (Hagerstown, Maryland: Washington County Board of Education).

[2]Clifford G. Erickson and Hymen M. Chausow, *Chicago's TV College: Final Report of a Three Year Experiment* (Chicago: Chicago City Junior College, August 1960).

[3]Midwest Program on Airborne Television Instruction, *Local Evaluations Handbook,* Draft Copy for Use in Summer Workshops, 1961, pp. xviii–xix.

separately, the committee will want to answer the following: What information will we need to answer the question? How will this information be gathered? When must each item of information be gathered? Who is responsible for gathering it? Will we need to condense or analyze the information after it is gathered? If so, how?

5. If the committee has selected questions that reflect common concerns, and has planned carefully for the gathering of necessary information, it has completed at least the more difficult half of its job. The task now is to see that the "dirty work" of collecting information (for example, test scores, teacher judgments, cost figures, comments) proceeds according to plan. Given a moderate amount of good luck to accompany the detailed plans, the information will be gathered as intended and the group will be ready for the last step in the process.

6. The final step in the evaluation consists of bringing together the collected information, examining and analyzing it, determining which conclusions (or answers) are supported by the information, and then preparing the evaluation report. A full evaluation report will probably include the following major sections: (a) the general background of the study and the reasons for interest in it, (b) the specific questions that were studied, (c) the procedures used in conducting the study, (d) the results obtained, and (e) the conclusions that can be drawn from these results.

The important thing to remember is that the necessary information should be gathered systematically. It is easy to demonstrate that the careful definition of the problem (the questions to be answered) and the careful planning of procedures not only contribute to the execution of an adequate local evaluation effort but also simplify the job and can save hours of work.

Evaluation of ITV Presentations Whether or not a school TV project is formally evaluated, in one way or another individual television lessons

need to be assessed. Ideally, materials should be evaluated *before* they are presented to students, and again when they are *used in the classroom*. A third evaluation of a lesson might be its *validation* with students (Section 17.3), and a final evaluation of a telelesson might be in conjunction with the *project evaluation* which would consider the lesson in relation to all other lessons (as discussed above).

One of the most comprehensive lesson evaluation processes, which also became a model for program evaluation for other projects, was the process developed by MPATI to evaluate individual telelessons before distribution. The purpose of lesson evaluation at this stage is to spot any major problems with the materials before they are used in the classroom. They can be screened for subject matter errors, teaching effectiveness, production and engineering qualities. If substantial problems are noted, the telelessons can be redone before distribution and utilization.

A four-part evaluation scheme was worked out for evaluating MPATI lessons, using separate sets of rating scales for each part. The four aspects of the lessons evaluated were: (1) teaching methodology, (2) subject matter, (3) production, and (4) engineering. The forms for the first three of these are shown in figures 17-1, 17-2, and 17-3. These forms were devised specifically for MPATI lessons, which in the main were major resource presentations using an on-camera TV teacher. Also it should be noted that three or four different reviewers completed the forms. Some items, such as the use of visuals, appear on more than one form—but one sheet would be completed by a production specialist, and in another case an educationist would be marking the form.

Feedback Evaluation Forms Because TV is initially a one-way medium of communication, anything that allows teachers—and possibly students—to furnish input back to the teachers and administrators responsible for the TV lessons

17-1 MPATI Evaluation Form for Teaching Methodology

```
        MIDWEST PROGRAM ON AIRBORNE          TV TEACHER_____
        TELEVISION INSTRUCTION
                                             TAPE NUMBER_____

                                             SERIES TITLE_____

           TAPE REVIEW FORM - REVIEW OF TEACHING METHODOLOGY


        Reviewer's Name_____      Position_____

        Date Recorded_____     Date Viewed_____

        Directions:  Place check marks along given continuum.
```

vl. EFFECTIVENESS OF OPENING

```
1            2            3            4            5
"Captures         "Somewhat absorbing"        "Creates
immediate                                      apathy or
attention"                                     antagonism"
```

2. LEVEL OF LEARNING

```
1            2            3            4            5
"Telecast         "Majority of telecast       "Beyond grasp
will be           will be understood"         of students
understood"                                   (Level of un-
                                              derstanding)
```

3. EASE AND CLARITY OF EXPRESSION

```
1            2            3            4            5
"Feels 'at        "Relaxed but conscious      "Ill at ease,
home' with            of speech"              incoherent"
subject"
```

4. INTEREST AND ENTHUSIASM IN SUBJECT

```
1            2            3            4            5
"Thoroughly       "Average interest and       "Would rather
engrossed in          enthusiasm"             be doing some-
subject"                                      thing else"
```

5. TEACHING TECHNIQUES

```
1            2            3            4            5
"Thorough use     "Moderate use of            "Knows little
of teaching        teaching aids &            of teaching
aids &              procedures"               aids & pro-
methods"                                      cedures"
```

6. ORGANIZATION OF SINGLE TELECAST

```
1            2            3            4            5
"Logical &        "Some error but general     "Needs help
clear pre-            content accurate"        with content"
sentation"
```

MPATI Tape Reviewing Form - Teaching Methodology

7. <u>PACING</u>

1	2	3	4	5
"Balances subject, importance, & time"		"Touches major points"		"Gets 'Hung up' on trivia"

8. <u>APPROPRIATENESS</u> <u>OF</u> <u>VISUALS</u>

1	2	3	4	5
"Visual thoroughly enhances learning"		"Uses visuals"		"Poor or irrelevant use of visuals"

9. <u>PUPIL</u> <u>INTEREST</u>

1	2	3	4	5
"Will evoke strong interest"		"Interesting"		"Discourages interest"

10. <u>OVERALL</u> <u>RATING</u>

1	2	3	4	5
"Will fill the bill completely"		"Satisfactory"		"Little learning will occur" (Poor)

<u>SERIES</u> <u>PROGRESS</u> <u>INDICATOR</u>
Check One:

_____ Reaches MPATI standards

_____ Caution! Problem_____

_____

_____

will cause series trouble if

compounded by other or

similar problems.

_____ Below MPATI standards

17-2 MPATI Evaluation Form for Subject Matter Review

```
              MIDWEST PROGRAM ON AIRBORNE          TV TEACHER_____
                 TELEVISION INSTRUCTION
                                                   TAPE NUMBER_____

                                                   NAME OF SERIES_____

              TAPE REVIEW FORM - REVIEW OF SUBJECT MATTER


              Reviewer's Name_____Position_____

              Date Recorded_____Date Viewed_____

              Directions:  Place check marks along given continuum

    1.  CORRECTNESS AND ACCURACY OF TELECAST MATERIALS

        1_____2_____3_____4_____5
        "Free of                  "Occasional errors"              "Loose with
         errors"                                                    facts"

    2.  RELEVANCE OF MATERIAL

        1_____2_____3_____4_____5
        "Very                     "Some of material                "Has little
         pertinent &               is pertinent"                    bearing"
         informative"

    3.  SCOPE OF CONTENT

        1_____2_____3_____4_____5
        "Correct                  "Occasionally too many           "Too many
         number of                 (too few) concepts"              (too few)
         concepts"                                                   concepts"

    4.  PACING

        1_____2_____3_____4_____5
        "Moves at                 "Occasionally too fast           "Hung up
         optimal                    (too few) concepts"             on trivia"
         rate (your
         opinion)"

    5.  ORGANIZATION OF TELECAST

        1_____2_____3_____4_____5
        "Very well                "Some organization"              "Poorly
         organized"                                                 organized"

    6.  KNOWLEDGE OF LESSON CONTENT

        1_____2_____3_____4_____5
        "Content is               "Some error but                  "Needs help
         accurate"                 general content                  with content
                                   accurate"
```

MPATI Tape Review Form - Subject Matter (Revised)

7. OMISSIONS OF INFORMATION

1	2	3	4	5
"Seldom omits pertinent material"		"Occasional omissions"		"Frequently omits valuable information"

8. AGREEMENT BETWEEN ANNOUNCED OBJECTIVES AND WHAT WAS ACTUALLY TAUGHT

1	2	3	4	5
"Complete agreement"		"Occasional digression"		"Extraneous & unrelated"

9. CALIBRE (OR QUALITY) OF PRESENTATION

1	2	3	4	5
"ETV put to maximum use"		"Some value educationally"		"Entertainment, not education"

10. OVERALL EFFECTIVENESS

1	2	3	4	5
"Excellent"		"Good"		"Poor"

11. FOLLOW-UP POSSIBILITIES

"Sound base for future learning"	"Occasional discussion of principles"	"Shallow; soon forgotten"

SERIES PROGRESS INDICATOR
Check One:

_____ Reaches MPATI standards

_____ Caution! Problem_____

_____

_____

will cause series trouble

if compounded by other or

similar problems.

_____ Below MPATI standards

17-3 MPATI Evaluation Form for Production Techniques

```
          MIDWEST PROGRAM ON AIRBORNE       TV TEACHER_____
             TELEVISION INSTRUCTION
                                            TAPE NUMBER_____

                                            NAME OF SERIES_____

          TAPE REVIEW FORM - REVIEW OF TELECAST PRODUCTION

          Reviewer's Name_____Position_____

          Date Recorded_____Date Viewed_____

          Directions:  Place check marks along given continuum.
```

1. OPENING (STANDARD & TELECAST)

1	2	3	4	5
"Intriguing, visually stimulating"		"Can take it or leave it"		"Confusing, visually poor"

2. VISUALS & PROPS

    a. Quality:

1	2	3	4	5
"Good"		"Of moderate quality"		"Poor"

    b. Use:

1	2	3	4	5
"Used effectively"		"Of moderate effectiveness"		"Awkward"

3. TEACHER'S APPEARANCE

1	2	3	4	5
"Well-groomed & pleasant"		"Acceptable, but could be improved"		"Sloppy, ill-kept"

4. TEACHER'S MANNERISMS

1	2	3	4	5
"Smooth"		"Occasionally distracting"		"Offensive"

5. SET

1	2	3	4	5
"Contributes to total telecast"		"A few minor changes would improve it"		"Very distracting"

6. CAMERA TECHNIQUES

1	2	3	4	5
"Properly used, smooth"		"Occasional poor shots"		"Poor placement & distracting movement"

MPATI Tape Reviewing Form - Telecast Production

7. LIGHTING

1	2	3	4	5
"Sufficient illumina- tion"		"Fairly well balanced"		"Inadequate"

8. AUDIO

1	2	3	4	5
"Makes max- imum contri- bution"		"Fairly good with a few bad spots"		"Scratchy, too low or high"

9. MUSIC

1	2	3	4	5
"Totally contributes to content of telecast"		"O.K. except for most most discriminating viewing"		"Gets in way (volume) poor quality, wrong mood"

10. TEACHER'S MOTIVATED MOVEMENT

1	2	3	4	5
"Direct, for a pur- pose,smooth- ly accom- plished"		"Moderately effective"		"No reason for move; jerky; too fast; awk- ward; no move- ment"

11. TELECAST CLOSING

1	2	3	4	5
"Students want to see next lesson"		"Average closing"		"Loses view- er's interest"

12. OVERALL PRODUCTION

1	2	3	4	5
"Good 'production' contributes to the max- imum		"Fair, not particularly outstanding"		"Poor 'pro- duction' hinders learning"

CREDITS:
1.

2.

3.

4.

SERIES PROGRESS INDICATOR
Check One:

_____ Reaches MPATI standards

_____ Caution! Problem_____

_____

will cause series trouble if compounded by other or similar problems.

_____ Below MPATI standards

should be utilized. In order to facilitate this type of classroom feedback, several different methods have been used. The most widespread technique is the use of telelesson evaluation forms or feedback evaluation sheets. These forms, which are designed to be filled out by the classroom teacher after using a particular TV lesson, serve a dual purpose—first, they provide a valuable source of legitimate feedback to the STV producer or administrator; and, second, they provide an opportunity for the receiving teacher to participate to some extent in the total ITV evaluation/planning experience.

These forms are designed and used in a wide variety of different ways. Some consist primarily of rating scales and boxes to be checked, while others rely heavily upon open-ended questions and written responses. Some ask teachers to respond to every telelesson in a series, while others are designed to be used only when the teacher feels strongly that some comment needs to be made about a given program. Some forms are tailor-made for a specific series, while others are designed rather broadly so that they may be used with any number of different types of programs. Sometimes the forms are integrated into a formal feedback/evaluation project and teachers are strongly requested to participate; other situations simply allow a teacher to respond voluntarily.

An evaluation form should be no longer and no more complicated than is necessary to obtain the needed information. Figure 17-4 is a feedback form used by teachers in an ETV project on the Island of Jamaica. Notice the simple, direct questions that are asked of the classroom teachers.

Feedback forms are frequently put in the teacher's guides (Section 14.8) so they will be at hand when a question, concern, or compliment comes to mind during a TV presentation. One of the more successful of these was an open-ended format used by MPATI. This simple form, which was included in multiple copies in every guide, simply had spaces for comments by teachers as they viewed the telecasts (figure 17-5). The lines on the form allowed sufficient space

for writing (and also were calibrated to typewriter line spacing). The address of the project headquarters was on each form so that a teacher could just drop off the form at the school office for forwarding to MPATI. After the comments were read by the MPATI staff, they were filed in the lesson folder for consideration in any remakes of the lesson.

Also in each MPATI guide was an end-of-semester evaluation form that asked for comments in an unusual way. Rather than specific questions to be answered by classroom teachers, the form had only words and phrases that began sentences to be completed by the classroom teacher (figure 17-6). The form worked well, and the information obtained aided the MPATI staff in planning new lessons and in revising old series.

Evaluation of TV lessons by administrators and teachers can have limited usefulness. In fact, one research study showed a negative correlation between estimates by educators as to which instructional materials would work best with students and the actual relative success of the materials in the classroom. The judgments of the educators were all too often just the opposite of results from actual use in classrooms. Educators simply are not always good estimators of what will be effective with students.[4]

17.3   Validation of Instructional Telecommunications

Validation is a special type of evaluation that considers student performance to see if a lesson or a series is able to do what the designer said it would do. When properly done, it is like a Good Housekeeping Seal of Approval on the lessons—the user has a guarantee that the product can deliver what it claims to be able to deliver (Section 13.8).

Unfortunately, we cannot just take a videotape off the shelf and begin to validate it

---

[4]Warren F. Seibert, "Broadcasting and Education," *Educational Broadcasting Review*, June 1972, p. 149.

17-4 Jamaica ETV Telelesson Evaluation Report

```
          We welcome your opinion on the lessons.  Please return this report
          to us regularly every week.  Do not wait for them to accumulate.

          School.....................................Parish...............

          Lesson No.........Title...........................................

          Age of Class............Form...........Number Viewing...........

          1.   Quality of Reception:

               Good...........................Poor.........................

               Tell why............................................

               ...................................................

          2.   Suitability of material for age/grade of children:

               Satisfactory.......Too Difficult........Too Easy............

               Tell why...........................................

               ...................................................

          3.   How did you prepare children for the telecast?

               ...................................................

          4.   What parts of lesson were particularly effective?

               ...................................................

          5.   Did the teacher move through the lesson too quickly?

               ...................................................

          6.   What follow-up activities resulted?

               ...................................................

               ...................................................

          7.   List any difficulties the children encountered in following
               the lesson......................................

               ...................................................

          8.   General comments and suggestions for Studio Teacher:

               ...................................................

               ...................................................

          9.   General comments including projects, special local activity
               shown by the children

               ...................................................

               Signature of Teacher...................Date...............
```

17-5 MPATI Open-ended Telelesson Evaluation Form

```
        To improve MPATI telecast offerings or the resource guides,
        there needs to be information from those using them as to
        their effectiveness, and as to the ways they can be improved.
        Please use these forms to tell us about a particularly effec-
        tive telecast, or about a telecast which can be improved.
        ------------------------- CUT ON DOTTED LINE ----------------------

     TO: MPATI                    FROM:
         Memorial Center            Name_____
         Purdue University
         Lafayette, Indiana         Grade_____

                                    School_____

                                    City and State_____

     SUBJECT:   Telecast No._____, Course_____

     Comments/Suggestions:_____

     _____

     _____

     _____

     _____

     _____

     _____

     _____

     _____

     _____

     _____

     _____

     _____

     _____

     _____

     _____

     _____

                    (Continue on Reverse Side)
```

with students. There are a few questions that we first need to ask: For whom is the lesson intended? What was it designed to do? What must students do at the end of the lesson to demonstrate that they have learned what they were supposed to learn? What score, obtained in what length of time, will we consider to be minimally acceptable? How many in the class will have to get this score before we can say that overall the lesson is successful? When we have answers to these questions, it becomes a simple matter to validate TV instruction.[5]

Defining the Target Audience A number of other questions need to be asked as we attempt to answer the larger question, "For whom is this lesson intended?" (Section 13.5.) We need to be able to answer the following and similar questions: What will the students' grade level be? What will be the experience level of the students? Will they come from medium and high socioeconomic families? Do we assume that the students have had particular units of instruction (prerequisites) before they view the lesson in question? Is a certain intelligence/mental aptitude level necessary? Will the students have to be able to perform certain skills before they can view the lesson? How will we ensure that the students have these so-called "entry behaviors"? Will there be pretests that students must pass before they begin the lesson? When we have answers to all these questions, we can move on to the next step in the process of validation.

Specifying Behavioral Objectives We next must be able to state in precise terms the behavioral objectives for our lesson (Section 13.6). We must describe what the learner should be able to do in performance terms—what we

should see the student do so that we can say without any question that the learner has learned. What operations will the students perform when they demonstrate that the instruction has been successful? Under what conditions will these operations be performed? Within what time limits will the operations be completed if we are to say they were successfully performed?

Two books mentioned in Section 13.6 are very helpful to anyone specifying behavioral objectives in order to validate instruction. Each of these books has the imposing title of *The Taxonomy of Educational Objectives: The Classification of Educational Goals*. The first of the books, by Bloom and others, is subtitled *Handbook I: Cognitive Domain;*[6] it deals with objectives having to do with thinking, knowing, and problem solving. The second book, by Krathwohl and others, is subtitled *Handbook II: Affective Domain;* it is concerned with objectives dealing with attitudes, values, interest, appreciation, and social-emotional adjustments.[7]

The taxonomies, as their name suggests, are hierarchical classification systems. They list behaviors, starting with those that are very simply attained, then progressing to more and more complex behaviors—each of which implies that previously described behaviors are included in the next higher behavior. For example, in the cognitive domain, the first major heading is "Knowledge," and under that is "Recall of fact." The next higher major heading is "Comprehension," with a subheading "Ability to restate knowledge in new words." Clearly, before one can "restate the concept 'taxonomy' in one's own words," he must first be able to "state a definition of 'taxonomy.'" One could learn by rote a definition of "taxonomy" without really understanding

[5]Warren L. Wade, *How to Provide Instructional VideoTape/Film Accountability* (Washington, D.C.: National Association of Educational Broadcasters, undated, circa. 1968), pp. 13–14. Information also can be found in Wade, "Let's Program Instructional TV Programs," *NAEB Journal,* January-February 1967, pp. 78–84.

[6]Benjamin S. Bloom, Max D. Englehart, Walker H. Hill, Edward J. Furst, and David R. Krathwohl, *Taxonomy of Educational Objectives, Handbook I: Cognitive Domain* (New York: David McKay Company, 1956).

[7]David R. Krathwohl, Benjamin H. Bloom, and Bertram B. Masia, *Taxonomy of Educational Objectives, Handbook II: Affective Domain* (New York: David McKay Company, 1964).

17-6 MPATI End-of-Semester Evaluation Form

TO: MPATI   FROM: Name_____
  Memorial Center   Grade_____
  Purdue University   School_____
  Lafayette, Indiana   City and State_____

Please give us your reactions to the MPATI telecasts by completing sentences for each item below. Do not spend more than
a few minutes completing the sentences. Use separate forms for
each course being used.

My classes watched approximately_____telecasts of the MPATI
course_____taught by_____

The quality of the TV picture has been

The television teacher

The content of telecasts

The level of telecasts

The students

Classroom discipline

As a teacher, I enjoy

I dislike

With respect to educational TV, I now believe

I believe educational television's greatest strength

To improve the telecasts, I suggest

The thing I would most like to see changed

The teachers' guide

To improve the guide

Please

the word; comprehension is a higher order of learning, and therefore is higher in the taxonomy for the cognitive domain.[8]

Preparing the Criterion Test   Before we begin constructing our TV lesson, we should prepare the test that students will take in order to demonstrate that they can perform the behavior(s) described in the objective(s). The test can be a conventional paper and pencil test; it can be a manipulation (such as tying a particular knot); or it can be an act that will demonstrate a particular attitude ("The learner will volunteer to serve on a committee involved in preparations for Citizenship Day"). The test will not only ask the student to do something, it also will contain time limits, specify other constraints ("The learner will be able to use only the tables in Chapter One"), and an indication of the minimal level of performance that will be considered "acceptable" (Section 13.6).

Preparing and Validating the Instruction If an instructional designer has been able to do all the foregoing, it now becomes a relatively simple matter to prepare the TV lesson and test learners to see if the program is successful. The remarkable thing about validated instruction is that, when we are successful, we can predict with confidence that, if the lesson is utilized exactly as it was when it was validated, we know that it will again be successful. In other words, if the telelesson is given to students with the same entry behaviors, under the same viewing conditions as before, and the same criterion test is administered, the students should achieve just as the original learners did when the TV lesson was validated. Upon reflection, this emerges as a powerful, and an exciting, idea. Although videotape has been practical in the classroom for a decade or two, we are only beginning to grasp its real significance: the ability to *replicate* instruction—to be able to predict precisely what learning

will take place when a given instructional stimulus is presented to similar types of learners. To some this may seem like an impersonal, mechanistic idea, but it really is the opposite. True individualized learning becomes possible only when a teacher is able to select the right learning materials for a given student at just the moment the student needs that material (Section 12.6). With the help of a computer and a variety of validated learning materials, including television, the teacher of the future may be able to prescribe to individual students those learning programs which will be the most exciting and challenging experiences possible for each individual learner.

17.4   Research on Learning from Television

Most experts agree that television has been researched as has no other device that has ever been introduced into our schools. As Jack McBride stated, "It is possible that in the entire history of education no single facet has been subjected to as exhaustive testing and experimentation and research as has educational television."[9] Literally thousands of different ITV research projects have been conducted. Over the years several individuals and organizations have attempted to distill and synthesize this research. The best-known reviews of the research were published in 1953 by Finn,[10] 1956 by Kumata,[11] 1958 by Barrow and Westley,[12] 1959 by Holmes,[13]

---

[8]David R. Krathwohl, "Stating Objectives Appropriately for Program, for Curriculum, and for Instructional Materials Development," *Journal of Teacher Education,* March 1965, p. 87.

[9]Jack McBride, "The Twenty Elements of Instructional Television: A Summary," paper presented at the Seminar on Educational Television Sponsored by the Centre for Educational Television Overseas, Cairo, Egypt, February 6–12, 1966.

[10]James D. Finn, "Television and Education: A Review of Research," *Audiovisual Communication Review,* Spring 1953, pp. 106–126.

[11]Hideya Kumata, *An Inventory of Instructional Television Research* (Ann Arbor, Michigan: Educational Television and Radio Center, 1956).

[12]Lionel C. Barrow and Bruce H. Westley, *Television Effects: A Summary of the Literature and Proposed General Theory,* Research Bulletin No. 9 (Madison: University of Wisconsin Television Laboratory, 1958).

[13]Presley D. Holmes, Jr., *Television Research in the Teaching-Learning Process* (Detroit: Wayne State University, 1959).

1960 by Allen,[14] 1962 by Schramm,[15] 1967 by Reid and MacLennan,[16] and in 1968 by Chu and Schramm.[17] More than 500 studies have been published comparing televised instruction with conventional instruction or comparing one way of using TV for instruction with a different way. After all this research, and all these *reviews* of the research, what do we know about learning from television? Not much!

Results of most of these research studies indicate "no significant difference"—there is no statistical basis for concluding that television per se affects the learning situation, positively or negatively. In most studies, the reasons for the "NSD" results can be traced to faulty research design, improper use of research tools, or simply that the items being compared really were not testing significantly different factors (that is, comparing the impact of a mediocre teacher giving a dull lecture "live" in front of a classroom and the impact of the same teacher giving the same lecture on a TV tube really is not testing any difference at all in the educational situation). Problems in designing and conducting effective ETC research are discussed below (Section 17.5). However, when there have been statistically significant differences, more often than not the differences have been favorable to the use of television.

What We Do Know about Learning from Television   The best synthesis of the research is also, fortunately, the most recent comprehensive review of the research. *Learning from Television: What the Research Says* is a report on a project conducted by Godwin Chu and Wilbur Schramm, then with the Stanford Institute for Communications Research, which attempted to take a "wide-angle view of the field with low definition."[18] That is, Chu and Schramm tried to determine what the research was trying to say, rather than what the reports did say. We can here only summarize some of the major conclusions of Chu and Schramm.

Given favorable conditions, children learn efficiently from instructional television. Students learn from any experience that seems relevant to them, and TV is good at providing vicarious experience. Much of the research leading to this conclusion compared TV instruction with conventional instruction—which presents difficulties both for the researchers and for the synthesizers of the research (see below). In his 1962 study, Schramm looked at nearly 400 experimental comparisons of television and classroom teaching. "He reported that 255 of these comparisons showed no significant differences, 83 were significantly in favor of televised teaching, and 55 significantly in favor of conventional teaching."[19] All the research on which the conclusion that students learn efficiently from TV is based used television completely to teach a part or all of a lesson. As Chu and Schramm state, "this is an unreal comparison, because almost nowhere in the world is television being used in the classrooms without being built into a learning context managed by the classroom teacher."[20]

[14]William H. Allen, "Audio-Visual Communication" (containing a review of ITV research findings), *in* C. Harris (ed.), *Encyclopedia of Educational Research* (New York: Macmillan Company, 1960).

[15]Wilbur Schramm, "What We Know About Learning From Instructional Television," *in* Wilbur Schramm (ed.), *Educational Television: The Next Ten Years* (Stanford, California: Institute for Communication Research, Stanford University, 1962), pp. 52–76.

[16]J. Christopher Reid and Donald W. MacLennan, *Research in Instructional Television and Film*, U.S. Department of Health, Education, and Welfare publication OE-34041 (Washington, D.C.: U.S. Government Printing Office, 1967).

[17]Godwin C. Chu and Wilbur Schramm, *Learning From Television: What the Research Says* (Washington, D.C.: National Association of Educational Broadcasters, 1968). In addition to these compilations, various doctoral dissertations have reviewed research findings for one reason or another. Particularly pertinent is David W. Stickell, "A Critical Review of the Methodology and Results of Research Comparing Televised and Face-to-Face Instruction" (unpublished doctoral dissertation, Pennsylvania State University, 1963).

[18]Chu and Schramm, *Learning from Television*, p. v.

[19]*Ibid.*, p. 5.

[20]*Ibid.*, p. 6.

Indeed, some of the most successful uses seem to depend on the studio teacher and the classroom teacher working as a team, toward the same learning goals. Therefore, the finding of "no significant differences" seems to mean that television can do its part in this combination, and one goal of future research and practice is to find what *combinations* will be more efficient than *either* classroom teaching or television teaching alone.[21]

By and large, instructional television can more easily be used effectively for primary and secondary school students than for college students. Moreover, the logistics of using television make it easier to use STV in elementary grades than in secondary grades. Chu and Schramm speculate that television becomes less effective as students reach higher grade levels because of characteristics of the medium (such as a lack of feedback) or characteristics of students (younger children basically like TV). Another possibility is that TV has been better used in the lower grades and so it has been more effective and better liked than unimaginative TV lectures viewed by college students.

As far as we can tell from present evidence, television can be used efficiently to teach any subject matter for which one-way communication will contribute to learning. Television is not subject-bound. As far as content is concerned, there seems to be no discipline that TV cannot teach, providing immediate feedback is not required. TV can teach a sailor how to tie a half hitch; it cannot correct his mistakes as he practices tying the knot. TV can demonstrate good posture; it cannot correct bad posture. Those subject areas in which there tends to be a great deal of demonstration, such as the physical sciences, lend themselves best to TV instruction using an on-camera teacher. Those subjects

that describe places or events outside of the classroom, such as the social studies, are able to use voice-over narration and "actuality" television techniques. Subjects involving psychomotor skills, such as typing, are able to capitalize on TV's ability to show close-ups from the point of view of the learner—the subjective camera angle. To the extent that these kinds of teaching are involved in a given subject area, TV can be used in the instruction if other factors justify its use.

Television is most effective as a tool for learning when used in a suitable context of learning activities at the receiving end. This is further explication of the point made earlier: TV is best used when it is a part of a total learning experience that combines classroom activities with TV and other media—on both a total planned basis and on a spur-of-the-moment basis relying upon decision making by skilled classroom teachers as they perceive learning difficulties by individuals and groups in the classroom.

These are some of the most important of the Chu and Schramm conclusions. In all, they state 60 conclusions under six broad headings:

1. Do pupils learn from instructional television?

2. What have we learned about the efficient use of instructional television in a school system?

3. What have we learned about the treatment, situation, and pupil variables?
    a. Physical variations
    b. Pedagogical variations
    c. Viewing conditions
    d. Lack of two-way communication
    e. Student's response

4. Attitudes toward instructional television

5. Learning from television in developing regions

6. Learning from television: Learning from other media

---

[21] *Ibid.*

For administrators faced with a decision whether or not to use TV, the conclusions about attitudes can be helpful. It appears that predispositions toward TV as a teaching tool are important at all levels. Administrators who are favorably disposed toward STV will encourage teachers to use it and will remove impediments to effective utilization of TV. Teachers who are enthusiastic about TV find that students like TV instruction; teachers who are negative about the medium tend to find their students do not like television. This is not only evident in the research, it can be readily verified by any STV utilization coordinator or administrator.[22]

In the conclusion to their report, Chu and Schramm ask a rhetorical question: "What kind of guidelines can we extract from this body of research?"

For one thing, it has become clear that there is no longer any reason to raise the question whether instructional television can serve as an efficient tool of learning. This is not to say that it always *does*. But the evidence is now overwhelming that it *can,* and, under favorable circumstances, does. This evidence now comes from many countries, from studies of all age levels from preschool to adults, and from a great variety of subject matter and learning objectives. The questions worth asking are no longer whether students learn from it, but rather, (1) does the situation call for it? and (2) how, in the given situation, can it be used effectively?[23]

Other questions worth asking are how can we have better research in the future, and how can we have better utilization of the specific research findings we now have? Areas of needed future research are explored in Section 17.6.

17.5   Designing and Conducting ITC Research

In order to consider what directions ITC research should and may take in the future, we need to review the ways research is typically done and look at some of the problems of conducting research in education. A list of steps normally followed in conducting research—an example of the scientific method referred to in our definition of research given above—would likely include the following:

1. *Definition of the Problem:* Isolate the problem from other, related problems. Review previous research to see if/how this problem has been approached in the past. Consider related theories to see if they correspond to the specific problem. Define all constructs (such as "motivation") so that we will know them when we see them.

2. *Stating the Research Purpose:* Phrase the problem in such a way that possible explanations for the problem are stated. Formulate hypotheses or alternative solutions that will be tested by the research.

3. *Developing a Research Design:* Decide on the exact procedure that will be followed. Design tests, observations, interview schedules, questionnaires, apparatus, experiments, comparisons, and activities as required.

4. *Collecting the Data:* Record your facts and figures, display them so they can be interpreted, and analyze them.

5. *Generalizing from the Data and the Analysis:* Arrive at conclusions that extend beyond the individuals or groups studied. Confirm research hypotheses, or suggest implications when research hypotheses could not be supported by the findings.

---

[22]Both authors have experienced this phenomenon of negative teacher predisposition affecting students' attitudes. One MPATI evaluation form submitted by a dissatisfied teacher noted that students did not like ITV even though the teacher *pretended* to like it. The teacher was sure her initial negative feelings were concealed from the pupils. Not very likely!

[23]Chu and Schramm, *Learning from Television,* p. 98.

6. *Reporting Results and Applying Findings:* If it is appropriate, publish the findings so others can benefit from your efforts. Recommend other research to continue investigation of the problem. Suggest ways the findings can be applied to practices. Apply the research findings in your own activities.

These steps can be followed whether one is conducting a carefully controlled experiment or is comparing a new way of doing things with some alternative ways of doing the same thing. In either case, an objective attitude and a systematic approach such as that described will lead to justifiable, logical, effective decision making.

Problems in Conducting and Applying ITV Research   Anyone who attempts to analyze research in the media or to explore the needs for research in mediated instruction, must consider a number of factors that have obstructed research in the past and will continue to do so. We have taken these problems and drawbacks from a number of thoughtful discussions of the situation, and added some concerns of our own.[24]

The Scientific Tradition   Educational researchers in general have for years tried to emulate the researchers in the physical sciences, both at the conceptual and the methodological levels. Consequently, they have designed their research to follow certain rules of the game that make it very difficult to find "significant differences"

except when there is so much control of the situation that it becomes unrealistic, with limited utility and generalizability. Even the researchers sometimes forget, or never understand, that a finding of "no significant difference" does not mean that no differences were found in the study—only that the differences were not sufficiently great, and/or that the dispersion of scores was so great, that the differences could not be called statistically "significant" according to the rules that scientists go by (the testing of null hypotheses with .05 levels of significance, two-tailed tests, data meeting the assumptions of interval data when means and correlations are computed, and so on).

The Ambiguity of Research Findings   It has been said that anyone can prove anything he wants with the scientific method just by the way the research is approached. The conflicting studies in ITV would bear this out. If one wants to find out the relation of eye contact (with the TV camera) to learning by TV, results of some studies indicate that it matters whether the teacher looks at the camera; at least one study found that it does not matter (and that study had a consistent difference in favor of *no* eye contact, although certain methodological problems may account for this).[25]

The Failure of Research to Address Itself to Timely Questions   Researchers tend to avoid the critical and timely questions that are faced by practitioners, partly because they often are not easily fitted to the rules of scholarly research, and partly because there is a fascination with the *original* idea; and practical questions are seldom unique and original.

The Pressure to Do Research   Probably too much research is done on learning and television by people who never become really proficient in educational research. This problem has

[24]Egon G. Guba, "The Theory-Practice Dichotomy in Instructional Television," unpublished paper read at the Region II Seminar, National Association of Educational Broadcasters, Miami University, Oxford, Ohio, March 22, 1965; Lee S. Dreyfus, "Closing the Gap—Research and Practice," unpublished paper read at NAEB Region III, supra.; Siebert, "Broadcasting and Education," pp. 139–150; and George L. Hall, "Letter from Wingspread," report on a seminar on improving research in public broadcasting, *Public Telecommunications Review,* March/April 1975. The first three problems, and two others, are excerpted directly from the Guba paper, which reflects the concerns of one who is a researcher and who was at the time the director of the Bureau of Educational Research and Service, The Ohio State University.

[25]Bruce H. Westley and J. B. Mobius, "The Effects of 'Eye-Contact' in Televised Instruction" (Madison: University of Wisconsin Television Laboratory, 1960).

arisen in part because we force graduate students to conduct scholarly research—which is often experimental research because it is less "messy" than case studies and the loosely structured investigations that might be more useful to practitioners. It also is possible that too much money has been indiscriminately used for research about utilizing TV in education and that the grants have too often been given to those with relatively little experience with research techniques.

PR Masquerading as Research Many so-called research reports are little more than public relations pieces for projects that have been funded by grants. As the grant begins to run out, a project may take funds designated for evaluation reports and publish selected findings that say to one and all, "Look at all we have accomplished—and see why we ought to receive more funds." This is not necessarily always a deliberately dishonest misuse of research techniques and traditions; when individuals have invested a year or two of their energies in a project, then begin to see how much more could be done in their area of interest, they are human enough to want a new grant to enable them to continue the investigation.

The Tendency of Decision Makers to Ignore What They Do Know of Research Findings What little *is* known about learning from television is often ignored by administrators, producers, directors, and TV teachers. They often have not been sufficiently concerned about research findings to pay much attention to them, or they find it more convenient to maintain the status quo when it is challenged by a particular research finding or theory. Even organizations that employ researchers often ignore findings of their own research teams. As George Hall says, "Those of us on the management side *do* indeed tend to regard research findings . . . as the Mafiosi do their madonnas: decorations to be shut away in the parlor, safe from the dangerous hustle-bustle of making a living."[26]

The Stereotype of Local Autonomy Egon Guba, director of the Bureau of Educational Research and Service, The Ohio State University, points out that the tradition of local autonomy, tenaciously held by educators, prevents a researcher from suggesting the far-reaching implications of a particular finding. Research results must have application at the local level, or they almost certainly will be rejected. In Chapter 12 we suggested that sometime in the future education may involve nationally produced media materials that are validated according to nationally agreed-upon performance objectives. This possibility is controversial, but it deserves consideration and investigation. Could research on these possibilities be conducted; and, if the research indicated it would benefit students to operate in such fashion at the national level, would the research findings be accepted? Guba indicates that we should not be optimistic:

> Attempts to produce applications on a regional, state, or national level, or, indeed, in any way that does not actively involve the classroom teacher or other practitioner, are rejected on the grounds that such activities subvert local autonomy. Researchers as individuals or as a group are not exceptions to this lockout; if they were to take seriously the expectations that they, the researchers, should make applications from their own research they would likely find their applications rejected as insidious and patronizing.[27]

Vested Interest in Maintaining the Status Quo We have already indicated that administrators often find it advantageous to defend the status quo. We also should acknowledge that there are many who for one reason or another have a vested interest in maintaining "the old way." There are others who simply are incapable of thinking beyond "the way we always have done it." Among those who may feel the need to hold

[26]Hall, "Letter from Wingspread."

[27]Guba, "Theory-Practice Dichotomy," p. 3.

to the status quo are textbook publishers that have a financial stake in keeping textbooks as the "major resource" in classrooms, librarians who fear disruption of their traditional domain, teachers' unions that see a threat to their power; school systems that have large investments in textbooks and other materials; taxpayers who are suspicious of new fads and frills; and teachers who have invested time and possibly money in developing files of materials they can refer to and use in their classrooms. It is no wonder that researchers have found that it takes about fifty years for a new idea or device to become accepted in the classroom.[28]

Lack of Time, Resources, and Competence on the Part of Practitioners   Since the local teacher and administrator are expected to make use of ITV (and other) research findings, it is not surprising that little application is made of research. As Guba says:

> Who has time or energy for engaging in developmental activities after a full day of coping with the turmoil of the classroom or the principal's office? Where will the money come from to pay for the materials that will have to be produced? Where will the extra staff come from that will be needed to relieve those persons whom we might charge with this responsibility? How will we find the special competencies that will be needed: the developers who will create, program, and evaluate the applications of research; the disseminators who will report and interpret the development; or the demonstrators who will produce and stage convincing and credible illustrations of the development in action? . . . To expect [such things from practitioners] is simply to be unreasonable and unrealistic.[29]

---

[28]*Ibid.*
[29]*Ibid.*

Despite these problems, research continues, as well it should. And—in order to cope with the increasingly frustrating problem of trying to deal with an ever-growing number of research reports—the government is trying to make it easier for researchers and practitioners alike with ERIC.

Educational Research Information Center (ERIC)   In the years following passage of the National Defense Education Act (Section 4.1), vast amounts of research were conducted in specialized areas of education such as reading, teaching the disadvantaged student, counseling, linguistics, and educational media. In all these areas (and others), the amount of research became so great that the U. S. Office of Education set up centers to catalog and disseminate information on research in their designated subject areas. There now are nineteen such national centers scattered throughout the United States.

Research on instructional television, radio, and recorded audio-visual material—plus information on programed instruction, computer-assisted or computer-managed instruction, and other audio-visual means of instruction—is collected and disseminated by the ERIC Clearinghouse on Educational Media and Technology at Stanford University. The ERIC clearinghouse collects current publications and documents of importance that are not available in principal journals or textbooks. These are reviewed, abstracted, coded for easy reference, and sent to a computer center where they are stored on computer tape and prepared for further reference use.

The Stanford center now has on computer tape hundreds of research studies and other writings on ITV from 1945 to the present time, from foreign as well as U.S. sources, and at least subject entries from chief journals as well as less common journals with occasional articles on ITV. It is possible to query the computer file to find what research has been done on such a specific topic—to offer one example—as the teaching of mathematics by television in the elementary school.

New research reports that come to the attention of the ERIC staff are abstracted and given reference codes when they come in; and these are forwarded to the U.S. Office of Education for inclusion in the next monthly issue of *Research in Education,* which lists documents indexed by all ERIC centers. When one notes a desired research report in *Research in Education,* or in a computer search, the information from ERIC will tell how to obtain the report. If it is not readily available in a journal, the document probably can be ordered from the ERIC Document Reproduction Service (EDRS), which can supply a microcard of the report or a slightly reduced photocopy (which tends to be rather expensive). Many libraries are ERIC repositories which automatically receive a microfiche copy of every document sent to ERIC that is not otherwise readily available.

Periodically ERIC commissions a review of the research studies on file. In 1973, the second paper summarizing the ITC research abstracted in the computer was published. The author of both summaries, Warren Seibert of Purdue University, noted an encouraging trend as he reflected on the early ITV research compared with the more recent investigations:

> For so many years, educational projects employing television or other modern media were peripheral, impromptu, short-term, and tentative. These projects were rarely granted time or resources enough to refine and stabilize their efforts or to show cumulative effects. When they succeeded, almost no one cared and when they failed, almost no one was surprised—or even knew. . . .
>
> Now, projects and programs in instructional television provide a different impression. Television has begun to matter and, as

some reports clearly show, in individual instances it now matters greatly.

There is no great victory to be seen here, however. There is only growing realization of promise. Just as there is a long way to go before the best can be realized—consistently—from traditional forms and media, so there is at least as far to go before the best is seen from the still-new medium of television. But to scan the current uses of television is to gain the belief that many educators are committed to the long haul and to the fullest realization of the promise.[30]

17.6   Needed Research in Instructional Telecommunications

It has been suggested above that many areas do not need any further research (for example, experimental studies comparing learning situations that use ITC to those that do not use electronic media). There are, however, numerous areas in which research concerning educational applications of ITC could be fruitfully conducted. We need to investigate how radio, television, and recorded audio and video can best be utilized in different kinds of learning situations. We need to know how to combine telecommunications with other media and learning techniques to teach a given subject at a given level more effectively. We need to learn how to adapt various teaching techniques for different kinds of learners. The research we need might best be called "systems" research since it broadly considers the role of instructional telecommunications as a part of a total system that is occurring in the classroom at

---

[30]Warren F. Seibert, "Instructional Television: The Best of ERIC," ERIC Clearinghouse on Media and Technology, Stanford University (Lafayette, Indiana: Measurement and Research Center, Purdue University, October 1973), p. 1.

a given moment of time.[31] Many aspects of the total system could profitably be examined in future research involving ITC.

Aptitude-Treatment Interaction Research

An approach to media research based loosely upon medical research techniques is beginning to help explain why some students, and some situations, seem better suited to television than others:

> Interaction hypotheses are not recent phenomena nor are they the exclusive domain of educational researchers. Early medical researchers who discovered the curative benefits of blood transfusions soon learned that there was considerable error in their new procedure. Although some patients immediately improved, others became more seriously ill, and a few died. It wasn't until the hypotheses were advanced concerning possible interactions between blood types that transfusions became beneficial to the greatest numbers. Interaction hypotheses were also suggested by the first generation of media researchers to explain the limited results of experiments which sought the one best medium for instruction. Freeman (1924)[32] conducted what appears to be the first systematic series of experiments with motion films in instruction and concluded that "the relative effectiveness of verbal instruction as contrasted with the various

forms of concrete or realistic material in visual media depends on the nature of the instruction to be given and the character of the learner's previous experience with objective materials" (p. 116).[33]

It was not until the early 1970's that aptitude-treatment interaction research really began to excite educational researchers.[34] Studies were undertaken which began with hypotheses concerning interaction effects. Those who had summarized research by other approaches reanalyzed collections of research to look for interaction results. One distiller of research, William Allen of the University of Southern California, looked at intellectual ability in relation to viewing lessons on television and films, and arrived at a set of "prescriptions" for designing ITV materials for students with different learning abilities. For example, Allen recommends:

> Instructional material designed for learners of high mental ability should employ the following design techniques: . . . higher information density, pictorial and conceptual complexity, and richness in images, ideas, and relationships, . . . rapid rate of development of information and concepts being communicated, . . . a format that places requirements on the learner to organize, hypothesize, abstract, and manipulate the stimuli mentally in order to extract meaning from it . . .[35]

As useful as such prescriptions are, the greater utility of the aptitude-treatment interaction research approach is in generating hypotheses

[31]C. Ray Carpenter, "Approaches to Promising Areas of Research in the Field of Instructional Television" *in New Teaching Aids for the American Classroom* (Stanford, California: Institute for Communication Research, 1960), pp. 73–94. *See also* Schramm ("What We Know About Learning," p. 71), who quotes Carpenter and further discusses systems research. A paper by Stowe compares the systems approach with research in education: Richard A. Stowe, "Research and the Systems Approach as Methodologies for Education," *AV Communication Review,* Summer 1973, pp. 165–175.

[32]F. N. Freeman (ed.), *Visual Education* (Chicago: University of Chicago Press, 1924). Citation is by Clark (see below) who further cites as the original reference Paul Saettler, "Design and Selection Factors," *Review of Educational Research,* 1968.

[33]Richard E. Clark, "Adapting Aptitude-Treatment Interaction Methodology to Instructional Media Research," *AV Communication Review,* Summer 1975, p. 133.

[34]The term "aptitude-treatment interaction" was coined by Lee J. Cronbach and Richard E. Snow in a 1969 technical paper for the U.S. Office of Education (USOE Contract No. OEC 4-6-061269-1217, Stanford University, 1969).

[35]William H. Allen, "Intellectual Abilities and Instructional Media Design," *AV Communication Review,* Summer 1975, pp. 163–164.

that attempt to answer the fundamental question: What kinds of students, in what kinds of situations and in what kinds of courses, learn better from television and audio messages as compared with other ways of receiving the information?

Examining Fundamental Characteristics of Television Despite all the research that has been done on teaching by television, we still know very little about the way in which television viewers become involved with what they see on the tube. McLuhan has said that TV is a "cool" medium that tends to involve us in ways that films do not; the viewer must participate in filling in the picture.[36] Others have said that we accept people we see on TV as being more real than people we see on film because when we are young we think that little people live inside the TV set! We have talked about the "three I's" of television—intimacy, involvement, and immediacy—all of which suggest that TV has a "realness" quality to it that makes everything we see on it a little more real than if we saw the same thing on film. Whatever the explanation or hypothesis, the fact remains that we know comparatively little about how people relate to what they see on television.

There also needs to be more investigation about basic characteristics of the perception of various video and audio elements—for example, color versus black-and-white presentations. We know that monochrome television can teach as well as or better than color TV except when color differentiation is critical to the learning task at hand (such as being able to watch litmus paper turn red in a test for acids). However, we know little about the ways students conditioned to color TV in the home (in some parts of the nation, three-fourths or more of the homes have color TV sets) react emotionally to presentations in black and white. Most STV programs are now broadcast in color, but most sets in public school classrooms are monochrome. Does this matter? Do students who view telelessons in black and white perceive the TV teacher, the subject matter, or the importance of the presentations, any differently from those who view the programs in color? How about sound? Does high-fidelity audio improve presentations, and do students who receive ITV programs in hi-fi sound react to elements in the course differently?

Tutorial Telecommunications We need additional research projects to investigate how to make recorded TV and recorded audio presentations individualized for different student needs, and how these individualized learning programs can be integrated into a management system that directs the student to the right presentation using the best communications medium at the proper time.

Both recorded audio and recorded video lectures have been used in tutorial style to individualize instruction and to make learning materials available at all times and at rates to which learners of varying needs could more readily adapt. When designed in a total systems approach to instruction, these tutorial systems (often using cassettes) have been successful in terms of student learning—although not always accepted enthusiastically by students. We need to analyze successes and failures of these attempts to use audio recordings and television in systematically designed learning projects. We need to go further than these pilot efforts to use telecommunications more systematically; we need to examine how these systems can be accepted and integrated into complete learning programs.

Lewis Rhodes' example of the tendency to use TV as we did elevators during their early stages of application (Section 12.2)—merely to get from one place to another more rapidly, rather than thinking about what this new technology permits us to do *beyond* what we are presently doing—suggests that we need research on using TV-based instructional systems to do things we

[36]Marshall McLuhan, *Understanding Media: The Extensions of Man* (New York: McGraw-Hill Book Company, 1964), pp. 27–28.

presently do not do in our schools. Maybe we can use telecommunications media to teach foreign languages to kindergarteners or computer programing to high school freshmen or real estate law to liberal arts college students (assuming these are legitimately recognized needs—that there is agreement that it is useful for kindergarteners to know Spanish, ninth-graders to know how to talk with a computer, or college students to know real estate law).

New Patterns of Instruction If we are able to design instructional systems that use media to tutor learners on a one-on-one basis, and if we find that we can use TV and audio recordings to teach subjects and material that we previously have not been able to teach, we then need to investigate how schooling patterns can best accommodate the new media and techniques. Some of these questions were raised in Chapter 12. Should students always have to go to school to receive instruction? Should schools consist primarily of classrooms? What new or revised roles for teachers should be supported? How are teachers going to learn their new roles? Should we continue to compartmentalize subjects, or should we attempt to use media in more of a core curriculum approach to learning, at least in the elementary grades? If we compartmentalize knowledge, should we expand the range of subjects available in secondary schools?

Student Feedback and Performance Much research indicates that students learn better from mediated instruction (or from any kind of instruction, for that matter) if they can actively respond to the material being presented. Experiments have shown that learning can be improved simply by using the practice of stopping a presentation periodically to ask the students a question. Yet little has been done to determine how interpolated questions can best be incorporated in a lesson, and how student responses can be recorded in order to continually validate the instruction.

We also do not know how covert responses (silent answers given only in the learn-

er's mind) can best be integrated into a lesson. Good instructional TV producers and on-camera teachers tend to ask rhetorical questions—intuitively leading the learner's mind forward, questioning and responding silently all the time the lesson is on the screen. However, there are few researched guidelines as to how this can be accomplished consistently in ITC productions.

Research also is needed on repeat viewing of telelessons of various kinds. We know that students learn new things from the same TV lesson each time they see it; but how much more—for different kinds of presentations and subject areas or levels—is not known.

In our investigation of student feedback, we also should research more systematically the validation of presentations. With improved student feedback mechanisms, it should be possible to design lessons that can be validated with differing groups, validated on local levels, and improved and revalidated as feedback data become available. Thus, it is theoretically possible to have lessons so carefully validated that one could give a group of students a pretest and, by then referring to a chart, make precise predictions about how that group of students will learn from a series of presentations.

Technical Advances Several aspects of hardware technology deserve considerable research emphasis. The video disc represents a great deal more than a new way to reproduce TV programs. The ability of laser-based disc systems to still-frame pictures and to retrieve specific frames of information opens up new worlds for applying the principles of programed instruction to video formats—particularly "branching" and "looping" to accommodate different rates and comprehension in learning given material. Research is needed on how best to use the new capabilities of video discs. We need to design adaptations of home video-disc units that will be more suitable for use in learning spaces.

Similarly, we need to research the possibilities of using satellites to do more than to relay demonstration programs from one place to another or to simply extend conventional ITV programing into remote areas of the world. Can satellites be used to let students in one area of the country learn from people in a different area? Can high school government classes visit with their congressmen, or sit in on congressional debates, by means of satellites?

Even the conventional, inexpensive, slant-track video recorder has not been fully exploited. For example, only a few projects have been concerned with the opportunities for exchanging student-produced videotapes—notably the "Video Explorations" project of the Division of Telecommunications of the Association for Educational Communications and Technology.[37] Action research is needed on ways to facilitate the exchange of student-prepared "pen-pal" messages between classes representing different cultures and/or environments.

Simulation and Role Playing   Television has been used in education primarily as a channel for delivering instructional content. Little has been done to use TV as a *simulation* device. Yet, television's ability to make the viewer feel that what is on the screen is actually happening *now* (the characteristic of immediacy) makes it useful as a simulation mechanism. One excellent example was a lesson from a series on flying that simulated a cross-country flight. Even without a verbal audio track, the learner was made to feel he was flying a plane, controlling the speed and direction of the aircraft.

A series of videotapes recorded in a classroom, in which simulated instructional problems suddenly occur, was very effective in teacher training classes at the University of West Florida. With video discs it should be possible to develop role-playing materials showing simulated class-room activities that provide for an education student to respond to what is seen on the screen. The appropriate consequence can then be selected by the supervising critic teacher to show what probably would be the response of the simulated class following the particular action selected by the education student role-playing the teacher of the class.

ITC and World Challenges   Finally, research is needed that will increase the effectiveness of ITC in areas of high illiteracy in the world. In these regions there often are so many different languages and dialects that it becomes virtually impossible to use any verbal instructional messages. In such areas, the pictorial element of television—combined with television's ability to involve us in events and processes with little or no verbal explanations—can be a powerful means for communicating simple and, sometimes, even very complex messages. Because we can begin with visual-only messages, TV becomes an effective way to initiate literacy training.[38] The research needed in this area ranges from basic research in visual literacy for the verbal illiterate to operations research on establishing ITC programs for literacy training.

It is clear that those ITC practitioners and theorists who are willing and able to conduct research on learning from instructional telecommunications will not want for questions worthy of study—or for reasons for undertaking the research. The need for quality research on mediated and media-supported instruction is greater than ever.

17.7   A Summing Up

It is altogether fitting and appropriate that this book close with a section on evaluation and research. In the instructional design model (figure 13-2), the act of evaluation is indicated

---

[37]This project was originated by Mrs. Kathy Busick in Hawaii. The current address of Mrs. Busick and the present coordinator of the project can be obtained from the Division of Telecommunications, AECT, 1201 Sixteenth Street, N. W., Washington, D. C. 20036.

[38]See Presley D. Holmes, "On Understanding Television: Significant Differences?" *AV Communication Review*, July-August 1962, p. 261.

as the final step of the system. However, as has been pointed out, the process of evaluation and assessment actually leads back to a modification of the system—a revision of objectives and procedures, a reexamination of our way of looking at the process, a new beginning.

So it is with this book. We hope that this work will lead the reader to a reevaluation of the ways in which we have traditionally examined the phenomena of noncommercial radio and television and related "telecommunications used for educational purposes"—the ways in which we define instructional telecommunications, school TV, public broadcasting, and related terms; the way we perceive the relationships between public broadcasting and ITC; the ways we examine the history of noncommercial electronic media; the ways we analyze the structure of educational telecommunications; the way we interpret current programing and financial issues in light of decisions made a quarter of a century ago; the way we organize formal schooling uses of ETC; the way we structure practical aspects of a functional educational program; and the way we predict future directions in ITC and schooling patterns.

In conclusion, it would be safe to state that educational telecommunications has indeed come a long way since the first meeting of the Association of College and University Broadcasting Stations more than half a century ago. It has faced many grueling challenges, received unexpected support from various quarters, and experienced heartbreaking setbacks in its struggle to become a recognized self-sufficient, viable force in our educational and cultural milieu.

The peculiar place that noncommercial media hold in America's educational system is still ambiguous and ill defined. To some extent, this is but a reflection of America's educational pluralism. Many questions have yet to be answered about the role of informal adult education in our society, the purposes of formal schooling in our institutions, the general cultural needs of our citizens, and the overall issue of the role of free media in our unique political system. Until these issues are resolved, the roles of public broadcasting and instructional media never can be determined with any degree of finality.

This book has but outlined some of the developments of the first half-century of educational telecommunications. It is up to the readers of this book to determine how we chart the course of ETC for the next half-century.

abcde

**Selected List
of ETC-Related
Organizations**

This appendix provides an alphabetical listing, by initials, of *some* of the key organizations, agencies, and institutions involved directly or indirectly in educational telecommunications. The year of the beginning of the organization is given in parentheses (where pertinent), and relationships with other agencies are indicated (in some instances).

ACER Advisory Committee on Education by Radio (1929).

ACNO Advisory Council of National Organizations (to the CPB).

ACUBS Association of College and University Broadcasting Stations (1925). Became the NAEB in 1934.

AECT Association for Educational Communications and Technology. Originally the DVI, then DAVI. Became the AECT in 1970.

AER(T) Association for Education by Radio (and, later, Television), (1940). Merged with the NAEB in the mid-1950's.

AIT Agency for Instructional Television. Originally part of NITL, became NCSCT, then NIT(C). Reconstituted as AIT in 1973.

APRS Association of Public Radio Stations (1973).

BFA Broadcasting Foundation of America.

CEN Central Educational Network (1967).

CPB Corporation for Public Broadcasting (1967).

CTW Children's Television Workshop (1967).

DAVI Department of Audio-Visual Instruction (1947), a department of NEA. Formed originally as DVI. Reconstituted as the independent AECT in 1970.

DOT Division of Telecommunications, of the AECT.

DVI Department of Visual Instruction (1923), of the NEA. Became DAVI, then AECT.

EEN Eastern Educational (Television) Network (1960).

ERIC Educational Research Information Center (Stanford University).

ERN Educational Radio Network (1962), a division of NETRC.

ETRC Educational Television and Radio Center (1952). Became NETRC in 1959, then NET.

ETS Educational Television Stations (1963), a division of the NAEB. Merged with the "new" PBS In 1973.

FAE (1) Fund for Adult Education (1951). Created by the Ford Foundation. (2) Fund for the Advancement of Education (1951). Also created by the Ford Foundation.

FCI Family Communications Incorporated, a production unit at station WQED.

FCC Federal Communications Commission (1934). Superseded the FRC.

FRC Federal Radio Commission (1927). Incorporated into the FCC.

FREC Federal Radio Education Committee (1935).

GPNITL Great Plains National Instructional Television Library (1962). Originally part of the NITL.

HEW (Federal Department of) Health, Education, and Welfare.

IER(T) Institute for Education by Radio (and, later, Television), (1930).

IHETS Indiana Higher Education Telecommunication System.

JCET Joint Council on Educational Telecommunications (1950). Originally the Joint Committee on Educational Television, later the Joint Council on Educational Broadcasting, now the JCET.

MET Midwestern Educational Television (network), (1961).

MPATI Midwest Program on Airborne Television Instruction (1959). Incorporated into GPNITL in 1971.

NACRE National Advisory Council on Radio in Education (1938).

NAEB National Association of Educational Broadcasters. Formed as the ACUBS. Became the NAEB in 1934.

NAVI National Academy of Visual Instruction. Merged into DVI in 1932.

NCCET National Citizens Committee for Educational Television (1952).

NCCGBC National Coordinating Committee of Governing Board Chairmen (of PTV stations), (1972). Merged with the "new" PBS in 1973.

NCER National Committee on Education by Radio (1931).

NCET National Center for Experiments in TV (1969), a unit of Station KQED.

NCSCT National Center for School and College Television. Originally part of NITL. Became the NCSCT in 1965, then NIT(C), then AIT.

NEA National Education Association.

NER National Educational Radio, a division of the NAEB until 1973.

NET National Educational Television. Originally the ETRC. Became NETRC, then NET in 1963. Merged with Station WNDT (now WNET) in 1969.

NETRC National Educational Television and Radio Center. Originally the ETRC. Became NETRC in 1959, NET in 1963.

NFPB National Friends of Public Broadcasting.

NIT(C) National Instructional Television Center. Originally part of NITL. Became the NCSCT, then NIT in 1968, and AIT in 1973.

NITL National Instructional Television Library (1962). Originally a division of NETRC.

NPACT National Public Affairs Center for Television (1971). Became a unit of Station WETA in 1974.

NPR National Public Radio (1970).

OTP Office of Telecommunications Policy (1970), part of the White House.

PBL Public Broadcast Laboratory (1967), a one-year project of NET.

PBS Public Broadcasting Service (1969). The "new" PBS, incorporating ETS and NCCGBC, was formed in 1973.

PCEB Preliminary Committee on Educational Broadcasting (1927).

PSSC Public Service Satellite Consortium (1975).

PTL Public Television Library (1973). Originally the ETS Program Service (formed in 1965).

RETAC (Los Angeles) Regional Educational Television Advisory Council (1959).

RMPBN Rocky Mountain Public Broadcasting Network (1969).

SCETVC South Carolina Educational Television Commission (1960).

SECA	Southern Educational Communications Association (1967).
SUN	State University of Nebraska.
TEMP	Texas Educational Microwave Program.
USOE	United States Office of Education (part of HEW).
VIAA	Visual Instruction Association of America. Merged into DVI in 1932.
WEN	Western Educational Network (1968).
WEST	Western Educational Society for Telecommunications.

The Twelve Recommendations of the Carnegie Commission

Listed below are the twelve recommendations of the Carnegie Commission as excerpted from *Public Television: A Program for Action* (New York: Bantam Books, Inc., 1967), pages 33 to 80.

1. We recommend concerted efforts at the federal, state, and local levels to improve the facilities and to provide for the adequate support of the individual educational television stations and to increase their number.

2. We recommend that Congress act promptly to authorize and to establish a federally chartered, nonprofit, nongovernmental corporation, to be known as the "Corporation for Public Television." The Corporation should be empowered to receive and disburse governmental and private funds in order to extend and improve Public Television programming. The Commission considers the creation of the Corporation fundamental to its proposal and would be most reluctant to recommend other parts of its plan unless the corporate entity is brought into being.

3. We recommend that the Corporation support at least two national production centers, and that it be free to contract with independent producers to prepare Public Television programs for educational television stations.

4. We recommend that the Corporation support, by appropriate grants and contracts, the production of Public Television programs by local stations for more-than-local use.

5. We recommend that the Corporation on appropriate occasions help support local programming by local stations.

6. We recommend that the Corporation provide the educational television system as expeditiously as possible with facilities for live interconnection by conventional means, and that it be enabled to benefit from advances in technology as domestic communications satellites are brought into being. The Commission further recommends that Congress act to permit the granting of preferential rates for educational television for the use of interconnection facilities, or to permit their free use, to the extent that this may not be possible under existing law.

7. We recommend that the Corporation encourage and support research and development lead-

ing to the improvement of programming and program production.

8. We recommend that the Corporation support technical experimentation designed to improve the present television technology.

9. We recommend that the Corporation undertake to provide means by which technical, artistic, and specialized personnel may be recruited and trained.

10. We recommend that Congress provide the federal funds required by the Corporation through a manufacturer's excise tax on television sets (beginning at 2 percent and rising to a ceiling of 5 percent). The revenues should be made available to the Corporation through a trust fund.

11. We recommend new legislation to enable the Department of Health, Education, and Welfare to provide adequate facilities for stations now in existence, to assist in increasing the number of stations to achieve nationwide coverage, to help support the basic operations of all stations, and to enlarge the support of instructional television programming.

12. We recommend that federal, state, local, and private educational agencies sponsor extensive and innovative studies intended to develop better insights into the use of television in formal and informal education.

The Public Broadcasting Act of 1967

**Public Law 90–129
90th Congress, S. 1160
November 7, 1967**

To amend the Communications Act of 1934 by extending and improving the provisions thereof relating to grants for construction of educational television broadcasting facilities, by authorizing assistance in the construction of noncommercial educational radio broadcasting facilities, by establishing a nonprofit corporation to assist in establishing innovative educational programs, to facilitate educational program availability, and to aid the operation of educational broadcasting facilities; and to authorize a comprehensive study of instructional television and radio; and for other purposes.

Be it enacted by the Senate and House of Representatives of the United States of America in Congress assembled. That this Act may be cited as the "Public Broadcasting Act of 1967".

Title I—Construction of Facilities

Extension of Duration of Construction Grants for Educational Broadcasting

Sec. 101 (a) Section 391 of the Communications Act of 1934 (47 U.S.C. 391) is amended by inserting after the first sentence the following new sentence: "There are also authorized to be appropriated for carrying out the purposes of such section, $10,500,000 for the fiscal year ending June 30, 1968, $12,500,000 for the fiscal year ending June 30, 1969, and $15,000,000 for the fiscal year ending June 30, 1970."

(b) The last sentence of such section is amended by striking out "July 1, 1968" and inserting in lieu thereof "July 1, 1971".

Maximum on Grants in Any State

Sec. 102 Effective with respect to grants made from appropriations for any fiscal year beginning after June 30, 1967, subsection (b) of section 392 of the Communications Act of 1934 (47 U.S.C. 392 (b)) is amended to read as follows: "(b) The total of the grants made under this part from the appropriation for any fiscal year for the construction of noncommercial educational television broadcasting facilities and noncommercial educational radio broadcasting facilities in any State may not exceed 8½ per centum of such appropriation."

Noncommercial Educational Radio Broadcasting Facilities

Sec. 103 (a) Section 390 of the Communications Act of 1934 (47 U.S.C. 390) is amended by inserting "noncommercial" before "educational" and by inserting "or radio" after "television".

(b) Subsection (a) of section 392 of the Communications Act of 1934 (47 U.S.C. 392 (a)) is amended by—(1) inserting "noncommercial" before "educational" and by inserting "or radio" after "television" in so much thereof as precedes paragraph (1); (2) striking out clause (B) of such paragraph and inserting in lieu thereof "(B) in the case of a project for television facilities, the State noncommercial educational television agency or, in the case of a project for radio facilities, the State educational radio agency,"; (3) inserting "(i) in the case of a project for television facilities," after "(D)" and "noncommercial" before "educational" in paragraph (1) (D) and by inserting before the semicolon at the end of such paragraph ", or (ii) in the case of a project for radio facilities, a nonprofit foundation, corporation, or association which is organized primarily to engage in or encourage noncommercial educational radio broadcasting and is eligible to receive a license from the Federal Communications Commission; or meets the requirements of clause (i) and is also organized to engage in or encourage such radio broadcasting and is eligible for such a license for such a radio station"; (4) striking out "or" immediately preceding "(D)" in paragraph (1), and by striking out the semicolon at the end of such paragraph and inserting in lieu thereof the following: ", or (E) a municipality which owns and operates a broadcasting facility transmitting only noncommercial programs;"; (5) striking out "television" in paragraphs (2), (3), and (4) of such subsection; (6) striking out "and" at the end of paragraph (3), striking out the period at the end of paragraph (4) and inserting in lieu thereof "; and", and inserting after paragraph (4) the following new paragraph:

"(5) that, in the case of an application with respect to radio broadcasting facilities, there has been comprehensive planning for educational broadcasting facilities and services in the area the applicant proposes to serve and the applicant has participated in such planning, and the applicant will make the most efficient use of the frequency assignment."

(c) Subsection (c) of such section is amended by inserting "(1)" after "(c)" and "noncommercial" before "educational television broadcasting facilities," and by inserting at the end thereof the following new paragraph: "(2) In order to assure proper coordination of construction of noncommercial educational radio

broadcasting facilities within each State which has established a State educational radio agency, each applicant for a grant under this section for a project for construction of such facilities in such State, other than such agency, shall notify such agency of each application for such a grant which is submitted by it to the Secretary, and the Secretary shall advise such agency with respect to the disposition of each such application."

(d) Subsection (d) of such section is amended by inserting "noncommercial" before "educational television" and inserting "or noncommercial educational radio broadcasting facilities, as the case may be," after "educational television broadcasting facilities" in clauses (2) and (3).

(e) Subsection (f) of such section is amended by inserting "or radio" after "television" in the part thereof which precedes paragraph (1), by inserting "noncommercial" before "educational television purposes" in paragraph (2) thereof, and by inserting "or noncommercial educational radio purposes, as the case may be" after "educational television purposes" in such paragraph (2).

(f) (1) Paragraph (2) of section 394 of such Act (47 U.S.C. 394) is amended by inserting "or educational radio broadcasting facilities" after "educational television broadcasting facilities," and by inserting "or radio broadcasting, as the case may be" after "necessary for television broadcasting". (2) Paragraph (4) of such section is amended by striking out "The term 'State educational television agency' means" and inserting in lieu thereof "The terms 'State educational television agency' and 'State educational radio agency' mean, with respect to television broadcasting and radio broadcasting, respectively," and by striking out "educational television" in clauses (A) and (C) and inserting in lieu thereof "such broadcasting".

(g) Section 397 of such Act (47 U.S.C. 397) is amended by inserting "or radio" after "television" in clause (2).

Federal Share of Cost of Construction

Sec. 104. Subsection (e) of section 392 of the Communications Act of 1934 (47 U.S.C. 392(e) is amended to read as follows: "(e) Upon approving any application under this section with respect to any project, the Secretary shall make a grant to the applicant in the amount determined by him, but not exceeding 75 per centum of the amount determined by the Secretary to be the reasonable and necessary cost of such project. The Secretary shall pay such amount from the sum available therefor, in advance or by way of reimbursement, and in such installments consistent with construction progress, as he may determine."

Inclusion of Territories

Sec. 105 (a) Paragraph (1) of section 394 of the Communications Act of 1934 is amended by striking out "and" and inserting a comma in lieu thereof, and by inserting before the period at the end thereof ", the Virgin Islands, Guam, American Samoa, and the Trust Territory of the Pacific Islands".

(b) Paragraph (4) of such section is amended by inserting "and, in the case of the Trust Territory of the Pacific Islands, means the High Commissioner thereof" before the period at the end thereof.

Inclusion of Costs of Planning

Sec. 106 Paragraph (2) of section 394 of the Communications Act of 1934 is further amended by inserting at the end thereof the following: "In the case of apparatus the acquisition and installation of which is so included, such term also includes planning therefor."

Title II—Establishment of Nonprofit Educational Broadcasting Corporation

Sec. 201 Part IV of title III of the Communications Act of 1934 is further amended by—(1) inserting

"Subpart A—Grants for Facilities"

immediately above the heading of section 390; (2) striking out "part" and inserting in lieu thereof "subpart in sections 390, 393, 395, and 396; (3) redesignating section 397 as section 398, and redesignating section 394 as section 397 and inserting it before such section 398, and inserting immediately above its heading the following:

"Subpart C—General"

(4) redesignating section 396 as section 394 and inserting it immediately after section 393; (5) inserting after "broadcasting" the first time it appears in clause (2) of the section of such part IV redesignated herein as section 398 ", or over the Corporation of any of its grantees or contractors, or over the charter or bylaws of the Corporation"; (6) inserting in the section of such part IV herein redesignated as section 397 the following new paragraphs:

"(6) The term 'Corporation' means the Corporation authorized to be established by subpart B of this part.

"(7) The term 'noncommercial educational broadcast station' means a television or radio broadcast station, which (A) under the rules and regulations of the Federal Communications Commission in effect on the date of enactment of the Public Broadcasting Act of 1967, is eligible to be licensed or is licensed by the Commission as a noncommercial educational radio or television broadcast station and which is owned and operated by a public agency or nonprofit private foundation, corporation, or association or (B) is owned and operated by a municipality and which transmits only noncommercial programs for educational purposes.

"(8) The term 'interconnection' means the use of microwave equipment, boosters, translators, repeaters, communication space satellites, or other apparatus or equipment for the transmission and distribution of television or radio programs to noncommercial educational television or radio broadcast stations.

"(9) The term 'educational television or radio programs' means programs which are primarily designed for educational or cultural purposes."

(7) striking out the heading of such part IV and inserting in lieu thereof the following:

"Part IV—Grants for Noncommercial Educational Broadcasting Facilities; Corporation for Public Broadcasting"

(8) inserting immediately after the section herein redesignated as section 398 the following:

"Editorializing and Support of Political Candidates Prohibited

"*Sec. 399* No noncommercial educational broadcasting station may engage in editorializing or may support or oppose any candidate for political office."

(9) inserting after section 395 the following new subpart:

"Subpart B—Corporation for Public Broadcasting

"Congressional Declaration of Policy

"*Sec. 396* (a) The Congress hereby finds and declares—(1) that it is in the public interest to encourage the growth and development of noncommercial educational radio and television broadcasting, including the use of such media for instructional purposes; (2) that expansion and development of noncommercial educational radio and television broadcasting and of diversity of its programing depend on freedom, imagination, and initiative on both the local and national levels; (3) that the encouragement and support of noncommercial educational radio and television broadcasting, while matters of importance for private and local development, are also of appropriate and important concern to the Federal Government; (4) that it furthers the general welfare to encourage noncommercial educational radio and television broadcast programing which will be responsive to the interests of people both in particular localities and throughout the United States, and which will constitute an expression of diversity and excellence; (5) that it is necessary and appropriate for the Federal Government to complement, assist, and support a national policy that will most effectively make noncommercial educational radio and television service available to all the citizens of the United States; (6) that a private corporation should be created to facilitate the development of educational radio and television broadcasting and to afford maximum protection to such broadcasting from extraneous interference and control.

"Corporation Established

"(b) There is authorized to be established a nonprofit corporation, to be known as the 'Corporation for Public Broadcasting', which will not be an agency or establishment of the United States Government. The Corporation shall be subject to the provisions of this section, and, to the extent consistent with this section, to the District of Columbia Nonprofit Corporation Act.

"Board of Directors

"(c) (1) The Corporation shall have a Board of Directors (hereinafter in this section referred to as the 'Board'), consisting of fifteen members appointed by the President, by and with the advice and consent of the Senate. Not more than eight members of the Board may be members of the same political party. (2) The members of the Board (A) shall be selected from among citizens of the United States (not regular fulltime employees of the United States) who are eminent in such fields as education, cultural and civic affairs, or the arts, including radio and television; (B) shall be selected so as to provide as nearly as practicable a broad repre-

sentation of various regions of the country, various professions and occupations, and various kinds of talent and experience appropriate to the functions and responsibilities of the Corporation. (3) The members of the initial Board of Directors shall serve as incorporators and shall take whatever actions are necessary to establish the Corporation under the District of Columbia Nonprofit Corporation Act. (4) The term of office of each member of the Board shall be six years; except that (A) any member appointed to fill a vacancy occurring prior to the expiration of the term for which his predecessor was appointed shall be appointed for the remainder of such term; and (B) the terms of office of members first taking office shall begin on the date of incorporation and shall expire, as designated at the time of their appointment, five at the end of two years, five at the end of four years, and five at the end of six years. No member shall be eligible to serve in excess of two consecutive terms of six years each. Not withstanding the preceding provisions of this paragraph, a member whose term has expired may serve until his successor has qualified. (5) Any vacancy in the Board shall not affect its power, but shall be filled in the manner in which the original appointments were made.

"Election of Chairman; Compensation

"(d) (1) The President shall designate one of the members first appointed to the Board as Chairman; thereafter the members of the Board shall annually elect one of their number as Chairman. The members of the Board shall also elect one or more of them as a Vice Chairman or Vice Chairmen. (2) The members of the Board shall not, by reason of such membership, be deemed to be employees of the United States. They shall, while attending meetings of the Board or while engaged in duties related to such meetings or in other activities of the Board pursuant to this subpart be entitled to receive compensation at the rate of $100 per day including travel time, and while away from their homes or regular places of business they may be allowed travel expenses, including per diem in lieu of subsistence, equal to that authorized by law (5 U.S.C. 5703) for persons in the Government service employed intermittently.

"Officers and Employees

"(e) (1) The Corporation shall have a President, and such other officers as may be named and appointed by the Board for terms and at rates of compensation fixed by the Board. No individual other than a citizen of the United States may be an officer of the Corporation. No officer of the Corporation, other than the Chairman and any Vice Chairman, may receive any salary or other compensation from any source other than the Corporation during the period of his employment by the Corporation. All officers shall serve at the pleasure of the Board. (2) Except as provided in the second sentence of subsection (c) (1) of this section, no political test or qualification shall be used in selecting, appointing, promoting, or taking other personnel actions with respect to officers, agents, and employees of the Corporation.

"Nonprofit and Nonpolitical Nature of the Corporation

"(f) (1) The Corporation shall have no power to issue any shares of stock, or to declare or pay any dividends. (2) No part of the income or assets of the Corporation shall insure to the benefit of any director, officer, employee, or any other individual except as salary or reasonable compensation for services. (3) The Corporation may not contribute to or otherwise support any political party or candidate for elective public office.

"Purposes and Activities of the Corporation

"(g) (1) In order to achieve the objectives and to carry out the purposes of this subpart, as set out in subsection (a), the Corporation is authorized to—

"(A) facilitate the full development of educational broadcasting in which programs of high quality, obtained from diverse sources, will be made available to noncommercial educational television or radio broadcast stations, with strict adherence to objectivity and balance in all programs or series of programs of a controversial nature; (B) assist in the establishment and development of one or more systems of interconnection to be used for the distribution of educational television or radio programs so that all noncommercial educational television or radio broadcast stations that wish to may broadcast the programs at times chosen by the stations; (C) assist in the establishment and development of one or more systems of noncommercial educational television or radio broadcast stations throughout the United States; (D) carry out its purposes and functions and engage in its activities in ways that will most effectively assure the maximum freedom of the noncommercial educational television or radio broadcast systems and

local stations from interference with or control or program content or other activities.

"(2) Included in the activities of the Corporation authorized for accomplishment of the purposes set forth in subsection (a) of this section, are, among others not specifically named—

"(A) to obtain grants from and to make contracts with individuals and with private, State, and Federal agencies, organizations, and institutions; (B) to contract with or make grants to program production entities, individuals, and selected noncommercial educational broadcast stations for the production of, and otherwise to procure, educational television or radio programs for national or regional distribution to noncommercial educational broadcast stations; (C) to make payments to existing and new noncommercial educational broadcast stations to aid in financing local educational television or radio programing costs of such stations, particularly innovative approaches thereto, and other costs of operation of such stations; (D) to establish and maintain a library and archives of noncommercial educational television or radio programs and related materials and develop public awareness of and disseminate information about noncommercial educational television or radio broadcasting by various means, including the publication of a journal; (E) to arrange, by grant or contract with appropriate public or private agencies, organizations, or institutions, for interconnection facilities suitable for distribution and transmission of educational television or radio programs to noncommercial educational broadcast stations; (F) to hire or accept the voluntary services of consultants, experts, advisory boards, and panels to aid the Corporation in carrying out the purposes of this section; (G) to encourage the creation of new noncommercial educational broadcast stations in order to enhance such service on a local, State, regional, and national basis; (H) conduct (directly or through grants or contracts) research, demonstrations, or training in matters related to noncommercial educational television or radio broadcasting.

"(3) To carry out the foregoing purposes and engage in the foregoing activities, the Corporation shall have the usual powers conferred upon a nonprofit corporation by the District of Columbia Nonprofit Corporation Act, except that the Corporation may not own or operate any television or radio broadcast station, system, or network, community antenna television system, or interconnection or program production facility.

"Authorization for Free or Reduced Rate Interconnection Service

"(h) Nothing in the Communications Act of 1934, as amended, or in any other provision of law shall be construed to prevent United States communications common carriers from rendering free or reduced rate communications interconnection services for noncommercial educational television or radio services, subject to such rules and regulations as the Federal Communications Commission may prescribe.

"Report to Congress

"(i) The Corporation shall submit an annual report for the preceding fiscal year ending June 30 to the President for transmittal to the Congress on or before the 31st day of December of each year. The report shall include a comprehensive and detailed report of the Corporation's operations, activities, financial condition, and accomplishments under this section and may include such recommendations as the Corporation deems appropriate.

"Right To Repeal, Alter, or Amend

"(j) The right to repeal, alter, or amend this section at any time is expressly reserved.

"Financing

"(k) (1) There are authorized to be appropriated for expenses of the Corporation for the fiscal year ending June 30, 1968, the sum of $9,000,000, to remain available until expended. (2) Notwithstanding the preceding provisions of this section, no grant or contract pursuant to this section may provide for payment from the appropriation for the fiscal year ending June 30, 1968, for any one project or to any one station of more than $250,000.

"Records and Audit

"(l) (1) (A) The accounts of the Corporation shall be audited annually in accordance with generally accepted auditing standards by independent certified public accountants or independent licensed public accountants certified or licensed by a regulatory authority of a State or other political subdivision of the United States. The audits shall be conducted at the place or places where the accounts of the Corporation are normally kept. All books, accounts, financial records, reports, files, and all other papers, things, or property belonging to or in use by the Corporation and necessary to facilitate the audits shall be made available to the person or persons conducting the audits; and full facilities for verifying transactions with the balances or securities held by depositories, fiscal agents and custodians shall be afforded to such person or persons. (B) The report of each such independent audit shall be included in the annual report required by subsection (i) of this section. The audit report shall set forth the scope of the audit and include such statements as are necessary to present fairly the Corporation's assets and liabilities, surplus or deficit, with an analysis of the changes therein during the year, supplemented in reasonable detail by a statement of the Corporation's income and expenses during the year, and a statement of the sources and application of funds, together with the independent auditor's opinion of those statements.

"(2) (A) The financial transactions of the Corporation for any fiscal year during which Federal funds are available to finance any portion of its operations may be audited by the General Accounting Office in accordance with the principles and procedures applicable to commercial corporate transactions and under such rules and regulations as may be prescribed by the Comptroller General of the United States. Any such audit shall be conducted at the place or places where accounts of the Corporation are normally kept. The representative of the General Accounting Office shall have access to all books, accounts, records, reports, files, and all other papers, things, or property belonging to or in use by the Corporation pertaining to its financial transactions and necessary to facilitate the audit, and they shall be afforded full facilities for verifying transactions with the balances or securities held by depositories, fiscal agents, and custodians. All such books, accounts, records, reports, files, papers and property of the Corporation shall remain in possession and custody of the Corporation. (B) A report of each such audit shall be made by the Comptroller General to the Congress. The report to the Congress shall contain such comments and information as the Comptroller General may deem necessary to inform Congress of the financial operations and condition of the Corporation, together with such recommendations with respect thereto as he may deem advisable. The report shall also show specifically any program, expenditure, or other financial transaction or undertaking observed in the course of the audit, which, in the opinion of the Comptroller General, has been carried on or made without authority of law. A copy of each report shall be furnished to the President, to the Secretary, and to the Corporation at the time submitted to the Congress.

"(3) (A) Each recipient of assistance by grant or contract, other than a fixed price contract awarded pursuant to competitive bidding procedures, under this section shall keep such records as may be reasonably necessary to fully disclose the amount and the disposition by such recipient of the proceeds of such assistance, the total cost of the project or undertaking in connection with which such assistance is given or used, and the amount and nature of that portion of the cost of the project or undertaking supplied by other sources, and such other records as will facilitate an effective audit. (B) The Corporation or any of its duly authorized representatives, shall have access for the purpose of audit and examination to any books, documents, papers, and records of the recipient that are pertinent to assistance received under this section. The Comptroller General of the United States or any of his duly authorized representatives shall also have access thereto for such purpose during any fiscal year for which Federal funds are available to the Corporation."

Title III—Study of Educational and Instructional Broadcasting

Study Authorized

Sec. 301   The Secretary of Health, Education, and Welfare is authorized to conduct, directly or by contract, and in consultation with other interested Federal agencies, a comprehensive study of instructional television and radio (including broadcast, closed circuit, community antenna television, and instructional television fixed services and two-way communication of data links and computers) and their relationship to each other and to instructional materials such as videotapes, films, discs, computers, and other educational materials or devices, and such other aspects thereof as may be of assistance in determining whether and what Federal aid should be provided for instructional radio and television and the form that aid should take, and which may aid communities, institutions, or agencies in determining whether and to what extent such activities should be used.

Duration of Study

Sec. 302   The study authorized by this title shall be submitted to the President for transmittal to the Congress on or before June 30, 1969.

Appropriation

Sec. 303   There are authorized to be appropriated for the study authorized by this title such sums, not exceeding $500,000, as may be necessary.

Approved November 7, 1967.

Noncommercial Public TV Stations, 1975

The following list of noncommercial television stations represents those stations that were on the air in late 1975, according to the *1975 NAEB Directory of Public Telecommunications* (Washington, D.C.: National Association of Educational Broadcasters, 1975) pages 106–120.

Alabama
Birmingham ..... WBIQ (Ch. 10)
Demopolis ...... WIIQ (Ch. 41)
Dozier .... WDIQ (Ch. 2)
Florence .... WFIQ (Ch. 36)
Huntsville .... WHIQ (Ch. 25)
Louisville .... WGIQ (Ch. 43)
Mobile .... WEIQ (Ch. 42)
Montgomery ..... WAIQ (Ch. 26)
Mount Cheaha ... WCIQ (Ch. 7)

Alaska
Bethel .... KYUK (Ch. 4)
Fairbanks .... KUAC (Ch. 9)

Arizona
Tempe/Phoenix .. KAET (Ch. 8)
Tucson .... KUAT (Ch. 6)

Arkansas
Little Rock ...... KETS (Ch. 2)

California
Eureka .... KEET (Ch. 13)
Huntington Beach KOCE (Ch. 50)
Los Angeles ..... KCET (Ch. 28)
Los Angeles ..... KLCS (Ch. 58)
Los Angeles ..... KVST (Ch. 68)
Redding .... KIXE (Ch. 9)
Sacramento ..... KVIE (Ch. 6)
San Bernardino .. KVCR (Ch. 24)
San Diego .... KPBS (Ch. 15)
San Francisco ... KQED (Ch. 9)
San Francisco ... KQEC (Ch. 32)
San Jose .... KTEH (Ch. 54)
San Mateo ...... KCSM (Ch. 14)

Colorado
Denver .... KRMA (Ch. 6)
Pueblo .... KTSC (Ch. 8)

Connecticut
Bridgeport .... WEDW (Ch. 49)
Hartford .... WEDH (Ch. 24)
New Haven ...... WEDY (Ch. 65)
Norwich .... WEDN (Ch. 53)

Delaware
Wilmington ...... WHYY (Ch. 12)

District of Columbia
Washington ..... WETA (Ch. 26)

Florida
Gainesville ...... WUFT (Ch. 5)
Jacksonville ..... WJCT (Ch. 7)
Miami .... WPBT (Ch. 2)
Miami .... WTHS (Ch. 2)
Miami .... WLRN (Ch. 17)
Orlando .... WMFE (Ch. 24)
Pensacola .... WSRE (Ch. 23)
Tallahassee ..... WFSU (Ch. 11)
Tampa .... WEDU (Ch. 3)
Tampa .... WUSF (Ch. 16)

Georgia
Athens .... WGTV (Ch. 8)
Atlanta .... WETV (Ch. 30)
Chatsworth ...... WCLP (Ch. 18)
Cochran .... WDCO (Ch. 15)
Columbus .... WJSP (Ch. 28)
Dawson .... WACS (Ch. 25)
Pelham .... WABW (Ch. 14)
Savannah .... WVAN (Ch. 9)
Waycross .... WXGA (Ch. 8)
Wrens .... WCES (Ch. 20)

Hawaii
Honolulu .... KHET (Ch. 11)
Wailuku .... KMEB (Ch. 10)

Idaho
Boise .... KAID (Ch. 4)
Moscow .... KUID (Ch. 12)
Pocatello .... KBGL (Ch. 10)

Illinois
Carbondale ...... WSIU (Ch. 8)
Chicago .... WTTW (Ch. 11)
Chicago .... WXXW (Ch. 20)
Olney .... WUSI (Ch. 16)
Peoria .... WTVP (Ch. 47)
Urbana .... WILL (Ch. 12)

Indiana
Bloomington ..... WTIU (Ch. 30)
Elkhart .... WNIT (Ch. 34)
Evansville .... WNIN (Ch. 9)
Indianapolis ..... WFYI (Ch. 20)
Muncie .... WIPB (Ch. 49)
St. John .... WCAE (Ch. 50)
Vincennes .... WVUT (Ch. 22)

Iowa
Council Bluffs ... KBIN (Ch. 32)
Des Moines ..... KDIN (Ch. 11)
Iowa City .... KIIN (Ch. 12)
Red Oak .... KHIN (Ch. 36)
Sioux City ...... KSIN (Ch. 27)
Waterloo .... KRIN (Ch. 32)

Kansas
Topeka .... KTWU (Ch. 11)
Wichita .... KPTS (Ch. 8)

Kentucky
Ashland .... WKAS (Ch. 25)
Bowling Green ... WKGB (Ch. 53)
Covington .... WCVN (Ch. 54)
Elizabethtown ... WKZT (Ch. 23)
Hazard .... WKHA (Ch. 35)
Lexington .... WKLE (Ch. 46)
Louisville .... WKMJ (Ch. 68)
Louisville .... WKPC (Ch. 15)
Madisonville ..... WKMA (Ch. 35)
Morehead .... WKMR (Ch. 38)
Murray .... WKMU (Ch. 21)
Owenton .... WKON (Ch. 52)
Pikeville .... WKPI (Ch. 22)
Somerset .... WKSO (Ch. 29)

Louisiana
Baton Rouge .... WLPB (Ch. 27)
Biddeford .... WMEG (Ch. 26)
New Orleans .... WYES (Ch. 12)

Maine
Augusta .... WCBB (Ch. 10)
Calais .... WMED (Ch. 13)
Orono .... WMEB (Ch. 12)
Presque Isle ..... WMEM (Ch. 10)

Maryland
Baltimore .... WMPB (Ch. 67)
Hagerstown ..... WWPB (Ch. 31)
Salisbury .... WCPB (Ch. 28)

Massachusetts
Boston .... WGBH (Ch. 2)
Boston .... WGBX (Ch. 44)
Springfield ...... WGBY (Ch. 57)

Michigan
Alpena .... WCML (Ch. 6)
Detroit .... WTVS (Ch. 56)
East Lansing .... WKAR (Ch. 23)
Grand Rapids ... WGVC (Ch. 35)
Marquette .... WNPB (Ch. 13)
Mount Pleasant .. WCMU (Ch. 14)
University Center WUCM (Ch. 19)

Minnesota
Appleton .... KWCM (Ch. 10)
Austin .... KAVT (Ch. 15)
Duluth .... WDSE (Ch. 8)
St. Paul-
  Minneapolis .... KTCA (Ch. 2)
St. Paul-
  Minneapolis .... KTCI (Ch. 17)

Mississippi
Ackerman .... WMAB (Ch. 2)
Booneville .... WMAE (Ch. 12)
Bude .... WMAU (Ch. 17)
Inverness .... WMAO (Ch. 23)
Jackson .... WMAA (Ch. 29)
McHenry .... WMAH (Ch. 19)
Oxford .... WMAV (Ch. 18)
Rose Hill .... WMAW (Ch. 14)

Missouri
Kansas City ..... KCPT (Ch. 19)
St. Louis .... KETC (Ch. 9)

Nebraska
Alliance .... KTNE (Ch. 13)
Bassett .... KMNE (Ch. 7)
Hastings .... KHNE (Ch. 29)
Lexington .... KLNE (Ch. 3)
Lincoln .... KUON (Ch. 12)
Merriman .... KRNE (Ch. 12)
Norfolk .... KXNE (Ch. 19)
North Platte ..... KPNE (Ch. 9)
Omaha .... KYNE (Ch. 26)

Nevada
Las Vegas ...... KLVX (Ch. 10)

New Hampshire
Berlin .... WEDB (Ch. 40)
Durham .... WENH (Ch. 11)
Hanover .... WHED (Ch. 15)
Keene .... WEKW (Ch. 52)
Littleton .... WLED (Ch. 49)

New Jersey
Camden .... WNJS (Ch. 23)
Montclair .... WNJM (Ch. 50)
New Brunswick .. WNJB (Ch. 58)
Trenton .... WNJT (Ch. 52)

New Mexico
Albuquerque .... KNME (Ch. 5)
Las Cruces ...... KRWG (Ch. 22)
Portales .... KENW (Ch. 3)

New York
Binghamton ..... WSKG (Ch. 46)
Buffalo .... WNED (Ch. 17)
Garden City ..... WLIW (Ch. 21)
New York .... WNET (Ch. 13)
New York .... WNYE (Ch. 25)
New York .... WNYC (Ch. 31)
Norwood .... WNPI (Ch. 18)
Rochester .... WXXI (Ch. 21)
Schenectady .... WMHT (Ch. 17)
Syracuse .... WCNY (Ch. 24)
Watertown ...... WNPE (Ch. 16)

North Carolina
Asheville .... WUNF (Ch. 33)
Chapel Hill ...... WUNC (Ch. 4)
Charlotte .... WTVI (Ch. 42)
Concord .... WUNG (Ch. 58)
Columbia .... WUND (Ch. 2)
Greenville .... WUNK (Ch. 25)
Linville .... WUNE (Ch. 17)
Wilmington ...... WUNJ (Ch. 39)
Winston-Salem .. WUNL (Ch. 26)

North Dakota
Fargo .... KFME (Ch. 13)
Grand Forks ..... KGFE (Ch. 2)

Ohio
Alliance .... WNEO (Ch. 45)
Athens .... WOUB (Ch. 20)
Bowling Green ... WBGU (Ch. 57)
Cambridge ...... WOUC (Ch. 44)

Cincinnati .... WCET (Ch. 48)
Cleveland .... WVIZ (Ch. 25)
Columbus .... WOSU (Ch. 34)
Kettering .... WOET (Ch. 16)
Newark .... WGSF (Ch. 31)
Oxford .... WMUB (Ch. 14)
Portsmouth ...... WPBO (Ch. 42)
Toledo .... WGTE (Ch. 30)

Oklahoma
Oklahoma City ... KETA (Ch. 13)
Oklahoma City ... KOKH (Ch. 25)
Tulsa .... KOED (Ch. 11)

Oregon
Corvallis .... KOAC (Ch. 7)
Portland .... KOAP (Ch. 10)

Pennsylvania
Allentown .... WLVT (Ch. 39)
Clearfield .... WPSX (Ch. 3)
Erie .... WQLN (Ch. 54)
Hershey .... WCTF (Ch. 33)
Philadelphia ..... WUHY (Ch. 35)
Pittsburgh .... WQED (Ch. 13)
Pittsburgh .... WQEX (Ch. 16)
Scranton .... WVIA (Ch. 44)

Rhode Island
Providence ...... WSBE (Ch. 36)

South Carolina
Allendale .... WEBA (Ch. 14)
Beaufort .... WJWJ (Ch. 16)
Charleston ...... WITV (Ch. 7)
Columbia .... WRLK (Ch. 35)
Florence .... WJPM (Ch. 33)
Greenville .... WNTV (Ch. 29)
Sumter .... WRJA (Ch. 27)

South Dakota
Aberdeen .... KDSD (Ch. 16)
Brookings .... KESD (Ch. 8)
Eagle Butte ..... KPSD (Ch. 13)
Lowry .... KLSD (Ch. 11)
Pierre .... KTSD (Ch. 10)
Rapid City .... KBHE (Ch. 9)
Vermillion .... KUSD (Ch. 2)

Tennessee
Chattanooga .... WTIC (Ch. 45)
Knoxville .... WSJK (Ch. 2)
Lexington .... WLJT (Ch. 11)
Memphis .... WKNO (Ch. 10)
Nashville .... WDCN (Ch. 8)

Texas
Austin-
  San Antonio .... KLRN (Ch. 9)
College Station .. KAMU (Ch. 15)
Corpus Christi ... KEDT (Ch. 16)
Dallas .... KERA (Ch. 13)
Houston .... KUHT (Ch. 8)
Killeen .... KNCT (Ch. 46)
Lubbock .... KTXT (Ch. 5)

Utah
Provo .... KBYU (Ch. 11)
Salt Lake City ... KUED (Ch. 7)

Vermont
Burlington .... WETK (Ch. 33)
St. Johnsbury ... WVTB (Ch. 20)
Rutland .... WVER (Ch. 28)
Windsor .... WVTA (Ch. 41)

Virginia
Annandale ...... WNVT (Ch. 53)
Harrisonburg .... WVPT (Ch. 51)
Norfolk .... WHRO (Ch. 15)
Norton .... WSVN (Ch. 47)
Richmond .... WCVE (Ch. 23)
Richmond .... WCVM (Ch. 57)
Roanoke .... WBRA (Ch. 15)

Washington
Lakewood Center KPEC (Ch. 56)
Pullman .... KWSU (Ch. 10)
Seattle .... KCTS (Ch. 9)
Spokane .... KSPS (Ch. 7)
Tacoma .... KTPS (Ch. 62)
Yakima .... KYVE (Ch. 47)

West Virginia
Grandview ...... WSWP (Ch. 9)
Huntington ...... WMUL (Ch. 33)
Morgantown ..... WWVU (Ch. 24)

Wisconsin
Green Bay ...... WPNE (Ch. 38)
La Crosse .... WHLA (Ch. 31)
Madison .... WHA (Ch. 21)
Menomonie ..... WHWC (Ch. 28)
Milwaukee ...... WMVS (Ch. 10)
Milwaukee ...... WMVT (Ch. 36)

American Samoa
Pago Pago ...... KVZK (Chs. 2,
                4, 5, 8, 10, 12)

Guam
Agana .... KGTF (Ch. 12)

Puerto Rico
Mayaguez .... WIPM (Ch. 3)
San Juan .... WIPR (Ch. 6)

Virgin Islands
St. Thomas ...... WTJX (Ch. 12)

National Public Radio Member Stations, 1975

The following list of non-commercial radio stations are those educational stations listed as full-service "NPR member" stations as of February 1976 (membership list provided by NPR).

Alaska
College .... KUAC (FM)
Kotzebue .... KOTZ (AM)

Arizona
Phoenix .... KMCR (FM)
Tucson .... KUAT (AM-FM)
Yuma .... KAWC (AM)

Arkansas
Jonesboro .... KASU (FM)

California
Long Beach ..... KLON (FM)
Los Angeles ..... KUSC (FM)
Northridge .... KCSN (FM)
Pasadena .... KPCS (FM)
San Bernardino .. KVCR (FM)
San Diego .... KPBS (FM)
San Francisco ... KALW (FM)
San Francisco ... KQED (FM)
San Mateo ...... KCSM (FM)
Santa Monica ... KCRW (FM)
Santa Rosa ..... KBBF (FM)
Stockton .... KUOP (FM)

Colorado
Denver .... KCFR (FM)
Greeley .... KUNC (FM)

District of Columbia
Washington ..... WAMU (FM)
Washington ..... WETA (FM)

Florida
Boynton Beach .. WHRS (FM)
Jacksonville ..... WJCT (FM)
Miami .... WLRN (FM)
Tallahassee ..... WFSU (FM)
Tampa .... WUSF (FM)

Georgia
Atlanta .... WABE (FM)

Illinois
Carbondale ...... WSIU (FM)
Chicago .... WBEZ (FM)
DeKalb .... WNIU (FM)
Edwardsville ..... WSIE (FM)
Peoria .... WCBU (FM)
Springfield ...... WSSR (FM)
Urbana .... WILL (AM-FM)

Indiana
Bloomington ..... WFIU (FM)
Indianapolis ..... WIAN (FM)
West Lafayette .. WBAA (AM)
Vincennes .... WVUB (FM)

Iowa
Ames .... WOI (AM-FM)
Cedar Falls ...... KHKE (FM)
Cedar Falls ...... KUNI (FM)
Cedar Rapids ... KCCK (FM)
Iowa City .... WSUI (AM)

Kansas
Lawrence ,...... KANU (FM)
Manhattan ...... KSAC (AM)
Wichita .... KMUW (FM)

Kentucky
Lexington .... WBKY (FM)
Louisville .... WFPK (FM)
Louisville .... WFPL (FM)
Morehead .... WMKY (FM)
Murray .... WKMS (FM)

Louisiana
New Orleans .... WWNO (FM)

Maine
Bangor .... WMEH (FM)
Portland .... WMEA (FM)

Maryland
Baltimore .... WBJC (FM)
Takoma Park .... WGTS (FM)

Massachusetts
Amherst .... WFCR (FM)
Boston .... WBUR (FM)
Boston .... WGBH (FM)

358

Michigan

Ann Arbor . . . . WUOM (FM)
Berrien Springs . . WAUS (FM)
Detroit . . . . WDET (FM)
East Lansing . . . . WKAR (AM-FM)
Flint . . . . WFBE (FM)
Houghton . . . . WGGL (FM)
Interlochen . . . . . . WIAA (FM)
Kalamazoo . . . . . . WMUK (FM)
Marquette . . . . WNMU (FM)
Mount Pleasant . . WCMU (FM)

Minnesota

Collegeville . . . . . . KSJR (FM)
Duluth . . . . WSCD (FM)
Minneapolis/
St. Paul . . . . KSJN (FM)
Minneapolis . . . . . KUOM (AM)
Moorhead . . . . KCCM (FM)
Northfield . . . . WCAL (FM)
Pipestone . . . . KRSW (FM)

Mississippi

Senatobia . . . . WNJC (FM)

Missouri

Buffalo . . . . KBFL (FM)
Columbia . . . . KBIA (FM)
Jefferson City . . . KLUM (FM)
Kansas City . . . . . KCUR (FM)
Maryville . . . . KXCV (FM)
Point Lookout . . . KSOZ (FM)
Rolla . . . . KUMR (FM)
Springfield . . . . . . KSMU (FM)
St. Louis . . . . KWMU (FM)
Warrensburg . . . . KCMW (FM)

Montana

Missoula . . . . KUFM (FM)

Nebraska

Omaha . . . . KIOS (FM)

New Mexico

Albuquerque . . . . KIPC (FM)
Las Cruces . . . . . . KRWG (FM)
Ramah . . . . KTDB (FM)

New York

Albany . . . . WAMC (FM)
Buffalo . . . . WBFO (FM)
Canton . . . . WSLU (FM)
Binghamton . . . . . WSKG (FM)
New York . . . . WNYC (AM-FM)
Oswego . . . . WRVO (FM)
Rochester . . . . WXXI (FM)
Schenectady . . . . WMHT (FM)
Syracuse . . . . WCNY (FM)

North Carolina

Chapel Hill . . . . . . WUNC (FM)
Winston-Salem . . WFDD (FM)

North Dakota

Belcourt . . . . KEYA (FM)
Fargo . . . . KDSU (FM)
Grand Forks . . . . . KFJM (AM)

Ohio

Athens . . . . WOUB (AM-FM)
Bowling Green . . . WBGU (FM)
Cincinnati . . . . WGUC (FM)
Columbus . . . . WCBE (FM)
Columbus . . . . WOSU (AM-FM)
Kent . . . . WKSU (FM)
Oxford . . . . WMUB (FM)
Toledo . . . . WGTE (FM)
Wilberforce . . . . . . WCSU (FM)
Yellow Springs . . WYSO (FM)
Youngstown . . . . . WYSU (FM)

Oklahoma

Stillwater . . . . KOSU (FM)

Oregon

Corvallis . . . . KOAC (AM)
Eugene . . . . KLCC (FM)
Eugene . . . . KWAX (FM)
Portland . . . . KBPS (AM)
Portland . . . . KOAP (FM)
Portland . . . . KBOO (FM)

Pennsylvania

Erie . . . . WQLN (FM)
Hershey . . . . WITF (FM)
Philadelphia . . . . . WUHY (FM)
Pittsburgh . . . . WDUQ (FM)
Pittsburgh . . . . WQED (FM)
Pittsburgh . . . . WYEP (FM)
Scranton . . . . WVIA (FM)

South Carolina

Charleston . . . . . . WSCI (FM)
Greenville . . . . WEPR (FM)

South Dakota

Brookings . . . . KESD (FM)
Vermillion . . . . KUSD (AM)

Tennessee

Collegedale . . . . . WSMC (FM)
Johnson City . . . . WETS (FM)
Knoxville . . . . WUOT (FM)
Memphis . . . . WKNO (FM)
Murfreesboro . . . . WMOT (FM)
Nashville . . . . WPLN (FM)

Texas

Austin . . . . KUT (FM)
Beaumont . . . . KVLU (FM)
Commerce . . . . . . KETR (FM)
Dallas . . . . KERA (FM)
El Paso . . . . KTEP (FM)
Killeen . . . . KNCT (FM)

Utah

Logan . . . . KUSU (FM)
Provo . . . . KBYU (FM)
Salt Lake City . . . KUER (FM)

Virginia

Harrisonburg . . . . WMRA (FM)
Norfolk . . . . WTGM (FM)
Richmond . . . . WRFK (FM)
Roanoke . . . . WVWR (FM)

Washington

Pullman . . . . KWSU (AM)
Seattle . . . . KRAB (FM)
Seattle . . . . KUOW (FM)
Tacoma . . . . KTOY (FM)

West Virginia

Beckley . . . . WVPB (FM)
Buckhannon . . . . . WVWC (FM)

Wisconsin

La Crosse . . . . WLSU (FM)
Madison . . . . WERN (FM)
Madison . . . . WHA (AM)
Milwaukee . . . . . . WUWM (FM)

Puerto Rico

Hato Rey . . . . WIPR (AM)